BIOLOGICAL METHODS OF PROSPECTING FOR MINERALS

BIOLOGICAL METHODS OF PROSPECTING FOR MINERALS

R. R. BROOKS

A Wiley-Interscience Publication

JOHN WILEY & SONS

New York   Chichester   Brisbane   Toronto   Singapore

Grateful acknowledgment is made to Harper and Row Publishers, Inc.,
for permission to reprint a substantial quantity of material from
Geobotany and Biogeochemistry in Mineral Exploration,
Copyright © 1972 by R. R. Brooks.

Library of Congress Cataloging in Publication Data:

Brooks, R. R.
   Biological methods of prospecting for minerals.

   "A Wiley-Interscience publication."
   Bibliography: p.
   Includes index.
   1. Biogeochemical prospecting.   2. Geobotanical
prospecting.   I. Title.
TN270.B828   1983   622′.12   82-21819
ISBN 0-471-87400-0

Printed in the United States of America

10 9 8 7 6 5 4 3 2 1

PREFACE

Biological methods of prospecting for minerals include geobotany (visual examination of the vegetation cover), biogeochemistry (chemical analysis of vegetation), and geozoology (use of animals for mineral prospecting), and have been used for many years. Until the 1950s, botanical techniques were used almost exclusively in the Soviet Union, but they have now been adopted in nearly every country in the world.

The first comprehensive English-language book on botanical methods was *Biogeochemical Methods of Prospecting,* a 1964 English translation of a Russian book by the late D. P. Malyuga. This work suffered to some extent from an overemphasis of the Russian experience and physical environment, with poor coverage of Western work. It was, nevertheless, an important pioneering text, the first in the English language. Another translation of a Russian book, *Biogeochemical Exploration for Mineral Deposits* by A. L. Kovalevsky, appeared in 1979. This relatively short work (136 pages) is a welcome addition to the literature on this subject and is more quantitative than Malyuga's book. Once again, however, it focuses predominantly on Russian research. Western work is represented by only 21 out of 104 references, none of which postdates 1971.

In 1972, I wrote a book entitled *Geobotany and Biogeochemistry in Mineral Exploration* which is now out of print. The friendly reception accorded this publication has encouraged me to produce a successor, the present work. This book coincides with a renewed interest in biological methods of prospecting and describes for the first time geozoological techniques. This renewed interest is a result of a number of factors. In the first place, most of the world's easily discoverable mineral deposits have now been delineated so that more sophisticated methods are required to find those that remain. The second factor is that biological methods of prospecting have become much more sophisticated and quantitative due to the development of statistical techniques facilitated by the advent of computers. Early botanical techniques for mineral exploration were essentially empirical because of a lack of adequate facilities to carry out such statistical procedures, and the field fell into disfavor largely because of this.

The present work is not merely an update of my previous book, although about 25% of the text is based on the previous publication. It is a new concept with a more quantitative approach and with a much greater coverage (966 references) of the literature. There is an expanded section on statistical methods for interpretation of biogeochemical data, as well as a description and discussion of aerial biogeochemical prospecting; a new exciting development that is likely to enhance greatly the potentiality of the art in the foreseeable future. The section on geobotany has been greatly expanded compared with my earlier book, and includes details of a number of important case histories. Much greater emphasis has also been placed on the important topic of remote sensing of the environment, an important up-and-coming tool in the field of geobotany.

This book is addressed to several classes of reader. It is addressed not only to field workers in exploration companies, but also to students of biogeography, applied geochemistry, economic geology, terrain science, geology, geophysics, chemistry, biochemistry, and phytochemistry. It is indeed addressed to all workers in the wide range of disciplines that overlap into the field of biological methods of exploration.

The foundation of this book is the pioneering work of many scientists such as Hiram Wild (formerly of the University of Zimbabwe), Helen Cannon and Hansford Shacklette of the U.S. Geological Survey, Harry Warren of British Columbia, and Alexander Kovalevsky of the Soviet Union. Other workers too numerous to mention also have been a useful source of inspiration and material for the compilation of this work.

I am particularly indebted to my many friends, colleagues, research students, and technicians who have contributed so much. In particular, I would like to thank G. L. Lyon, D. Nicolas, J. S. Nielsen, M. H. Timperley, N. E. Whitehead, P. J. Peterson, G. W. Butler, B. F. Quin, B. C. Severne, N. J. Marshall, C. R. Boswell, T. E. Yates, T. Jaffré, J. Lee, N. I. Ward, R. D. Reeves, M. T. Lyons, P. C. Kelly, S. Dilli, E. D. Wither, J. A. McCleave, F. Malaisse, M. J. B. Sedborough, E. K. Schofield, R. S. Morrison, C. C. Radford, J. M. Trow, W. J. Kersten, O. Vergnano Gambi, H. M. Crookes, B. Fredskild, J. R. Press, J.-M. Veillon, A. Asensi Marfil, S. Shaw, R. M. McFarlane, J. Holzbecher, D. E. Ryan, A. J. Kelton, T. E. Dudley, and J. R. Liddle. I dedicate this book to them.

Thanks are also due to the many funding agencies and exploration companies who provided funds, expertise, and logistic support for field work in Australia, Sri Lanka, Canada, New Caledonia, and New Zealand. These include Australian Selection (Pty.) Ltd., Carpentaria Exploration Co. (Pty.) Ltd., The Consolidated Silver Co. of New Zealand, Kennecott Explorations (Australia) Pty. Ltd., Lime and Marble Ltd., McIntyre Mines (N.Z.) Ltd.,

Norpac Mining Ltd., Western Mining Corp., Aquitaine Co. of Canada, the Natural Science and Engineering Council of Canada, and the Mineral Resources Sub-committee of the New Zealand University Grants Committee. I also thank Professors R. D. Batt, L. E. Smythe, and D. E. Ryan for providing excellent research facilities in New Zealand, Australia, and Canada, respectively.

The stated aim of this book is to attempt to bridge the gap between the various disciplines involved in biological methods of prospecting for minerals. If I have succeeded in persuading geologists to drop their hammers for a moment and seriously contemplate the vegetation around them, or if I have succeeded in making the analyst and field worker more aware of their mutual problems, then I will be well satisfied.

R. R. Brooks

Palmerston North, New Zealand
April 1983

CONTENTS

11. Insects as Indicators of Mineralization **101**

PART 3—BIOGEOCHEMISTRY IN MINERAL EXPLORATION

12. An Introduction to Biogeochemical Prospecting **109**

13. Soils and Their Formation **112**

14. Accumulation of Elements by Plants **127**

15. Biogeochemical Parameters and Their Significance for Mineral Prospecting **153**

CHAPTER
1
GENERAL INTRODUCTION

1.1. THE ROLE OF BIOLOGICAL METHODS IN MINERAL PROSPECTING

There is at present an unprecedented surge in prospecting for minerals throughout the world because of the increasing demands placed on this planet's dwindling resources by expanding economies and burgeoning populations. Most of the world's easily discoverable deposits were found many years ago. Therefore those that remain tend to occur either beneath the sea or in inaccessible parts of the world, perhaps covered by ice caps or by dense forest and bush where outcrops are hard to detect.

To discover these increasingly hard-to-detect mineral deposits, exploration geochemists have a wide array of methods at their disposal, ranging from aerial geophysics to various forms of geochemical procedure involving the analysis of air, waters, stream sediments, soils, rocks, plants, animals, and bogs. These methods have become increasingly sophisticated and it has now become a recognizable fact of life that no one technique by itself can give a complete delineation of mineralization. The judicious application of two or three different procedures can give a cumulative result that is far superior to that obtained by any single procedure alone. This is perhaps one of the reasons there recently has been renewed interest in biological methods of prospecting for minerals. Biological methods include biogeochemistry (chemical analysis of vegetation), geobotany (visual study of the plant cover), and geozoology (the use of animals in prospecting).

Biogeochemical prospecting has been stimulated by the much greater ease with which samples can now be analyzed by such powerful methods as atomic absorption spectrometry, and inductively coupled plasma emis-

1

sion spectrometry (see Chapter 18) coupled with new methods of aerial sampling. Geobotany has gotten a tremendous boost from the sophisticated methods of remote sensing of the environment that have become available during the past decade (see Chapter 6). Both it and biogeochemistry have greatly benefited from the development of sophisticated computer-assisted statistical techniques which significantly improve the precision and reliability of both procedures.

Vegetation as an exploratory tool has not been accepted as readily as have rocks, soils, waters, and stream sediments. This may be due to the fact that botanical methods sometimes demand a greater skill in execution and interpretation and may require some sort of preliminary orientation survey in a new area. To some field workers vegetation is a "nuisance," an impediment to progress that must be ruthlessly bulldozed out of the way before other operations can proceed. Fortunately this attitude is changing and many workers in mineral exploration are beginning to realize that vegetation can be an asset rather than a hindrance. The distribution of species and the elemental content of the plant cover present a complex picture that can provide important information to the prospectors if they have the motivation to try to interpret it.

The extent to which plants may be used in the search for minerals will depend of course on the nature and extent of the vegetation cover. About two thirds of the world's land surface is covered with vegetation of which 42% comprises forest, 24% grassland, and 21% consists of desert shrubs and grasses in semiarid terrain. Since developed areas with less natural vegetation already will have been explored more thoroughly than remoter regions, and because deserts will have obvious outcrops, it is likely that most of the world's remaining economic mineral deposits will be hidden beneath vegetation. For this reason, there should be increasing scope for the future use of biological methods in mineral exploration.

1.2. THE LITERATURE OF BIOLOGICAL PROSPECTING METHODS

It is very difficult to calculate the number of papers appearing each year in all of the various fields of biological prospecting because so much of this work is published in the Soviet Union and is often not available in the West. Entries in *Chemical Abstracts, Biological Abstracts* and even *Referativny Zhurnal* tend to be selective and do not cater completely to such a multidisciplinary field. According to A. L. Kovalevsky [447], about 460 Russian papers were published in all the fields during the period 1938–1970. The sum total of references for biological prospecting found in the above three sources was about 150. I have recently had a computer search made on the above world literature for the period 1965–1980 which turned up 333 references including only 61 Russian ones. By selecting the work of a well-

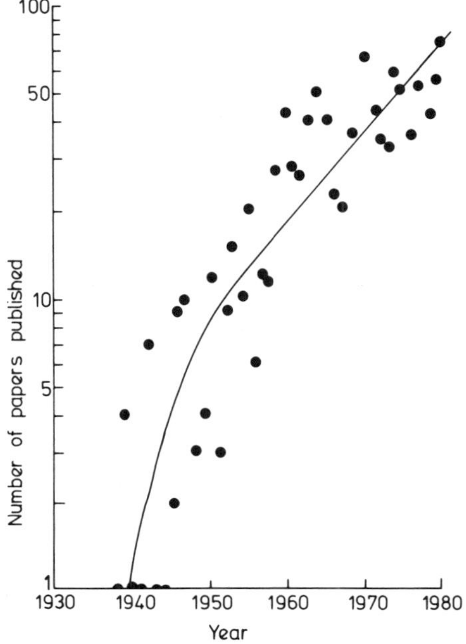

FIGURE 1.1. Number of papers published annually (log scale) in the field of biological methods of prospecting during the period 1938–1980.

known Russian biogeochemist (A. L. Kovalevsky) whose complete bibliography was in my possession, it appeared that the computer search revealed only about one quarter of his work. Applying the same procedure to my own publications, only 60% of these were identified. Making allowances for all these deficiencies, Figure 1.1 shows the total number of biological prospecting papers published since 1938 (the year of the pioneering biogeochemical paper by S. M. Tkalich [812]. These figures have been obtained by multiplying by four the number of Russian papers appearing in the abstracting journals and by using a factor of 1.7 for the Western papers. This gives a total of about 1000 papers for the period, of which 18% are geobotanical, 3% are geozoological, and 79% are biogeochemical. About a quarter of the geobotanical papers were concerned with remote sensing.

Table 1.1 lists sources of scientific papers for the period 1965–1980 and is based on the computer search which greatly underestimated Russian work as well as giving somewhat lower values for the Western publications. The Russian papers have been grouped together because of their wide dispersal among a large number of different journals. Perhaps the largest number of these Russian works was to be found in *Doklady Akademii Nauk SSSR*.

After allowing for underrepresentation in abstracting journals, about 60% of all biological prospecting papers during this period were in the English language (including English translations of Russian papers), 36% were in Russian, and about 1% were in each of German, French, and Spanish. All other languages together accounted for another 1% of the total.

TABLE 1.1 Journals that Published Papers on Biological Prospecting Methods 1965–1980[a]

Source	Number of Papers
Russian journals	73
U.S. Geological Survey publications	33
Proceedings of conferences and symposia	31
Journal of Geochemical Exploration	25
Economic Geology	17
Masterate and doctorate theses	13
Geological Survey of Canada publications	10
Books and chapters in books	9
International Atomic Energy Agency publications	6
Transactions of the Institution of Mining and Metallurgy	5
Proceedings of the Australasian Institute of Mining and Metallurgy	3
Bulletin of the Geological Society of Finland	3
Other publications	105

[a]Based on computer search and greatly under-representing Soviet literature.

1.3. CENTERS OF RESEARCH

The data in Table 1.1 give some information on the centers of research in the field of biological prospecting methods. In the West it is probable that a large percentage of the work is still carried out at the U.S. Geological Survey (mainly biogeochemical). The main centers of geobotany are Bedford College and the Remote Sensing Laboratory of Stanford University. The rest of the work tends to be fragmented among various universities and government research institutes, such as the Geological Survey of Canada. Except in organizations such as the U.S. Geological Survey where biological methods are well established, there is a tendency for the importance of various centers of this field to wax and wane as individual researchers retire or move elsewhere.

In the Soviet Union, centers of research tend to retain their identity even when well-known scientists move or retire. Institutions that are well known in the field include: The Institute of Geology, Ulan Ude (A. L. Kovalevsky); The Vernadsky Institute of Geochemistry and Analytical Chemistry, Moscow (V. V. Koval'sky); The Institute of Geology and Geophysics, Tashkent (R. M. Talipov); The University of Leningrad (M. D. Skarlygina-Ufimtseva), The Institute of Ore Geology, Moscow (F. V. Chukhrov). This short listing, which is far from complete, again shows the predominance of the Soviet Union in biological methods of prospecting.

1.4. THE SCOPE AND AIMS OF THIS BOOK

The purpose of this book is to bring together, in a single volume, the elements of the diversity of scientific disciplines (botany, zoology, geology, analytical chemistry, geophysics, biogeography, phytochemistry, plant physiology, and ecology) which are integral components of biological methods of prospecting for minerals. The lack of a broad overview is one of the continuing problems of current research in many fields of scientific endeavor. The old adage that an "expert" is someone who knows more and more about less and less until he finally knows everything about nothing, is more than just a facetious comment. It has an undeniable ring of truth about it. Until this unfortunate trend can be reversed and more interdisciplinary study encouraged, some scientists will continue to be ignorant of other disciplines. To no other field does this probably apply more than biological methods of prospecting. To break down these artificial barriers is the driving force behind this book, and I hope that this aim will be achieved.

1
GEOBOTANY IN MINERAL EXPLORATION

2

AN INTRODUCTION TO GEOBOTANY IN MINERAL EXPLORATION

Geobotanical methods of prospecting are the most simple and yet the most complex of all methods of biological prospecting. The apparent paradox in this statement is explained by the observation that the basic equipment needed is merely a pair of human eyes, whereas a good deal of experience is needed to interpret that which has been seen.

It has been known since Roman times that natural vegetation reflects to some extent the nature of its substrate. We find Vitruvius in the reign of Augustus Caesar (10 B.C.) observing that:

> The slender bulrush, the wild willow, the alder, the agnus castus, reeds, ivy and the like . . . usually grow in marshy places . . . . but water is to be sought in those regions and soils other than marshes, in which such trees are found naturally and not artificially planted.

Furthermore, the prospectors for ores in China and Europe have been aware for at least 400 years that there is a relationship between vegetation and its substrate. Though some of the early writings on plant indicators of ores and minerals contained fanciful or far-fetched material, many of these early observations are as valid today as they were several centuries ago.

In geobotanical methods of prospecting, the interpretation of the significance of the vegetation cover is carried out in a number of ways. These are: studying the nature and distribution of plant communities; studying the nature and distribution of individual indicator plants; observing morphological changes in vegetation (e.g., color changes); examining any of the above by aerovisual observations, by aerial photography, or by satellite imagery. The latter has become a very powerful tool in geobotany since the launching of the LANDSAT series of observation satellites (see Chapter 6).

One of the problems of geobotany as opposed to biogeochemistry (chemical analysis of vegetation), is that whereas the latter can be carried out quite easily by someone possessing only common sense and not necessarily having a botanical background, the same is not true to the same extent for geobotanical methods of prospecting. Preferably an expert geobotanist should have some knowledge of a wide range of disciplines such as biochemistry, biogeography, botany, chemistry, ecology, geology, and plant physiology. Very few persons will have expertise in more than a very few of these disciplines, so that sometimes a team would be desirable for field work. In spite of the extra cost and trouble involved in such an undertaking, the expenditure is far less than for most other exploration techniques, and the rewards are potentially very great. If persons with wide experience and expertise cannot be found, unskilled personnel can often provide useful data, so that geobotany should never be dismissed as being impracticable in a given area.

In the subsequent five chapters of this book, the broad outlines of geobotanical methods are presented and discussed. As far as possible an attempt has been made to give the theoretical basis of the concepts presented without delving too deeply into biochemistry or plant physiology. It is obvious that a field worker with some concept of the theoretical basis of his or her work will be more effective than a person who adopts a purely empirical approach without consideration of the principles involved.

In discussing the subject of geobotany in mineral exploration, an attempt has been made to avoid the danger of overstating the potential of the method. It is not claimed that geobotany by itself will necessarily give the whole picture of mineralization in a particular area, but it is certainly claimed that the technique can be of great value in completing the picture and in some cases can represent the difference between success and failure of an exploration project. In some cases indeed [214], mineral occurrences have been found by geobotany alone. The procedure is dependent above all on the skill of the operator, and unless skilled personnel can be employed, disenchantment may well follow. It is for this reason perhaps that geobotanical papers are a significant minority in the literature of biological methods of prospecting, and are in the main written by a small number of well-known researchers who have long specialized in this technique.

3

PLANT COMMUNITIES AS INDICATORS OF MINERALIZATION

3.1. INTRODUCTION

The Russian geologist Karpinsky [413] was one of the first to recognize that different plant associations exist on varying geological substrates such as sandstones, clays, limestones, and so on, and that this distribution might be used to characterize the geology of the area concerned. Karpinsky concluded that reliance should be placed on an examination of the whole community rather than on one or two characteristic plants within it. His classical work has ultimately led to the science of **indicator geobotany,** which is a division of botany concerned with the theoretical and practical aspects of vegetation and its component species as indicators of the conditions of the environment.

Indicator geobotany has been brought to a very high state of development by workers in the Soviet Union and elsewhere. Classical Soviet reviews of the field are those of Chikishev [190] and Viktorov et al. [849]. Indicator communities or **characteristic floras** will not in themselves necessarily indicate mineralization but will often serve to characterize regions where certain types of mineralization are likely to occur. Examples of this are the use of serpentine floras for locating chromite deposits [535] or the study of selenium floras [154] which can indirectly show the presence of uranium mineralization because of the geochemical association of these two elements. Linstow [515] in reviewing the field, has referred to such communities as **bodenanzeigende Pflanzen** (soil-indicating plants).

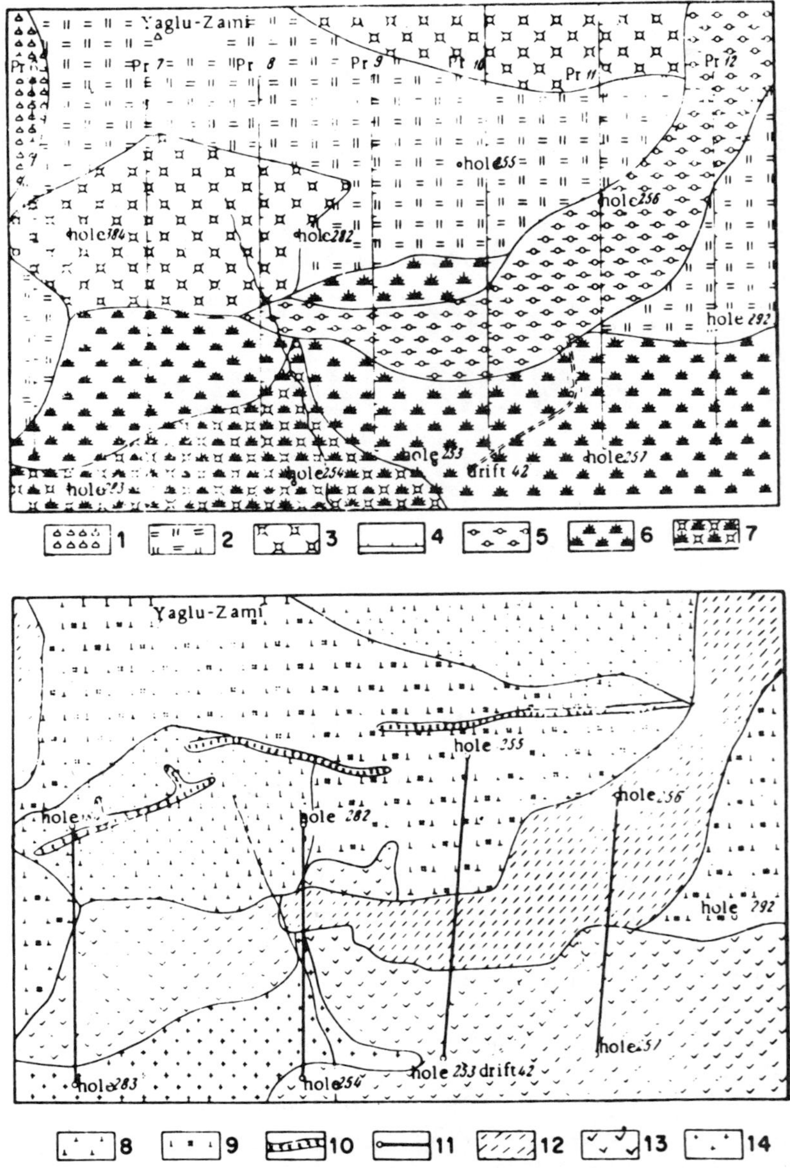

FIGURE 3.1. Geobotanical and geological maps of the Karmir-Karsky area of the Soviet Union. Upper map: 1—*Silene compacta* association; 2—grass association; 3—legume and grass association; 4—biogeochemical survey profiles; 5—*Lapsana communis* association; 6—thyme and tragacanth association; 7—thyme and tragacanth association with legumes and various grasses. Lower map: 8—weakly modified porphyrites; 9—strongly modified porphyrites; 10—granodiorite-porphyry dikes; 11—geological profiles between holes; 12—hornfels; 13—unmodified monzonites; 14—syenites. *Source:* Malyuga [563]. Copyright Plenum Press 1964. Reprinted by kind permission.

A distinction must be made between the classical geobotanical (plant sociological) studies of the European school and geobotanical studies oriented specifically to mineral exploration. The former are based on the pioneering work of Braun-Blanquet [90] and are hardly ever concerned with mineral exploration *per se,* though they are often carried out over mineralized areas because it is precisely in such places that the most extreme examples of community modification might be expected. Much of this plant sociological work has been summarized by Ernst [287].

There is a very extensive literature on the above type of geobotany but in the case of work oriented specifically to mineral exploration, the literature [211] is much more sparse. The relationship between plants and bedrock (mineralized or otherwise) is affected by a number of factors including climate, geomorphology, and edaphic variables such as soil pH and composition. It is obvious that a favorable climate will tend to nullify the adverse effects of toxic elements in the substrate. For this reason geobotany tends to be a more powerful tool in semiarid areas or in Mediterranean-type climates where there is a prolonged dry summer so that the elemental content of the soil moisture gradually increases to an unsupportable level. A further advantage of working in semiarid areas is that plant mapping is so much simpler in a region of, say, 10% vegetation cover than in thick bush or tropical forest.

The science of indicator geobotany as applied to mineral exploration has been stimulated by two very significant developments during the last decade or so. The first of these is satellite imagery (see Chapter 6) and the second is the development of sophisticated computer-assisted techniques (see Chapter 19) which allow for thorough statistical treatment of the raw data in order to evaluate very subtle variations in the composition of the plant community.

The most obvious advantage of geobotany is in regions where there is no surface expression of mineralization and where mineralization at depth is marked by a geochemical barrier such as calcrete. Characteristic plant communities may be small and dispersed or may cover very large areas. It has been reported [211] that such communities cover one of the largest iron ore deposits in the world and delineate the extensive phosphate deposits of Queensland [210].

An illustration of the potentially strong association of vegetation and geology is shown in Figure 3.1, which represents the Karmir-Karsky area near Kadzharan in Soviet Armenia [564] and it can readily be seen that there is a striking correlation between these two variables.

3.2. MAPPING TECHNIQUES IN INDICATOR GEOBOTANY

In many cases the distinction between two different plant communities is so pronounced that a mere visual observation is all that is needed to observe the geological boundaries. This is particularly the case for serpentine floras.

Even when the rock types are relatively similar, pronounced differences in vegetation can occur. This has been demonstrated [67] for vegetation growing over altered and nonaltered andesite in Nevada.

In cases where superficial observation is not sufficient, recourse must be made to geobotanical mapping. Such an operation should be ideally carried out by a skilled botanist or ecologist since the amateur can readily confuse plants which are superficially similar and might have difficulty in distinguishing different ecotypes (even if they were indeed distinguishable).

Plant mapping involves the selection of a number of sample plots known as **quadrats.** There is no general agreement on the best method of selection of these quadrats. Some workers believe that they should be selected in a random manner, whereas others believe that they should be chosen subjectively. This latter procedure, though frequently used, is open to criticism insofar as it makes the assumption that the vegetation associations are already known.

If the subjective approach is to be used, the following procedure should be adopted. Every plot should have the utmost uniformity it is possible to find in the area concerned; not only with regard to plant species, but also with consideration of such factors as aspect, slope, drainage, relief, and altitude. Particular care must be taken that the quadrats do not include two or more different associations as may occur with plots of nonuniform slope.

The size of the quadrat now has to be established. As a general rule, the size should be the minimum needed to include most of the plants of the association and will obviously be related to the homogeneity of the community. In assessing the size of the **minimal area,** the law of diminishing returns will obviously apply; that is, successive increases in size of the sample area will give successively smaller amounts of additional information. The concept of minimal area has been reviewed by Goodall [323]. A species-area plot at first rises sharply and then becomes flatter although never completely horizontal because the whole area would have to be included in the quadrat to be sure of including every single species in the area. Greig-Smith [328] has discussed the problem of minimal area at some length, and there appears to be no general agreement on a universal criterion to determine this area. As a general rule, however, the following procedure may be adopted in the field.

Begin with a small quadrat of perhaps 5 sq. m. note the species within it and then increase the size of the plot progressively (10, 50, 100 sq. m, etc.) noting at each stage additional species encountered. When there is an appreciable drop in the rate of increase of new species found, the optimum quadrat size will have been found.

When the size and position of the test plots have been established, the next procedure will involve an evaluation of the **density** of individual species and their **spacing** (reciprocal of density). In determining density, direct counting or a scale of numbers may be used. The scale is somewhat arbitrary but can give good results in the hands of a good field worker. The system is as follows: 1—very rare, 2—rare, 3—infrequent, 4—abundant, 5—very abundant. One disadvantage of this system is that data are heavily dependent on the personal assessment of one individual and are not always comparable with data collected by other workers. The use of this scale is nevertheless justified on the grounds of speed and practicality. It is also a useful system when the vegetation cover is dense so that counting of absolute numbers of individuals would have been impractical. A geobotanical map of an area need contain nothing more than the density or spacing data enumerated above. If other parameters are added, a more meaningful map can be compiled.

The space demand of a species introduces another concept, that of **cover**. It is assumed that the entire shoot system of a plant is projected on the ground and that this area (equal to the area of shade if the sun were directly overhead), represents the cover. Cover is usually expressed as a percentage. The total of all species will often exceed 100% due to overlap.

A quick method of assessing cover involves measuring the total length of interception by plants on line transects. The proportion of the total length of the transect intercepted by a given species is a measure of its cover. The reader is referred to Greig-Smith [328] for further details of these and other means of measurement.

Some mention should also be made of the concept of **layering.** The vegetation may be considered not only laterally but also vertically. In the vertical concept, a number of distinct layers are recognized. These are the tree layer, shrub layer, herb layer and moss layer. Clearly, the denser the upper tree layer, the greater will have to be the tolerance towards reduced light intensity by the lower members of the community. Mosses, as might be expected, will tolerate the least light intensity.

There are a number of other criteria of plant communities which can also be employed in geobotanical mapping. These are however somewhat outside the scope of this book and will be only mentioned briefly.

Sociability expresses the space relationship of individual plants and can be expressed in terms of a simple scale [90] as follows: soc. 1 growing in one place singly; soc. 2 grouped or tufted; soc. 3 in troops small patches or cushions; soc. 4 in small colonies in extensive patches or forming carpets; soc. 5 in great crowds.

Vitality is a measure of how a plant prospers in the community and can be expressed by a number of conventional symbols [90] as follows: ●—well developed and regularly completing life cycle; ◉—strong and increasing but usually not completing life cycle; ◎—feeble but spreading and never completing life cycle; ○—occasionally germinating but not increasing.

Periodicity is a measure of the regularity or absence of rhythmic phenomena in plants such as flowering, fruiting, and so on. A study of this criterion involves continuous and systematic research and is outside the scope of this book.

If mean values for density, spacing, cover, and other parameters are determined for each species and averaged over several quadrats (preferably chosen randomly), it will be possible to characterize the plant associations

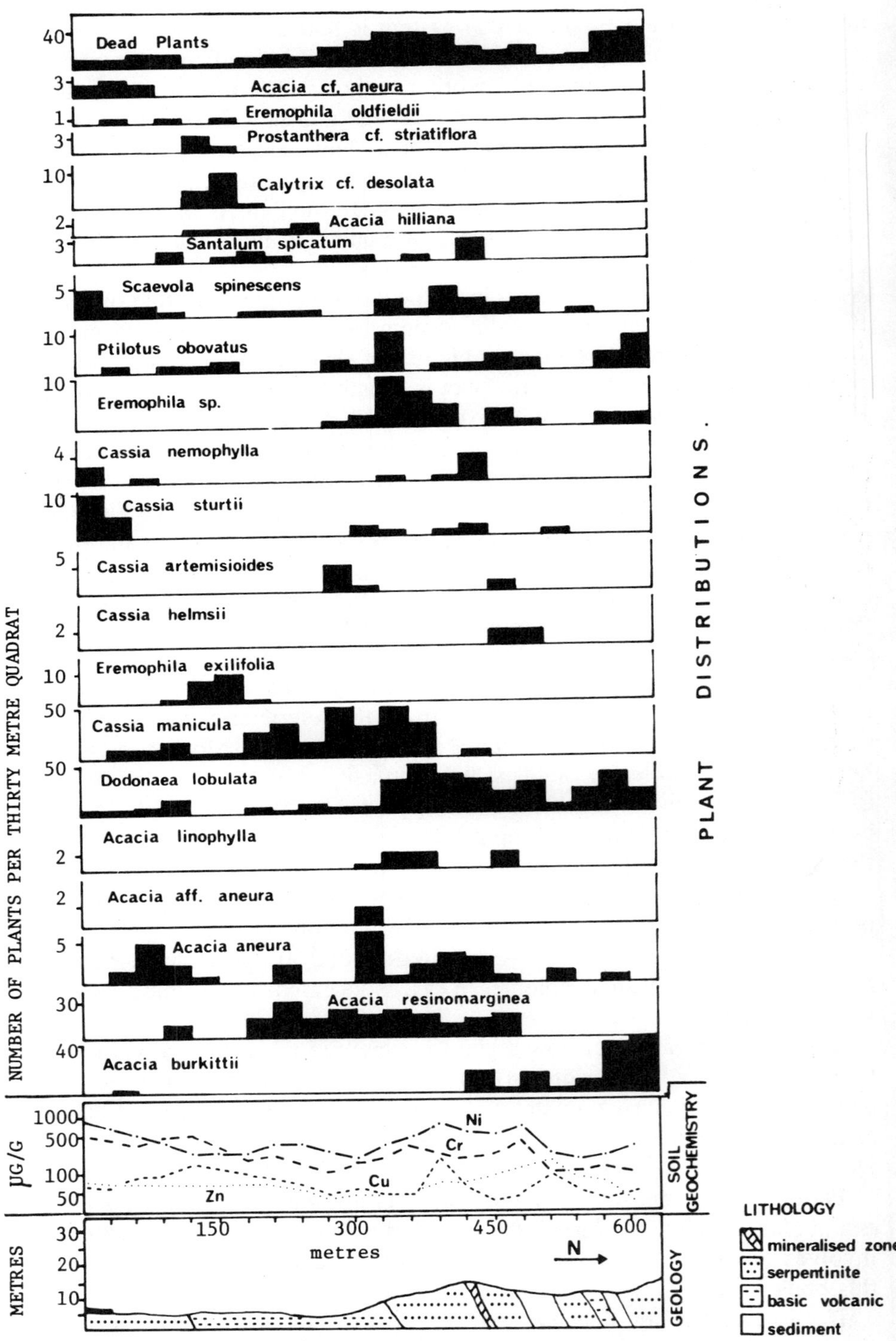

FIGURE 3.2. Belt transect across Marriott Prospect, Mt. Clifford, Leonora, Western Australia. *Source:* Severne [738].

in the test area and to attempt to correlate this with the geological environment.

Although the use of quadrats is the most usual approach in indicator geobotany, a different approach will be needed for plants growing over narrow ore bodies. In such cases, the use of **line transects** or **belt transects** is recommended. Line transects consist of parallel straight lines run through an area with the aid of tape measures and compasses, whereas belt transects consist of lines of continuous quadrats running across the profile of the area. Figure 3.2 shows data for a belt transect undertaken in western Australia.

Representation of plant densities in each quadrat can be made in various ways. A simple method is to record the number of individuals in each 30 m (a common dimension) quadrat or else to record each species as a percentage of the total number of individuals of all species in each quadrat.

As mentioned above, it is much easier to carry out plant mapping in arid areas than in thick bush. In the latter environment, the line transect is usually more appropriate. If a multistorey vegetation assemblage is involved. it is hard to evaluate the extent to which each plant or plant grouping intersects the line. In such cases an alternative method consists of laying a tape measure through the community and recording every individual of a specified size (e.g., trunk diameter) within a predetermined distance of the tape. There are, of course, many variants of this procedure that can be adapted to specific conditions.

Another type of plant mapping is more qualitative. Figure 3.3 represents work carried out on the copper-cobalt deposits of Shaba Province, Zaïre [553]. Representation is exaggerated, both in regard to altitude and the height of vegetation. This form of mapping nevertheless gives a better feeling of reality than does the more quantitative scheme.

In spite of earlier comments about the employment of skilled personnel

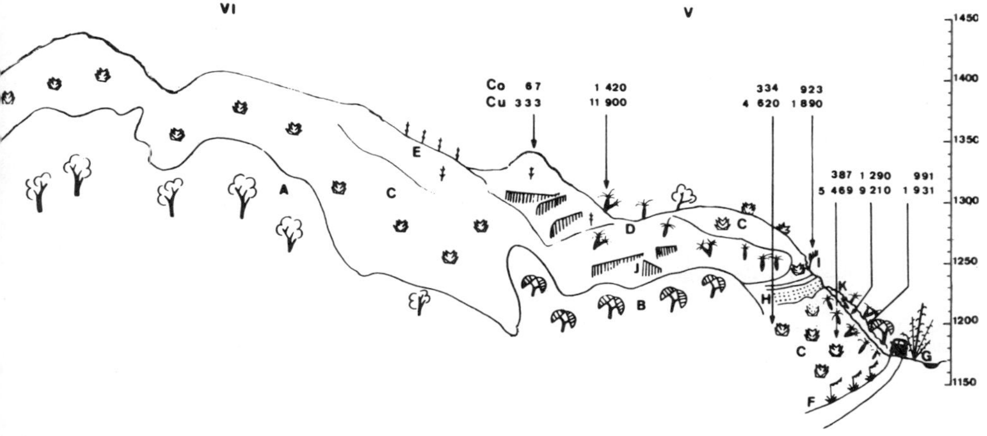

FIGURE 3.3. Representation of the vegetation of mineralized hillocks at Fungurume, Shaba Province, Zaïre. *Source:* Malaisse et al. [552].

TABLE 3.1. Characterization of Substrates by Discriminant Analysis (D^2 statistic) of Geobotanical Data from Spargoville, Western Australia

Species	D^2	Number of Correct Predictions		
		A (26 quad)	UB (11 quad)	A/UB (7 quad)
21	0.5	11	6	0
23	5.7	2	11	1
21, 23	6.2	10	6	1
21–23	9.4	15	3	1
6, 21, 23	7.4	16	7	1
2, 21, 23	8.3	21	7	1
2, 3, 21, 23	9.1	16	8	1
2, 4, 21, 23	12.1	17	8	1
2, 4, 14, 21, 23	21.1	19	9	1
2, 4, 14, 15, 21, 23	29.1	22	10	1
2, 4, 14–16, 21, 23	29.2	22	11	2
2, 4, 14–16, 21, 23, 25	29.4	20	10	2
2, 4, 14–16, 21, 23, 27	33.0	21	10	2
2, 4, 12, 14–16, 21, 23, 27	41.3	22	10	3
2, 4, 14–16, 21, 23, 24, 27	42.5	22	10	4
1, 2, 4, 12, 14–16, 21, 23, 24, 27	48.2	23	10	3
1, 2, 4, 5, 12, 14–16, 21, 23, 24, 27	50.4	24	10	3
1, 2, 4–6, 12, 14–16, 21, 23, 24, 27	51.8	24	10	3
1, 2, 5, 8, 12, 14–16, 21, 23, 24, 27	56.6	24	10	4
1, 2, 4, 5, 7, 8, 12, 14–16, 21, 23, 24, 27	58.7	24	10	4
1, 2, 4–8, 12, 14–17, 21, 23, 24, 27	67.8	24	10	4
1, 2, 4–8, 12, 14–17, 19, 21, 23, 24, 27	76.2	24	10	5
1, 2, 4–8, 12, 14–17, 19, 21, 23, 24, 27, 28	89.8	24	9	4
1, 2, 4–8, 12, 14–17, 19, 21, 23, 24, 27, 29	81.3	23	9	5
1, 2, 4–8, 12, 14–17, 19, 21, 23, 24, 27, 29	100.3	24	10	6
1, 2, 4–8, 12, 14–17, 19, 21, 23, 24, 27, 31	119.4	25	9	6
1–8, 12, 14–17, 19, 21, 23, 24, 27, 31, 32	137.5	25	10	6
1–9, 12, 14–17, 19, 21, 23, 24, 27, 31, 32	138.1	25	10	6
1–8, 10, 12, 14–17, 19, 21, 23, 24, 27, 31, 32	138.2	25	10	6
1–8, 11, 12, 14–17, 19, 21, 23, 24, 27, 31, 32	140.2	24	10	6
1–8, 12–17, 19, 21, 23, 24, 27, 31, 32	147.2	25	10	6
1–8, 12, 14–19, 21, 23, 24, 27, 31, 32	144.3	24	10	6
1–8, 12, 14–19, 21, 23, 24, 26, 27, 31, 32	142.8	24	10	6
1–8, 12, 14–19, 21, 23, 24, 27, 30–32	173.9	26	10	5

Source: Nielsen et al. [620]

KEY:

1—*Acacia acuminata*	12—*D. stenozyga*	23—*E. salubris*
2—*A, colletoides*	13—*Eremophila caerulea*	24—*E. torquata*
3—*Westringia cephalantha*	14—*E. dempsteri*	25—*Exocarpus aphyllus*
4—*Acacia erinacea*	15—*E. ionatha*	26—*Kochia pyramidata*
5—*A. graffiana*	16—*E. oppositifolia*	27—*Melaleuca sheathiana*
6—*Alyxia buxifolia*	17—*E. pachyphylla*	28—*Olearia muelleri*
7—*Cratystylis microphylla*	18—*E. sp. I*	29—*Pittosporum phillyraeoides*
8—*C. subspinescens*	19—*E. sp. II*	30—*Rhagodia sp.*
9—*Dodonea filifolia*	20—*Eucalyptus calycogona*	31—*Santalum spicatum*
10—*D. lobulata*	21—*E. lesouefii*	32—*Scaevola spinescens*
11—*D. microzyga*	22—*E. longicornis*	33—*Trymalium ledifolium*

NOTE:

A—amphibolites
UB—ultrasonic rocks
A/UB—mixed quadrats

for plant mapping, good results can be obtained by intelligent people lacking specialized knowledge. The only attribute required is an ability to recognize and rerecognize individual species. My own procedure is to place into a photograph album a specimen from each new species encountered and to give it a code number. At a later date the samples are sent to an herbarium for identification. In the interim, the code numbers are using in the preliminary graphic representation of the data. To assist in identification, flowers and/or fruits, as well as leaves, are desirable. However, the former may not always be available, and in the case of difficult genera such as the Acacias, flowers or fruits usually are essential. I well remember having had recourse to giving a daily watering to some *Acacia* species in western Australia to produce flowers (often within 10 days) for identification purposes.

In Figure 3.2 the response of specific plants to lithology is sufficiently well pronounced as to be readily appreciated by the naked eye. For example, it is clear that in this area at least, *Acacia burkittii* delineates serpentinite whereas *A. resinomarginea* is confined to gabbro. In some cases, however, statistical treatment of the data can greatly enhance very subtle changes in vegetation. This is illustrated in the following case history from the Spargoville area of central Western Australia, near Kalgoorlie [620]. Plant mapping involved 34 species sampled over 43 quadrats each measuring 30×15 m. Twenty-six of these quadrats were on amphibolites, 11 were on ultrabasic rocks, and 7 were transitional quadrats with elements of both rock types. Using discriminant analysis (see Chapter 19) a numerical score was assigned to each quadrat on the basis of an equation of the form:

$$z = l_1x_1 + l_2x_2 + l_3x_3 + \cdots + l_{34}x_{34} + c$$

where $x_1 - x_{34}$ are the number of individual plants of species 1–34 in each quadrant and $l_1 - l_{34}$ are coefficients so chosen as to maximize the differences between the scores for the three types of quadrat. These coefficients are chosen by computer calculations. C is a constant. Using this form of discrimination it was possible to predict the nature of the geological substrate in 26 out of 26 amphibolite, 10 out of 11 ultrabasic, and 6 out of 7 intermediate quadrats (Table 3.1). This degree of discrimination was achieved by use of only 20 of the 33 species observed. It is obvious that computer-assisted programs of this nature will serve to improve the potential of geobotanical mapping to a high degree.

3.3. CHARACTERISTIC FLORAS

3.3.1. Introduction

In the above generalized examples, it will have been noted that many mineralized areas apparently have a characteristic flora which may be unique to that particular locality or may be characteristic for all discrete areas of mineralization in that region. It is clear that not only mineralization but the rock itself will probably support a characteristic flora which may be readily recognizable to the naked eye, or which might require computer-assisted interpretation of very subtle changes of vegetation.

Most of the well-recognized characteristic floras are a reflection of rock type rather than of mineralization within it, but are nevertheless useful for the geologist in helping to identify the lithology, particularly if outcrops are sparse.

The ecology of a plant community will be greatly influenced by the pH of the soil and by the presence, excess, or deficiency of mineral nutrients. The **availability** of elements is affected by the pH of the substrate as can be seen from Figure 3.4, which represents the relative availability of several elements at various pH values. The pH of the soil can also affect the oxidation state of the elements since electrode potentials for most reactions are pH dependent. For example, in the reaction

$$Fe^{3+} + e = Fe^{2+}$$

the oxidation potential is 0.6 V at pH 4 and is -0.2 V at pH 8. For this reason, iron will tend to be oxidized at high pH values in which state it becomes less available to plants that in the reduced form, which is more stable at a low pH.

It is obvious from the above discussion that the Eh of the soil will also affect availability by controlling the oxidation state of an element at a particular pH value. If certain elements or groups of elements predominate in the substrate, they can affect the ecological balance either directly by their toxic effects or indirectly by their antagonistic effects on other ele-

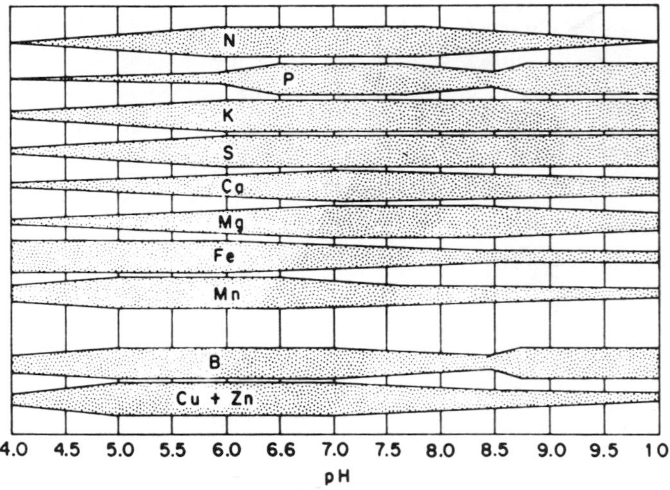

FIGURE 3.4.   Effect of pH on the availability of plant nutrients. *Source:* Ignatieff [376].

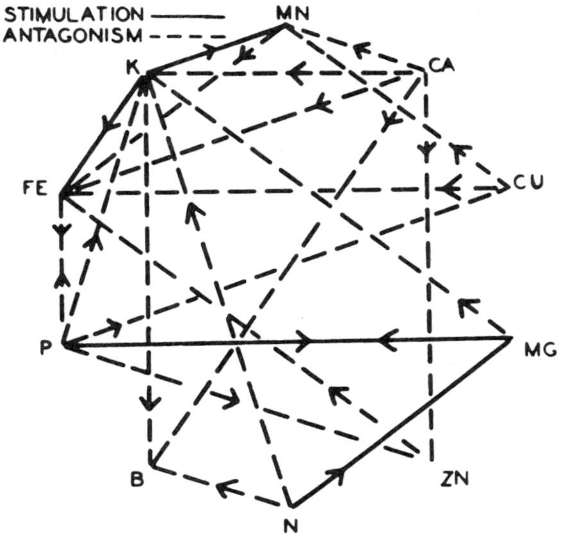

FIGURE 3.5. Stimulatory and antagonistic effects of various pairs of elements on their absorption by plants. *Source:* Mulder [608].

ments. At certain concentrations, these elements will be toxic to most plants, and only those species able to adapt to the hostile environment will survive. Figure 3.5 shows the antagonistic and stimulatory effects of various ions. For example, high copper or zinc levels in the soil should depress iron uptake and produce chlorosis in the vegetation.

The influence of nutrients on the ecology of plants is neatly summarized by the *Law of Relativity* [531] which may be expressed as follows: "The more nearly a factor is in minimum in relation to other factors acting upon the organism, the greater is the relative influence of a change of that factor upon the growth of the organism." In other words, as a factor increases in intensity, its relative effect on the organism decreases, and when the factor is in the region of its maximum, the effect of change on that organism is at a minimum.

It would be possible, no doubt, to find characteristic floras for nearly all types of substrate, but certain ecological groupings are well documented and will each be considered in turn in an attempt to present the basic factors influencing the development of these particular floras.

3.3.2. Calciphilous (Limestone) Floras

Calcium has a great effect on vegetation. It is not surprising, therefore, that calcareous rocks such as limestone and dolomite usually carry a characteristic flora which often renders the geological boundary easy to delineate.

The effect of calcium on plants depends more on its solubility than on the absolute amounts present. The solubility of limestone in water is about 0.1%, whereas dolomite is about half as soluble. Many of the effects of calcium are indirect. Lime has a favorable effect on the drainage and structure of the soil and is able to coagulate many of the colloidal compounds which make a soil heavy and poorly drained. Limestone soils, therefore, are well aerated and are good conductors of water and heat. The result of this conditioning effect is that many plants that thrive in any type of soil in warm and dry climates tend to confine themselves exclusively to calcareous soils at higher latitudes, since these are the only soils that can provide the conditions they require. A striking example of this is the distribution of the beech *Fagus sylvatica* which thrives everywhere in southern Europe, whereas in England it is found in the natural state only on limestones.

Plant communities that thrive on calcareous soils are said to be calciphilous, whereas those that have a specific requirement for calcium are calcicolous. Characteristic limestone communities therefore contain both types of plant and exclude calcifuge plants (such as the Ericaceae), which require a low pH in order to thrive.

Calciphilous plants have been described for well over a century [190, 271]. Because limestones do not support a peculiar or stunted flora, this poses problems in the precise identification of the substrate. By its very richness, the flora presents a problem in geobotanical mapping. One way around this difficulty is to look for certain species known to be characteristically calciphilous, such as the genera *Dianthus, Fagus, Bromus, Festuca, Linaria,* and so on. Another approach is to compare the richness of a limestone flora with the paucity of that of adjoining substrates.

3.3.3. Halophyte Floras

A characteristic plant association is found over saline soils containing sodium chloride, sodium carbonate, or sodium sulphate. Saline soils are found usually in areas of dry climate or near the sea. The ecology of **halophytes** has been studied by several workers [4, 762]. Extensive discussion of this particular plant association is beyond the scope of this book since much of the work has been carried out on estuarine species such as mangroves. In geobotanical mapping it is the halophytes of arid areas that have received the most attention.

Typical dry-climate halophytes are found in Australia (*Atriplex* spp.), the western United States, and the steppe regions of the Soviet Union. Halophytes usually have a high osmotic pressure in the cells and accumulate relatively high levels of salts. In the Soviet Union these plants have been studied extensively as a guide to subterranean water resources [190]. An interesting subgroup of the halophytes is the selenium floras, which are discussed below.

3.3.4. Selenium Floras

One of the most interesting cases of the use of a characteristic flora in mineral prospecting results from the discovery of the selenium floras of the western United States [52], Colombia, Canada, and Queensland [580]. Selenium communities indicate the presence of selenium in the soil, either because they have a specific requirement for this element or because they can tolerate large concentrations of it. Selenium indicators [822] include species of the genera *Astragalus, Stanleya, Aster,* and *Oryzopsis,* and are found among the characteristic shadscale (*Atriplex confertifolia*) associations. The Australian plants include *Neptunia amplexicaulis* and *Acacia cana.* Although selenium plants contain high concentrations of this element, they are probably capable of growth without it [758]. They are able to substitute selenium for sulphur in their metabolism without adverse effects [758].

Uranium deposits on the Colorado Plateau are primarily carnotite containing appreciable quantities of selenium. The presence of carnotite results in greater availability of selenium to plants. Cannon [154, 160] has been able to use the *Astragalus* species for indirect prospecting for uranium because the plants tend to grow in areas of maximum total or available selenium in the soils. Her classic work represents one of the most successful known applications of the geobotanical method, and will be discussed further in the next chapter.

Some of the elements of a selenium flora are able to absorb huge quantities of this element (up to 1% of the dry weight of some *Astragalus* species) so that the garliclike odor of volatile seleniferous compounds can often be detected in the plants themselves, sometimes even from a fast-moving vehicle. This has some connotations for the field of geozoology, since dogs have been trained to detect the faint odor of sulphide deposits (see Chapter 9), and it should be a much easier task for them to detect the very much stronger odor of aromatic plants.

3.3.5. Serpentine Floras

Of all morphological changes produced in vegetation by the substrate, those found in serpentine floras are certainly the most extreme. So great is the differentiation between serpentine floras and those of adjacent substrates that geological boundaries are readily observable. Plate 3.1 shows the boundary between serpentine and normal vegetation in the Dun Mountain serpentine occurrence of Nelson Province, New Zealand. The differences is very striking.

Extensive studies of serpentine floras have been carried out in several countries, for example, Finland [524], Italy [842], New Caledonia [391, 503], New Zealand [502, 535], Poland [723], Portugal [584, 652], Spain [322],

PLATE 3.1.   View of the Dun Mt. serpentine belt, Nelson, New Zealand. The sharp vegetation change between the serpentine and adjacent sedimentary formation is very obvious.

Soviet Union [377], Sweden [710], United States [487, 684, 700], and Zimbabwe [931]. Typical examples of this type of flora show a general sparseness of vegetation with a shortage of species as well as individuals. There are usually a few species that are endemic to a particular area, such as *Myosotis monroi* and *Pimelea suteri* in the Dun Mountain area of New Zealand.

Serpentine soils are considerably different from normal soils in that they are very rich in chromium, cobalt, iron, magnesium, and nickel, as well as being deficient in the nutrients calcium, molybdenum, nitrogen, phosphorus, and potassium.

Lounamaa [524] and Robinson et al. [700] concluded that the unusual serpentine flora results from excessive amounts of chromium, cobalt, and nickel. Other workers [593, 710] claimed that nickel was particularly responsible for the specialized vegetation. The low calcium content of serpentine soils led Kruckeberg [486], Walker [871], and Walker et al. [872] to conclude that the flora of these soils is unusually tolerant of low calcium/ magnesium ratios in the substrate. Yet another theory [483, 640, 723] is that survival of plants on serpentine soils depends on their ability to adapt, at least partially, to all the adverse edaphic factors of these soils. Extensive investigations on a New Zealand serpentine flora [502, 533, 535] have led to the conclusion that some plants have an ability to extract calcium to a greater extent than do other nonserpentine plants, and in this way counteract to some extent the deficiency of calcium in ultrabasic soils. It was also observed that serpentine-endemic species often have much higher nickel contents than do accompanying plants that are nonendemic. For example, the endemic *Pimelea suteri* contains over 400 µg/g nickel in its dried leaves [503].

Lee et al. [504] collected soil from the base of two serpentine-endemic plants of New Zealand (*P. suteri* and *Myosotis monroi*) and showed that nickel levels in these soils were substantially higher (3010 µg/g) than in those of other soils from this area selected on a random basis (<2000 µg/g). It may well be, therefore, that serpentine-endemic plants prefer such sites where, presumably, competition from other species is less severe.

Perhaps one of the most complete discussions of the serpentine problem, and easily the most comprehensive survey of the vegetation of a single ultrabasic region, is a study by Jaffré [391] of the serpentine flora of New Caledonia.

It is clear that the serpentine problem is far from being solved, and much work remains to be done before it can be established with certainty why this peculiar plant community is to be found over ultramafic rocks. It is even possible that the main factors influencing the development of a serpentine flora vary in different areas and that no universal factors exist.

3.3.6. Zinc (Galmei) Floras

Plant communities growing over soils high in copper, lead, or zinc have a certain similarity with serpentine floras: plant growth is retarded and stunted, broadleaf plants are absent, and endemic forms are often found if the area of mineralization is large enough [49, 704, 730, 732]. Some species such as *Agrostis tenuis, Silene cucubalus,* and *Campanula rotundifolia* produce ecotypes morphologically indistinguishable from those growing over serpentine soils [89, 663, 731].

In many cases it cannot be established whether or not the characteristic flora is due to one or all of the three base metals mentioned above, since all three usually are found together in areas of sulphide mineralization. There has been a tendency to classify such communities as zinc or galmei floras, since zinc is usually the main constituent.

The true zinc floras are found in western Germany and eastern Belgium, where the soils are rich in zinc and do not contain inordinately high levels

TABLE 3.2. Plant Communities on Mineralized Hillocks at Fungurume, Zaïre

Community	Dominant Species	Associated Species
A—open forest (up to 280 μg/g Cu in soil)	*Brachystegia bussei*	*Albizzia adianthifolia* *A. antunesiana* *Brachystegia spiciformis* *Cussonia arborea* *Dalbergia boehmii* *Diplorynchus condylocarpon* *Kirkia acuminata* *Ochna schweinfurthiana* *Pseudoachnostylis maprounei-folia* *Sterculia quinqueloba* *Steganotaenia araliacea* *Psychotria* sp. *Vitex madiensis* s. sp. *milanjensis*
B—steep thickets (250 μg/g Co and 350 μg/g Cu in soil)	*Uapaca robunsii*	*Loudetia superba* *Phragmanthera rufescens* var. *cornetii*
C—steppe-savanna (up to 350 μg/g Co and 5000 μg/g Cu in soil)		*Haumaniastrum rosulatum* *Loudetia simplex* *Polygala petitiana* *P. usafuensis*
D—Velloziaceae (up to 1500 μg/g Co and 12000 μg/g Cu in soil)	*Xerophyta retinervis* var. *equisetoides* *X. demeesmaekeriana*	*Crassula alba* *Lapeyrousiana erthyranthra* var. *welwitschii* *Moreae carsonii* *Pandiaka metallorum* *Spuriodaucus marthozianus*
E—Commelinaceae and Convolvulaceae steppe (700 μg/g Co and 25000 μg/g Cu in soil)	*Commelina zigzag* *Ipomoea alpina*	*Alectra sessiliflora* *Anisopappus davyi* *Bulbostylis abortiva* *B. mucronata* *Crotalaria cornetii* *Cryptosepalum dasycladum* *Cyanotis longifolia* *Hibiscus rhodanthus* *Sopubia drageana*
F—*Rendlia cupricola* swards	*Rendlia cupricola*	
G—*Oxytenanthera abyssinica* thickets	*Oxytenanthera abyssinica*	*Parkia filicoides*
H—*Haumaniastrum*	*Haumaniastrum robertii*	*Bulbostylis abortiva* *Eragrostis boehmii* *Rendlia cupricola*
I—loose thickets of *Euphorbia ingens*	*Euphorbia ingens* *Sarcostemma viminale*	*Annona senegalensis* *Cussonia arborea* *Dioscorea bulbifera* *Selaginella abyssinica* *Steganotaenia araliaceae* *Tacca leontopetaloides*

Community	Dominant Species	Associated Species
J—denuded vertical faces		
K—vegetation of outcrops of cellular siliceous rocks	Various	*Aeolanthus rosulifolius* *A. saxatile* *Aneimeia angolensis* *Anthoceros punctatus* *A. mandoni* *Euphorbia fanshawei* *Faroa acaulis* *Fossombronia* sp. *Gongylanthus ericetorum* *Mohria caffrorum* *Monadenium* sp. *Pellaea pectiniformis* *P. goudotii* *Plagiochasma eximium* *Riccia* sp. *Targionia hypophylla*

Source: Malaisse et al [553]

of copper or lead. *Galmei* floras have been known for well over a century, and early miners were guided to ore deposits by members of the galmei community such as *Viola calaminaria*. The capacity of zinc floras to accumulate this element is quite remarkable. Linstow [515] reports that *Thlaspi calaminare* contains ten times as much of this element in the leaves as in the roots. Recent studies [682] have shown that at least ten other species of *Thlaspi* contain comparable amount of zinc.

An excellent bibliography of zinc and other heavy metal floras has been compiled by Ernst [283] who lists 96 references. Further references are listed in his book on heavy-metal vegetation [287].

3.3.7. Other Characteristic Floras

A full discussion of other characteristic floras is beyond the scope of this work. For further information the reader is referred to Adriani [4]. In many instances evidence for some characteristic floras is sparse and based on the distribution of so few species that plants may be considered better as individual indicators (see Chapter 4) rather than as part of a specific ecological community. The communities of most interest to the exploration geochemist are usually so small and so widely dispersed that they often cannot be called characteristic floras in the true sense of the word. The reason for this is quite simple. For a characteristic flora to evolve depends largely on the two factors of time and space. Millions of years might be required for a plant community to evolve, and this is a requirement that usually can be satisfied. On the other hand, many, if not the majority, of mineralized areas usually have a surface expression of only a few hectares, so that the minimum area for community evolution is not achieved. However, there is one outstanding exception to this limitation of space. The copper-cobalt deposits of Zaïre/Zambia are dispersed in a total area of at least 30,000 sq. km and form one of the world's greatest metallogenic provinces. The individual mineralized areas (over 100 of them) often cover several square kilometers in some 10% of the total area of the province. The first detailed studies on this copper-cobalt flora of this region were by Robyns [704] and by Duvigneaud and his co-workers [253–255]. Later studies were carried out by Malaisse and others [551, 553]. A typical copper-cobalt flora from mineralized hillocks in the Fungurume area of Shaba Province, Zaïre, is described in Table 3.2. Although of far smaller extent than the deposits of Zaïre/Zambia, the copper deposits of Zimbabwe are sufficiently large to support a characteristic flora, which has been described by Wild [929].

3.3.8. Some Case Histories of Geobotanical Exploration

In spite of the very strong emphasis on botanical methods of exploration in the Soviet Union, geobotany in mineral exploration is less important than the biogeochemical method in that country. This is partly because so much of Soviet geobotany is concerned with identification of soil type and detection of subterranean water [190].

Table 3.3 gives examples of a number of geobotanical surveys carried out in various parts of the world. It does not include the classical European plant sociological work [287] since this was largely retrospective; that is, studies were carried out in the twentieth century on deposits that had been known in many cases since the Middle Ages. Furthermore, the data obtained are not likely to be of use elsewhere, because Europe has already been explored very thoroughly, and there cannot be many more ore deposits to be identified in that continent, except perhaps in Fennoscandia. Most of the Western work has been carried out in semiarid steppe or savanna where geobotanical anomalies are perhaps more pronounced because of the additional stress caused by aridity. These studies are of three types: retrospective studies, which are of potential value by determining mineral-indicating communities in one area of known ore occurrences and then applying the information for prospecting in adjacent unknown areas; detection of anomalous vegetation patterns over previously unknown mineralization; studies not specifically oriented to mineral exploration. Most of the studies reported in Table 3.3 are of the first kind. However, Cole and Le Roex [214] were able to detect copper mineralization beneath an overburden of sand and calcrete in the Witvlei area of South West Africa. The dominant species in

TABLE 3.3. Examples of the Use of Plant Communities in Geobotanical Exploration in Various Countries

Country	Elements Sought	Major Community Species	References
Australia	Copper, lead, zinc	*Polycarpaea glabra, Eriachne mucronata*	619
	Lead, zinc	*Polycarpaea synandra* var. *gracilis, Tephrosia* sp.	215
	Iron	*Acacia patens*	205
	Nickel	*Hybanthus floribundus*	209, 738, 740
		Acacia burkittii	738
Botswana	Copper	*Ecbolium lugardae*	214
New Caledonia	Nickel	*Phyllanthus serpentinus*	
		Homalium kanaliense, Hybanthus austrocaledonicus	391
Papua–New Guinea	Copper	*Albizzia* sp.	211
South West Africa	Copper	*Helichrysum leptolepis*	214
United States	Selenium, uranium	*Astragalus preussi*	154
Zaïre	Cobalt	*Crotalaria cobalticola, Silene cobalticola*	254
	Copper, cobalt	*Haumaniastrum robertii, Haumaniastrum katangense*	553
Zimbabwe	Copper	*Celosia trigyna, Becium homblei*	929
	Nickel	*Dicoma niccolifera*	931

this vegetation assemblage was the Composite *Helichrysum leptolepis*. Similarly, the same authors were able to predict the existence of copper mineralization beneath as much as 30 m of calcrete in Botswana, using an anomalous shrub layer dominated by *Ecbolium lugardae*.

Wild [929, 931] has worked extensively on heavy-metal plant communities controlled by chromium, copper, and nickel in mineralized areas in Zimbabwe. His work was not, however, oriented specifically to mineral exploration.

The extensive studies over the copper-cobalt deposits of Zaïre/Zambia [253–255, 551, 553] were largely retrospective in nature, since earlier workers had already used the well-known "copper flowers" (variously ascribed to *Becium homblei, Haumaniastrum katangese,* and *H. robertii*) to delineate mineralization. It is in this region that heavy-metal plant communities have their greatest concentration, multiplicity and diversity. The same area probably also has the greatest potential for future discoveries.

Extensive geobotanical investigations have been carried out by Jaffré and his co-workers [109, 110, 124, 391–394, 421, 422, 505] in New Caledonia. Plant communities indicative of nickel, chromium, cobalt, and manganese were identified and studied. New Caledonia is unique in possessing perhaps the world's richest and most diverse serpentine flora which includes a disproportionate number of hyperaccumulators (Chapter 21) of nickel and other elements.

Outside of Africa, perhaps the most extensive and successful geobotanical investigations of plant communities are those which have been carried out in Australia. Much of this work has been retrospective over previously known mineralization determined by other exploration methods, but so many useful data have now been accumulated, that there is a good chance that new discoveries will be made in virgin territory. It is probable that the most significant future progress in geobotany will be in the field of remote sensing of the environment (see Chapter 6).

4

INDICATOR PLANTS

4.1. INTRODUCTION

Indicator plants are individual species which are confined to a specific geological substrate which may be a given rock type or, more rarely, a specific type of mineralization. It has been known for many centuries that occurrence of certain plant species in a given area can indicate the existence of mineralization in the substrate. For example, the "kisplante" or pyrite plant (*Lychnis alpina*) was used by Scandinavian miners in the seventeenth century in the search for copper. Plants whose presence indicates a certain type of mineralization, rock type, or special condition in the substrate are known as **universal** or **local indicators.** Local indicators, as the name suggests, are purely local in their ability to indicate mineralization, whereas universal indicators may be used wherever they are found. When a plant gives a direct response to the element of interest, it is said to be a **primary indicator,** whereas if the response is indirect via another element geochemically associated with the former, it is said to be a **secondary indicator.**

There is fairly extensive literature on indicator plants and a number of reviews have appeared during the past two decades [97, 102, 119, 156, 161, 162, 563]. One of the most important of these reviews is by Cannon [161] who lists 122 indicators of ore deposits, including nearly all of the indicators given in the classical (pre-twentieth century) literature. A later review [102], however, listed far fewer indicators by eliminating some of the earlier doubtful sources. This same list is included in this chapter and is conservative in its coverage. This chapter also contains several references to indicator plants originally discovered by chemical analysis of herbarium material [see also Chapter 21]. Such analyses form part of the entirely different technique of biogeochemical prospecting but produce data that might be used for later geobotanical investigations.

4.2. THE MYTHOLOGY OF INDICATOR PLANTS

Numerous references to indicator plants have been made during the past 150 years, but the compiler of a table of these plants is immediately confronted with the problem of what to include and what to reject. It is not particularly important to adopt as a criterion the fact that a plant has or has not been used in prospecting, or that the results have or have not been successful, since if a plant is a true indicator of specific mineralization, its potential is always present. A far more serious problem is assessment of the credibility of an author's claims about a given species. Sometimes claims are lost in antiquity and are little more than folklore. Examples of folklore are stories about plant indicators of diamonds and other precious stones [248, 774].

Sometimes a plant can acquire the reputation of being an indicator on very slight evidence, and the myth can be perpetuated by successive reviews which never fail to include the species concerned. A very good example of this is the story of gold accumulation by *Equisetum* species (horsetails), a primitive plant sometimes known as the "scouring rush" because of its rough texture caused by its high silica content. It is also known in Arctic Canada as "goose grass" because it is frequently the food of geese during the spring.

The genus *Equisetum* includes about 25 species found throughout the world except Australasia [223]. Horsetails tends to favor damp areas and are commonly found on disturbed ground such as roadsides and tailing ponds around mining districts [770]. They commonly have very deep root systems; Malyuga [563] has reported lengths of up to 150 cm in specimens of *E. sylvaticum* from Siberia. The silica content of horsetails ranges up to 42% in the ash [950]. Bateman and Wells [48] found that *E. variegatum* was particularly tolerant of copper and contained up to 296 μg/g of this element in the ash.

Over 40 years ago, *Equisetum* acquired the reputation as an accumulator and indicator of gold as a result of dubious analytical work by Nemec et al. [613]. They reported 610 μg/g gold in the ash of *E. palustre* growing in an auriferous area at Oslany, Slovakia, in which the substrate contained only 0.1–0.2 μg/g gold. The method of analysis of these authors was essentially as follows: samples were ashed and the ash was dissolved in aqua regia. After boiling until nitrous fumes had disappeared, the samples were evaporated and then boiled with hydrochloric acid to precipitate silica which was filtered off. The dilute acid filtrate was then treated with hydrogen sulphide to precipitate acid-insoluble sulphides which were then assumed by the authors to represent entirely gold sulphide. The precipitates were dried and weighed or, in cases where they were too small, they were precipitated in the presence of pectin to give a colloidal sulphide solution which was determined turbidimetrically and compared with standards. This procedure clearly would not differentiate between gold and other sulphides such as those of copper and arsenic. There can be little doubt that the analytical data were in error and yet the findings were repeated in a whole string of subsequent articles and reviews [63, 135, 156, 530, 563, 804].

The myth has persisted in spite of attempts to disprove it. For example, Warren and Delavault [886] determined 0.17–0.34 µg/g gold in the ash of horsetails from British Columbia and Razin and Rozhkov [680] reported an average of only 0.4 µg/g in the ash of eight specimens of *E. pratense* from the Soviet Union. By far the best investigation of the myth was the work by Cannon et al. [167] who determined gold and 20 other elements in 26 horsetails from the United States. Gold values in the ash ranged from 0.4–1.0 µg/g. They concluded that

> . . . *Equisetum* consistently accumulates zinc in amounts that are greater than in the substrate; commonly these amounts are also greater than those in the plants that grow in the same soil. *Equisetum* is therefore an accumulator of zinc but contrary to other reports is not an accumulator of gold.

It might have been thought that the evidence of Cannon et al. [167] would have finally disproved the myth, yet Brussell [135] claimed to have discovered gold in the emission spectrum of *E. hyemale*. The lines cited were weak lines of gold (443 and 479 nm) and the absence of the strongest gold line at 267 nm would appear to indicate that these results were also in error. In a recent study in Nova Scotia, a number of specimens of *Equisetum* were analyzed for gold, arsenic, and antimony [108]. None of the samples contained measurable gold (i.e., >1 µ/g) but all the specimens collected from auriferous ground had high concentrations of arsenic ranging up to 738 µg/g in dry material. To a lesser extent this was also true for antimony. The evidence appears to indicate that the genus is tolerant of arseniferous ground and is also an arsenic accumulator. Because of the common geochemical association of gold and arsenic it is suggested that analysis of horestails for arsenic may be a useful prospecting tool where geological or other evidence indicates the presence of auriferous ground.

Although there are a number of early records of "lead plants," none of these has had its reputation confirmed by detailed investigations. Lidgey [513] has referred to *Amorpha canescens* said by prospectors in Michigan, Wisconsin, and Illinois to be most abundant in soils overlying galena. However, I have personally analyzed 36 specimens of this taxon supplied by the St. Louis Botanical Garden Herbarium and none of these specimens from widespread localities in the United States had an unusual accumulation of lead. Certainly this work does not exclude the possibility that factors other than ore elements may control the distribution of this species, but some elevated lead levels might have been expected in a plant said to favor ore deposits.

Yet another series of probable myths seems to originate from the work of W. Freise reported by Dorn [248]. The former mentions six indicators of gold in Brazil as well as two of tin. Documentation is poor in all cases and leads the reader to suspect that the findings may be unreliable. It is unwise, however, to assume that all poorly documented early work may be erroneous. For example Henwood [348] in discussing a copper bog in Merioneth, Wales, wrote:

> . . . Persons conversant with the copper turbaries consider the presence of metal in the soil indicated by the growth of the sea pink (Armeria maritima), which appears to flourish there with remarkable luxuriance . . . .

The reputation of *Armeria maritima* as a copper indicator originally was based only on this obscure paper of 1857 in which Henwood himself only quoted the opinion of the local people. Other workers [285, 290], however, have recently studied this plant growing in the selfsame bog and have confirmed its local exclusive occurrence in a copper-impregnated substrate.

4.3. SPECIFIC INDICATOR PLANTS

In Table 4.1 a number of well-known indicator plants have been tabulated and are further discussed below under the individual elements.

4.3.1. Aluminum

Although some reviews list at least three indicator plants of aluminum, there is in fact no evidence whatsoever of an association between these species and aluminum deposits. For example, *Ilex aquifolium* has been suggested as a possible indicator on the basis of a report by Pope Pius II [655] in which he described the discovery of an aluminum deposit in Italy by Giovanni di Castro in 1465:

> . . . per quos dum Innoes ambulat novam herbe faciem offendit: miratur, inquirst; deinde certior fit similem nasci herbam in montibus Asiae, qui Turcarum aerium alumine ditant. . . .

This unidentified plant was later suggested to be *Ilex aquifolium* by local people but was discredited by Targioni-Tozzetti [797] who showed that the holly tree grows equally well off and on aluminum deposits. The role of aluminum indicators has also been ascribed to various species of *Symplocos* which have been reported by Faber [289] as containing up to 7.23% of this element in dried leaves of *S. spicata*. However, there is no evidence that

TABLE 4.1. Plant Indicators of Mineral Deposits

Species	Family	Locality	References
Boron			
Eurotia ceratoides (L)	Chenopodiaceae	USSR	143
Limonium suffruticosum (L)	Plumbaginaceae	USSR	143
Salsola nitraria	Chenopodiaceae	USSR	143
Cobalt			
Crassula alba (L)	Crassulaceae	Zaïre	553
Crotalaria cobalticola (U)	Leguminosae	Zaïre	113, 254
Haumaniastrum robertii (U)	Labiatae	Zaïre	101
Silene cobalticola (U)	Caryophyllaceae	Zaïre	254
Copper			
Acalypha dikuluwensis (U)	Euphorbiaceae	Zaïre	255
Aeolanthus biformifolius (U)	Labiatae	Zaïre	552
Anisopappus hoffmanianus (U)	Compositae	Zaïre	255
Armeria maritima (L)	Plumbaginaceae	Wales	285, 290, 348
Ascolepsis metallorum (U)	Cyperaceae	Zaïre	255
Becium homblei (L)	Labiatae	Zaïre/Zambia	370
B. peschianum (U)	Labiatae	Zaïre	255
Bulbostylis barbata (U)	Cyperaceae	Australia	619
B. burchelli (L)	Cyperaceae	Australia	207
Commelina zigzag (U)	Commelinaceae	Zaïre	255
Crotalaria cobalticola (U)	Leguminosae	Zaïre	254
C. francoisiana (U)	Leguminosae	Zaïre	255
Cyanotis cupricola (U)	Commelinaceae	Zaïre	255
Ecbolium lugardae (L)	Acanthaceae	S. W. Africa	207, 214
Elsholtzia haichowensis (L)	Labiatae	China	735
Eschscholzia mexicana (L)	Papaveraceae	USA	177
Gladiolus actinomorphanthus (U)	Iridaceae	Zaïre	255
G. klattianus s. sp. *angustifolius* (U)	Iridaceae	Zaïre	255
G. peschianus (U)	Iridaceae	Zaïre	255
G. tshombeanus s. sp. *parviflorus* (U)	Iridaceae	Zaïre	255
Gutenbergia cuprophila (U)	Compositae	Zaïre	255
Gypsophila patrinii (L)	Caryophyllaceae	USSR	615
Haumaniastrum katangense (U)	Labiatae	Zaïre	255
H. robertii (U)	Labiatae	Zaïre	101, 255
Helichrysum leptolepis (L)	Compositae	S. W. Africa	207, 215
Impatiens balsamina (L)	Balsaminaceae	India	5
Lindernia damblonii (U)	Scrophulariaceae	Zaïre	255
L. perennis (U)	Scrophulariaceae	Zaïre	255
Lychnis alpina (L)	Caryophyllaceae	Fennoscandia	105, 122
Merceya latifolia (U)	Bryophyta[a]	Worldwide	649
Mielichhoferia mielichhoferi (U)	Bryophyta[a]	Worldwide	649
Minuartia verna (L)	Caryophyllaceae	UK	285
Oligotrichum hercynicum (U)	Bryophyta[a]	Alaska	161

TABLE 4.1. *(Continued)*

Species	Family	Locality	References
Copper			
Pandiaka metallorum (U)	Amaranthaceae	Zaïre	255
Polycarpaea corymbosa (L)	Caryophyllaceae	India	837
P. spirostylis (L)	Caryophyllaceae	Australia	117, 768
Rendlia cupricola (U)	Gramineae	Zaïre	255
Sopubia metallorum (U)	Scrophulariaceae	Zaïre	255
S. neptunii (U)	Scrophulariaceae	Zaïre	255
Sporobolus stelliger (U)	Gramineae	Zaïre	255
S. deschampsioides (U)	Gramineae	Zaïre	255
Tephrosia s. nov. (L)	Gramineae	Queensland	619
Vernonia cinerea (L)	Compositae	India	837
V. ledocteanus (U)	Compositae	Zaïre	255
Iron			
Acacia patens (L)	Leguminosae	Australia	205
Burtonia polyzyga (L)	Leguminosae	Australia	205
Calythrix longiflora (L)	Myrtaceae	Australia	205
Chenopodium rhadinostachyum (L)	Chenopodiaceae	Australia	205
Eriachne dominii (L)	Gramineae	Australia	205
Goodenia scaevolina (L)	Goodeniaceae	Australia	205
Lead			
Alyssum wulfenianum (U)	Cruciferae	Austria/Italy[b]	
Thlaspi rotundifolium var. *cepaeifolium* (U)	Cruciferae	Austria/Italy[b]	
Manganese			
Crotalaria florida var. *congolensis* (L)	Leguminosae	Zaïre	254
Maytenus bureauvianus (L)	Celastraceae	New Caledonia	389
Nickel			
Alyssum bertolonii (U)	Cruciferae	Italy	593
A. pintodasilvae (U)	Cruciferae	Portugal	584
A. spp. of Section Odontarrhena (U)	Cruciferae	S. Europe	115, 118
Hybanthus austrocaledonicus (U)	Violaceae	New Caledonia	109
H. floribundus (L)	Violaceae	Australia	209, 740
Lychnis alpina var. *serpentinicola* (L)	Caryophyllaceae	Fennoscandia	710
Selenium and Uranium			
Aster venusta (L)	Compositae	USA	154
Astragalus albulus (L)	Leguminosae	USA	154

TABLE 4.1. *(Continued)*

Species	Family	Locality	References
Selenium and Uranium			
A. *argillosus* (L)	Leguminosae	USA	154
A. *confertiflorus* (L)	Leguminosae	USA	154
A. *pattersoni* (U)	Leguminosae	USA	154
A. *preussi* (U)	Leguminosae	USA	154
A. *thompsonae*	Leguminosae	USA	154
Zinc			
Armeria halleri (L)	Plumbaginaceae	Pyrenees	636
Hutchinsia alpina (L)	Cruciferae	Pyrenees	636
Minuartia verna (L)	Caryophyllaceae	W. Europe	284
Thlaspi calaminare (U)	Cruciferae	W. Europe	284
Thlaspi ssp. (U)	Cruciferae	S. Europe	682
Viola calaminaria (U)	Violaceae	W. Europe	284

Key: L—local indicator.

U—universal indicator.

[a]Phylum

[b]R. D. Reeves and R. R. Brooks. *Environ. Pollut.* (in press).

any species of this genus has a distribution influenced by aluminum. A monumental survey of aluminum in the plant world was carried out by Chenery [185–188] who analyzed this element in several thousand herbarium specimens including every one of the families of dicotyledons. A good review of aluminum in plants has been furnished by Hutchinson [374].

4.3.2. Boron

Three plant indicators of boron are listed by Buyalov and Shvyryayeva [143]. As these have been used for prospecting, there seems little reason to doubt their authenticity. However, they are halophytes indicating a type of soil in which boron is found along with many other salts; they are therefore probably indirect or secondary indicators of boron.

4.3.3. Cobalt

All known indicators of cobalt are restricted to the copper-cobalt mineralized belt of Zaïre/Zambia, known as the Shaban Copper Arc in Zaïre and as the Copper Belt in Zambia. The problem with these indicators is that copper and cobalt invariably occur together in these deposits, though in varying proportions, and it is therfore difficult to decide to which element tolerance has been established. However, Brooks [101] has found inordinately high cobalt levels (up to 1% in dry leaves) in *Haumaniastrum robertii,* and Duvigneaud [254] has reported high concentrations in *Crotalaria co-balticola* and *Silene cobalticola*. If excessive uptake of an element indicates the tolerance of a plant towards that element, then all three species are clearly indicators of cobalt. With a very few exceptions, all cobalt indicators are endemic to cobalt deposits. There is a large number of other taxa confined to the copper-cobalt deposits [253–255, 553) but it is difficult to decide whether they indicate either or both of copper and cobalt. *Nyssa sylvatica* has been alleged to be an indicator of cobalt, and certainly my own studies [114] have indicated that all species of this genus are able to accumulate cobalt. As these taxa have never been used for mineral prospecting, however, they are not included in Table 4.1.

4.3.4. Copper

There are probably more indicators for copper than for any other element and the reputation of many of these indicators has been known for well over a century. The "kisplante" (*Lychnis alpina*; see Chapter 3) is a good example of this. *Armeria maritima,* which has already been mentioned [348], forms part of a heavy-metal community including *Minuartia verna* and *Agrostis tenuis* [285, 290]. *M. verna* is also found over old lead workings in Wales and Northern England, though it is not clear whether it indicates copper, lead, or zinc, or any combination of these three elements.

One of the best known of the copper indicators is the Australian *Polycarpaea spirostylis*. Its reputation as an indicator was first established by Skertchly [768] and later confirmed by other studies [117, 215, 224, 334, 619]. Brooks and Radford [117] analyzed 183 specimens of 12 species of *Polycarpaea* from Australian herbaria and noted over a dozen collection localities with anomalous levels of copper and/or zinc in the plants.

Easily the greatest concentration of copper indicators is found on the copper-cobalt deposits of Central Africa. Copper flowers extend as far south as Zimbabwe. The best researched of all of these copper flowers is *Becium homblei* [370]. Although this species is to be found on nonmineralized terrain, it is often found as the sole colonizer of soils with anomalous concentrations of copper. At least 26 other copper flowers have been reported from Central Africa. Although it is by no means certain whether copper and/or cobalt are indicated by these plants, the matter is somewhat academic because of the coexistence of both copper and cobalt in these deposits. The copper flowers are so numerous that it is possible to determine not only the presence of mineralization but also its degree [553].

An interesting sidelight is the remarkable ability of one of the copper flowers (*Haumaniastrum katangense*) to indicate archaeological sites. This is perhaps the first example of a new science which might be termed **phyto-archaeology.** It has been shown [241] that this species has escaped from its original habitat on cupriferous outcrops of the Shaban Copper Arc and has been able to colonize the base of ancient termite hills used as sites of precolonial smelting operations. Copper ores brought to the sites have formed artificial deposits favorable for such colonization. Excavation of these sites has revealed numerous artifacts of the fourteenth-century Kabambian culture, such as ceramics and copper crosses used as currency at that time.

In an investigation of herbarium specimens of the kisplanten *Lychnis alpina* and *Silene dioica,* Brooks et al. [122] analyzed about 700 specimens for copper, lead, and nickel and showed that most of the major serpentine and copper mining regions of Fennoscandia were reidentified. More locations with anomalous heavy-metal levels in vegetation are at present under investigation.

Another well-documented copper plant is *Gypsophila patrini,* which is found in the Soviet Union. Nesvetaylova [615] studied the distribution of this plant in soils of varying copper concentrations. The species apparently flourishes in soils averaging 0.1% base metals, with copper levels of 0.03–0.1%. It is totally absent at higher concentrations of these elements and has an infrequent distribution in soils with about 40 μg/g base metals (30 μg/g copper). At soil levels of copper below 30 μg/g, the plant is entirely absent.

A striking example of a copper plant is *Eschscholtzia mexicana* [177], which acts as a local indicator of copper in parts of Arizona.

4.3.5. Gold

As far as is known there are no primary indicators of gold. This is hardly surprising as gold concentrations in rocks and their associated soils can never approach levels sufficiently high so to influence the nature of the vegetation. Despite this obvious fact, there have been extravagant claims in the past (see above). Although there are no recorded cases of direct indicators of gold, certain plant communities can sometimes indicate the presence of rock types which can be potentially auriferous. An interesting case of this is recorded from Venezuela where Pagliuchi [635] described a "gold bird" (see Chapter 10) said to feed from the Mora tree which was alleged to colonize auriferous quartz reefs.

4.3.6. Iron

Although there are several classical references to so-called "iron plants" such as *Dammara ovata* [404], there is no evidence that such species do in fact indicate iron *per se*. Many of these "iron plants" are found over serpentinites and their distribution is more likely to be controlled by nickel, chromium, or magnesium than by iron. A well-researched work on iron plants is by Cole [205] who has shown that in Western Australia the Tertiary iron deposits are indicated by a relict Tertiary flora dominated by the species shown in Table 4.1. These species occur in varying combinations according to the degree of weathering and iron content of individual areas. A comparable situation has been discovered over haematite ore bodies in South America.

4.3.7. Lead

Until recently there was no firm evidence for the existence of any lead plant. Certainly there are many species characteristic of copper-lead-zinc mineralization, but it is not known with certainty whether each or all of these elements are instrumental in controlling the vegetation distribution. However, Nicolls et al. [619] have suggested that *Eriachne mucronata* growing over copper-lead-zinc mineralization in Queensland has a high tolerance to lead uptake. Even if some of these plants do indeed indicate lead, they are only local indicators and are not in any way universal indicators. Recently, however, we have observed (Reeves and Brooks, *Environ. Pollut.*, in press) the unusual hyperaccumulation of lead by *Alyssum wulfenianum* and *Thlaspi rotundifolium* s. sp. *cepaeifolium* growing over lead deposits near Raibl on the Italian–Austrian border. The former species accumulates up to 1000 μg/g lead whereas the *Thlaspi* accumulates up to 7000 μg/g (0.7%), a remarkable concentration for an element as toxic as lead. There is no concomitant undue accumulation of copper or zinc that leads us to the conclusion that these two taxa may be the first recorded examples of true indicators of lead.

4.3.8. Manganese

Apart from the usual dubious "manganese plants" of early literature, there is no firm indication of the existence of a true manganophyte with the possible exception of *Crolalaria florida* var. *congolensis*. This plant was found by Duvigneaud [254] in substrates at Kisenga (Zaïre) containing typically 1.6–16.7% manganese. The plant itself was not analyzed by this worker. Jaffré [389] has reported over 3.2% manganese in *Maytenus bureauvianus,* the uptake being independent of pH and probably related to the manganese content of the soil. Similarly, very high levels of manganese have been reported [124] for *Alyxia* species from New Caledonia.

4.3.9. Nickel

The situation of "nickel plants" is somewhat unusual. Strictly speaking, all plants endemic to serpentine substrates are nickel plants insofar as they indicate the presence of nickeliferous rocks. However, their distribution may be governed by magnesium rather than by nickel [504], and as there are many thousands of taxa endemic to ultrabasic substrates it is obviously impracticable to name them all.

Among serpentine endemics are plants with an inordinate capacity to accumulate very large concentrations of nickel. The first of these to be recorded was *Alyssum bertolonii* [593], which contains up to 1% nickel in dry leaves. Brooks [102] has listed 34 hyperaccumulators (> 1000 μg/g nickel in dry matter) and since that earlier review, a further 80 species (mostly in the genus *Alyssum*) have been discovered.

Perhaps the most intensively studied hyperaccumulator of nickel is *Hybanthus floribundus* [209, 738–740]. Hyperaccumulation of nickel in this species was first observed by M. M. Cole and reported in unpublished reports to a mining company. The first report in the open literature was by Severne and Brooks [740]. Cole [209] concluded that its presence was indicative of a nickeliferous environment, especially in regions of low pH and a high concentration of plant-available nickel in the substrate.

In studies on the flora of New Caledonia, which contains many hyperaccumulators of nickel, Lee et al. [505] showed that the relationship between the nickel content of *Hybanthus austrocaledonicus* (typically 1.4% in dry material) and the nickel content of the substrate was highly significant ($P < 0.005$). As the New Caledonian taxon is virtually restricted to nickeliferous substrates, it may be classified as a universal indicator. By contrast, *H. floribundus* is frequently found on non-nickeliferous substrates and is therefore only a local indicator.

Experiments on locations within New Zealand serpentinites which were particularly favored by endemics, showed that *Pimelea suteri* and *Myosotis monroi* were found predominantly in areas of high magnesium levels in soil rather than those of high nickel levels [504]. This element, rather than nickel, may therefore be the controlling factor.

Although Table 4.1 lists only a few hyperaccumulators of nickel in the genus *Alyssum,* it is almost certain that nearly all of the 44 hyperaccumulating *Alyssum* species reported by Brooks et al. [115], will be classified similarly if the necessary field work can be carried out.

Many lists of indicator plants include *Lychnis alpina* (*Viscaria alpina*) as a nickel plant. However, there is no certainty that nickel is indeed the controlling factor. The name kisplante (pyrite plant) implies that sulphide minerals including iron may be involved. All that can be safely said about *L. alpina* is that one form of it (var. *serpentinicola*) is tolerant of nickeliferous substrates [104].

4.3.10. Selenium and Uranium

Selenium indicators that indicate uranium have already been discussed in Chapter 3. Their use resulted in a very successful search for uranium in the Colorado Plateau of the United States. This work was carried out by H. L. Cannon and her co-workers [154]. The list of species in Table 4.1 is taken from Cannon [161].

4.3.11. Silver

There is only one taxon with a reputation for indicating silver. Lidgey [513] stated that ". . . in Montana experienced miners look for silver where *Eriogonum ovalifolium* flourishes . . ." Certainly this species is found over silver deposits in the United States but it is an extremely common plant found over a wide range of substrates. This species has recently been studied more fully by staff of the U.S. Geological Survey [330] who have found that it is the dominant ground cover in a number of barren areas of otherwise forested regions in Montana. However, these areas are mineralized with copper, lead, and zinc, as well as with silver. Since silver concentrations, even in ores, are relatively low, it is unlikely that the distribution of this species is controlled by silver alone.

4.3.12. Zinc

There are numerous plant species with a reputation of indicating zinc deposits. In some cases the claims are dubious because of the frequent association of zinc with copper and lead in sulfide deposits. Nevertheless there is little doubt that *Viola calaminaria* (a form of *V. lutea*) in Western Europe is an indicator of zinc deposits. This species is not only an indicator of zinc deposits but also accumulates significant concentrations of this element. The same is true of *Thlaspi alpestre* s. sp. *calaminare* (sometimes known as *T. calaminare*). Ernst [284] has reported 0.20% and 1.18% zinc, respectively, in *V. calaminaria* and *T. calaminare*. The same author has reported 0.47% zinc in *Minuartia verna* s. sp. *hercynica*.

Hyperaccumulation of zinc in the genus *Thlaspi* is not confined to *T. calaminare*. Reeves and Brooks [682] have reported that nearly all of the European species (27 out of 36) can contain over 1000 μg/g zinc in dry leaves and that a futher 8 of these taxa contain over 1% of this element. The genus also contains many hyperaccumulators of nickel.

4.3.13. Discussion

The list of indicators in Table 4.1 contains over 70 taxa. It is noteworthy that over a third of these belong to the three families Leguminosae, Caryophyllaceae, and Labiatae, which include the most successful copper indicators such as *Becium homblei* and *Polycarpaea spirostylis* as well as the selenium indicators of the genus *Astragalus*. Almost without exception, indicator plants are herbs or shrubs rather than trees. In many cases (e.g., *Polycarpaea spirostylis*) indicator plants tolerate mineralized ground by restricting uptake of the element indicated. In other cases (usually involving nonessential elements such as nickel) indicators appear to achieve their tolerance by accumulating large concentrations of the metal involved (e.g., *Homalium austrocaledonicum* [505].

Even though the plant indicators shown in Table 4.1 represent a selection from a much larger list of reputed indicators, further selection from this table may still be necessary to err on the conservative side. There are very few cases in which even a universal indicator is entirely restricted to a specific mineralization. For example, in the case of *Crotalaria cobalticola* and *Becium homblei,* it was always possible to find at least one isolated population that was not growing over mineralized ground.

In spite of the above reservations, there is little doubt that a thorough knowledge of the existence or potential presence of indicator plants over a given exploration target will do much to assist exploration geochemists in their work.

4.4. CLASSIFICATION OF INDICATOR PLANTS

To classify indicator plants as local or universal is a somewhat simplistic approach. It may well be superficially adequate for mineral exploration, but does not really serve to give the field worker a deep understanding of what indeed is indicated by the species concerned.

Duvigneaud and Denaeyer-De Smet [255] have devised a somewhat complicated system of classification for plants growing over copper-cobalt deposits in Zaïre. The system applies only to that particular area, though Wild [937] has found it generally applicable to Zimbabwe as well. The system is not perfect since there is overlap (i.e., species such as *Becium homblei* can occur in two separate classifications simultaneously) and moreover, because of the almost invariable occurrence of significant amounts of cobalt in all of the copper deposits, many of the taxa may be controlled by cobalt instead of by copper. Table 4.2 gives a representation of this system.

Duvigneaud and Denaeyer-De Smet [255] recognized 26 species or subspecies endemic to the cupriferous soils and 14 species (*oligocuprophytes*) that are almost endemic but which may have ecotypes on normal soils. These 40 species are therefore reliable indicators of copper and/or cobalt in Shaba Province and are of potential use in mineral exploration. A phytosociological classification of Central African metallophytes has been given by Ernst [287]. In an excellent review of heavy-metal plants in Central Africa, Wild [937] mentions the work of Jacobsen [384–386] who attempted to quantify metal tolerance by specific taxa. He used the term specific indicator value (s.i.v.), which was related to the highest and lowest concentrations of copper in which an individual species was found, and to the average copper level in the substrate. The expression is

$$\text{s.i.v.} = \frac{\text{highest copper level—lowest copper level}}{\text{average copper level}}$$

He suggested that good indicators should have an s.i.v. of 4 or less; that is, they are found over a small range of copper values of a high average concentration in the soil. Jacobsen's contribution is an important one since the classification of indicator plants had previously only been semiquantitative at best. The more that is known about a specific plant indicator, the greater will be its value for mineral exploration.

One unfortunate aspect about indicator geobotany is that even when a plant has an excellent reputation as a universal indicator, it always seems

TABLE 4.2. Classification of Vegetation in Zaïre in Relation to Copper Tolerance

Class	Description	Subclass	Description	Typical Species
I Cuprophytes	Plants growing on soils with highest copper values and often restricted to them (eucuprophytes).	1	Polycuprophytes found on soils with 0.5–1.0% copper.	Annual labiates, (*Haumaniastrum*), geophytes (*Icomum*), grasses (*Eragrostis*)
		2	Oligocuprophytes found on soils with 0.08–0.20% copper.	Perennial labiates (*Becium, Triumfetta, Thunbergia*) *Xerophyta, Ipomoea.*
		3	Eurycuprophytes found on wide range of copper levels.	
		4	Local cuprophytes restricted locally to copper soils.	*Becium homblei*
II Cuprophiles	Found on lower copper levels (0.05–0.10%) and not confined to cupriferous soils.			*Becium homblei, Uapaca robynsii, Olax obtusifolia*
III Cuproresistant species.	Ubiquitous species resistant to weak, medium, or strong copper concentrations.	1	Eucuproresistant tolerant to all copper ranges up 2%	*Monocymbidium ceresiiforme, Eragrostis boehmii, Crotalaria cornetii*
		2	Oligocuproresistant species of cupricolous steppe and rocky outcrops. Also adventives on mine waste.	*Andropogon filifolius, Gladiolus robiliartinus, Aeolanthus* spp., *Pennisetum* spp.
IV Cuprifuge	Species that never occur over copper.			*Hyparrhena* spp.

Source: Adapted from Wild [937] and based on data from Duvigneaud and Denaeyer-De Smet [255].

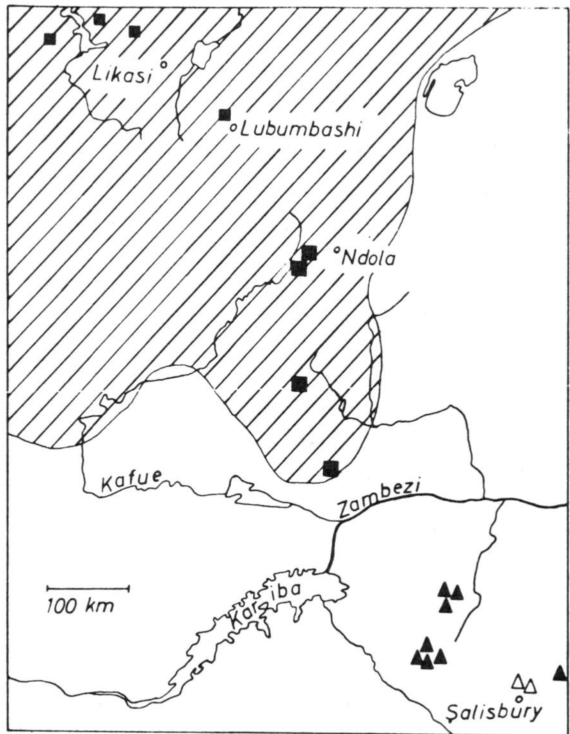

FIGURE 4.1. Distribution of *Becium homblei* in Central Africa. The main distribution is indicated by the shaded area and is in general cupriferous. Collections over mineralized ground are indicated by solid squares or solid triangles. Open triangles indicate a population on nonmineralized ground outside the main area. *Source:* Ernst [287].

to have been possible to find at least one isolated population growing over normal soil. *Becium homblei* already has been mentioned as an example of this. Until the work of Howard-Williams [370], it was thought to be confined exclusively to copper-rich soils. However, this author has now confirmed that the species is widely distributed in noncupriferous soils throughout Zambia and Zaïre, and is even found over barren granites near Salisbury, Zimbabwe (Figure 4.1). However, this does not diminish its great value as a local indicator in many parts of Central Africa.

4.5. THE EVOLUTION OF INDICATOR PLANTS

A common characteristic of many indicator plants is their highly disjunct distribution over widely separated islands of mineralization. It has been argued [729] that such species were once widely distributed before the Fourth Glacial Period (at least in more temperate parts of the globe) and are now glacial relics confined to mineralization where there is little competition from other species. Turrill [825] introduced the concept of *pa-*

laeoendemism and *neoendemism* to account for these disjunct populations. Palaeoendemics are species of formerly wide distribution that are now confined to mineralized islands; neoendemics are often ecotypes of common surrounding species that by natural selection have been able to colonize these mineralized sites. In support of this latter theory it will be noted that these indicator plants usually have a closely related nonmetallophyte that surrounds the mineral occurrence and yet does not colonize it. An example of this is *Becium homblei* and the closely related *B. obovatum*. In studies on *Alyssum,* Brooks et al. [115] found that in some subsections of Section Odontarrhena of this genus, all except one of the species were able to hyperaccumulate nickel (up to 2% in dry material) and had a narrow distribution on specific serpentine occurrences. The remaining member of the subsection such as *A. peltarioides* in Subsection Samarifera was widely distributed throughout the whole area containing the small and disjunct populations of eight different hyperaccumulators.

Available evidence points to the so-called "mine taxa" [27] as being neoendemics rather than palaeoendemics. We must of course differentiate between mining sites and areas of natural mineralization. The former may only be a century or so old, whereas the latter may be millions of years of age. The fact that metal-tolerant plants can evolve on mine sites points to the very great speed at which this evolution can occur. Malaisse and Brooks [549] have suggested a similar neoendemic evolutionary sequence for the Zaïrean plants *Silene burchelli* var. *angustifolia* from cupriferous outcrops and *S. cobalticola* from highly mineralized copper cobalt deposits. The former species is an ecotype of *S. burchelli.*

An interesting question is whether, in fact, an indicator plant has a specific physiological requirement for the element it indicates. All indicator plants appear to be able to grow quite well in normal soil, though it is my experience that they appear to have a poor tolerance to fungal and insect attack under such conditions. These plants, however, are stimulated by addition of small amounts of the relevant metal ion [282]. It has been suggested [27] that the correct interpretation of this is, that because of the plant's efficient tolerance mechanism in deactivating toxic elements, the external trace element requirement is higher than in normal plants.

Extensive further discussion of evolution is beyond the scope of this book. The reader is referred to standard texts [27, 287, 825, 937] for a fuller discussion of this topic.

4.6. MOSSES AND LICHENS AS INDICATORS OF MINERALIZATION

Bryophytes (mosses and liverworts) and lichens have an extraordinary ability to absorb trace elements from the substrate upon which they are found, and will often tolerate adverse edaphic conditions to a much greater degree than will vascular plants. The accumulation of trace elements by bryophytes is usually much greater than for other plant groups, as has been shown by extensive studies on mosses and lichens in Finland [526] and in North America [508,742,745–747]. Specialized types of bryophyte are the so-called copper mosses, which are reputed to grow only on substrates high in copper. The predilection of these plants for copper has been known ever since their discovery in the early nineteenth century, though a fuller report on them did not appear until 100 years later [606].

Copper mosses belong to mainly two genera: *Mielichhoferia* and *Merceya*. Mårtensson [575] and Persson [650] reported copper levels in substrates of *Mielichhoferia elongata* growing in Scandinavia and gave values ranging between 20–450 µg/g with most values in the higher part of the range. It has also been reported that specimens of *Mielichhoferia macrocarpa* and *M. mielichhoferi* in North America [728] grow on substrate rich in pyrite. The genus *Merceya* has been described by Noguchi [627] and Persson [649] and it is believed that all of their six species are cuprophile. There has been some controversy for several years as to whether the response of copper mosses is to copper itself or whether the plants are controlled by pH or the sulphide content of the substrate [724]. A paper by Hartman [342] has acribed all these variables to the ecology of *M. mielichhoferi* in Colorado. The low pH in copper deposits does not necessarily seem to be a significant controlling influence, because Persson [650] has reported a high copper concentration (320 µg/g) in the substrate of *Merceya latifolia* together with a high pH (7.6).

Shacklette [747] has carried out extensive analyses of Alaskan mosses and their substrates as well as substrates of *Mielichhoferia mielichhoferi* from Michigan. He also obtained data for *M. elongata* from Sweden and France. His findings strongly suggest copper as the controlling element and tend to disprove the role of sulphur in affecting the distribution of these bryophytes.

It is clear that it will never be a practical proposition to use lichens or copper mosses directly as prospecting guides because not only are they extremely rare, they are also very difficult to identify by anyone except a skilled bryologist. Persson [949, 650], however, has suggested a novel method of using these plants in mineral exploration. The procedure involves examining species of copper mosses from herbaria, noting the collection locality and then examining this locality by conventional explorations methods. Cannon [156] has stated that three copper deposits were located by this method. For a further discussion of the use of herbarium material in prospecting, the reader is referred to Chapter 21.

Lichens have also been used in prospecting for minerals. Viktorov [848] has reported that *Collema minor* and *Lecidea decipiens* are indicators of gypsum in desert areas of the Soviet Union. Alstrup and Hansen [16] have reported high metal tolerance of the lichens *Alectoria pubescens*, *Umbilicaria lyngei*, and *Lecanora polytropa*. These species were growing on a copper-rich rock on Disko Island, Greenland. The copper content of the

rock was 2%. A study of the relationship between lichens and the substrate was carried out in Northern Canada by Easton [264]. He found that ten rock types carried specific lichen assemblages. Of particular interest was the association of species such as *Lecidia dicksonii* with pyrite and pyrrhotite in gabbroic rocks. Because lichens give a characteristic color to specific rocks, they should form a ready method of identification of such rocks at a distance and might even be visible from aerial photographs or satellite imagery (see Chapter 6). Certainly the work of Easton is one of the most detailed and important studies of this nature ever carried out.

4.7. ASSESSMENT OF SUCCESS IN USE OF INDICATOR PLANTS

It is unfortunately true that much of the geobotanical work carried out by scientists unattached to exploration companies is not followed up by commercial enterprises. Very often such work is undertaken without the cooperation or even knowledge of the exploration companies. Furthermore, the tendency of some companies to restrict publication of data is often a discouragement for many scientists to become involved. In spite of this problem, there are many documented cases where follow-up drilling has justified amply the original expenditure of effort into the geobotanical survey.

One of the great success stories seems just as impressive today as it was 20 years ago. In work carried out in the Yellow Cat area of Utah, Cannon [160] was able to compare data on indicator plants with the results of drilling carried out over the whole of a 15 sq. km area. Her method of assessment of success is presented as an example of a procedure that might be copied advantageously.

Table 4.3 shows the distribution of indicator plants (*Astragalus pattersonii* and *A. preussi*) in the vicinity of drill holes driven into favorable and

TABLE 4.3. Association Selenium-Indicating Plants Plants with Presence or Absence of Uranium Mineralization as Determined by Drill Holes

	Number of Drill Holes	
Nature of Ground	Indicators Present	Indicators Absent
Geologically favorable	311 (206)	497 (602)
Geologically unfavorable	12 (117)	448 (343)
Ore-bearing	51 (17)	30 (64)
Mineralized	99 (46)	117 (170)
Nonmineralized	113 (200)	838 (751)

Source: Based on data from Cannon [160].
Note: Values in parentheses indicate expected distribution if there were no relationship between plants and mineralization.

TABLE 4.4. Comparison of Indicator Plant Distributions with Drilling Results at Various Depths

Depth (m)	Number of Holes	Percentage of Mineralized Holes in which Plants were Present	Percentage of Ore Holes in which Plants were Present	Percentage of Ore Holes in which Plants were Absent
0–3	16	54	60	54
3–7	35	78	33	70
7–10	32	75	100	81
10–16	32	41	60	46
16–22	33	38	56	42
22–33	34	44	44	43
33–38	36	28	78	42
38–50	35	29	81	45
50–56	31	36	41	38
56+	12	22	0	16
Total	296			
Average		46	62	50

Source: Cannon [160].
Note: Of 971 barren holes, only 14% had indicator plants growing near them.

nonfavorable ground. Of the holes driven into ground supporting indicator plants, 97% were later proved to have been driven into ground that was geologically favorable or semifavorable. Indicator plants were clearly extremely useful in locating favorable ground, though not necessarily the mineralization within it.

The next stage in Cannon's work was to establish the effectiveness of the indicators in the geologically favorable ground. The drill holes were classified as ore-bearing (more than 0.1%), mineralized (more than 0.02% uranium), and nonmineralized. The data are shown in Table 4.3. It will be noted that indicator plants were located near 63% of the ore holes, 46% of the mineralized holes, and only 12% of the nonmineralized drillings.

The above evaluation had not considered the depth at which mineralization was found. Clearly there must be some limit to the effective depth of the method in any particular areas. Table 4.4 shows data for the distribution of indicator plants in relation to results of drilling. Indicator plants favored ore holes even when the ore body was as deep as 50 m. The greatest effectiveness was for depths in the 7–10 m range.

Many other methods have been used to assess success in geobotanical surveys. In the Soviet Union, use has been made of the so-called **reliability index** and **validity index** [849], which involve not only a study of the distribution of indicator plants but also their uptake of the element that is sought.

5

MORPHOLOGICAL AND MUTATIONAL CHANGES INDUCED BY MINERALIZATION

5.1. INTRODUCTION

Changes in the morphology of plants and evidence of disease are useful aids in geobotanical prospecting, and have been used as field guides since the eighteenth century. Early workers had to rely on obvious changes such as *dwarfism* or variation in color, but with the increasing sophistication of modern science and greater knowledge of plant physiology, many other visual indications of mineralization have been noted and can be used in prospecting.

Morphological changes in plants under the influence of mineralization are very varied and include such factors as: *dwarfism, gigantism, mottling or chlorosis of leaves, abnormally shaped fruits, changes of color in flowers, disturbances in the rhythm of the flowering period, changes in growth form,* and a large number of other indications. Considerably more skill is required for recognizing morphological changes than is required for studying the distribution of indicator plants (see Chapter 4) unless the changes are very obvious. Usually a trained botanist or plant physiologist should be employed for this type of work. Some common morphological and mutational changes in plants are listed in Table 5.1 and will be considered in more detail below.

TABLE 5.1. Morphological and Mutational Changes in Plants Affected by Mineralization

Element or Mineral	Effect
Aluminum	Shortening of roots and leaf scorch [874].
Bitumen	Gigantism and early or second flowering [867].
Boron	Stunting, prostrate forms, deformation, blotching, and browning of leaves [143,873,940].
Chromium	Chlorosis of leaves [351].
Cobalt	Chlorosis or increase of chlorophyll [254,351].
Copper	Chlorosis of leaves and dwarfism [254,351,384], reduction of size of seeds and corollas [371].
Iron	Darkening of leaves [140].
Manganese	Chlorosis of leaves with white blotching [520,521].
Molybdenum	Abnormally colored shoots [910], chlorosis and vulnerability to insect attack [149].
Nickel	Chlorosis and necrosis of leaves [351].
Serpentine	Dwarfism, stunting, color change of flowers [483].
Uranium or Radioactivity	Variation in flower color, abnormal fruits, more chromosomes, stimulated growth [157,160,743,744].
Zinc	Chlorosis of leaves, symptoms of manganese deficiency [351].

Source: Brooks [97].

5.2. TOXIC EFFECTS OF MINERALS ON PLANTS

Most morphological and mutational changes in plants may be said to be due to the toxic effects of minerals in the soil. There is an extensive literature on the subject of toxicity of elements to plants and it is possible to make general conclusions concerning the effects of certain metals on plants and to establish basic reasons for the effects that are produced.

Bowen [79] has suggested that elements may be divided into three classes according to their toxicities.

1. *Very Toxic* toxicity symptoms appear at concentrations less than 1 µg/g in the soil. Such elements include copper and mercury.

2. *Moderately Toxic* toxicity symptoms appear at concentrations between 1–100 µg/g in the substrate. Examples of this are the transition elements and most of the elements of Groups III, IV, V, and VI of the Periodic Table.

3. *Scarcely Toxic* toxicity symptoms rarely appear at concentrations normally encountered in all but special substrates. Examples of such elements are the halogens, nitrogen, phosphorus, sulfur, titanium, the alkali metals, and the alkaline earths.

The above classification applies to vascular plants rather than to bryophytes and the data are based on the concentrations of the elements in nutrient solutions where availability to the plants is 100%. Where plants are found under natural conditions, substantially higher concentrations of these elements can be tolerated in the soil, provided that the availability is low. The most common mechanism of toxic action in plants are as follows.

5.2. Poisoning of Enzymes

The more electronegative metals such as copper, mercury, and silver, have a great affinity for sulfydryl groups which are reactive sites for many enzymes. The enzyme is therefore unable to function so that toxicity results. As might be expected, there is an approximate relationship between the toxicity of an element to a plant and its electronegativity [773]. For divalent metals, the order of electronegativity decreases in the sequence mercury, copper, tin, lead, nickel, cobalt, cadmium, iron, zinc, manganese, magnesium, calcium, strontium, and barium, so that mercury is the most toxic in the series and barium the least so. It must be emphasized that this rule is very approximate and differs with species.

5.2.2. Replacement of Essential Nutrients

Arsenate and chlorate can occupy sites that normally involve phosphate and nitrate, respectively [1]. Phosphorus and nitrogen deficiency symptoms can therefore appear in the plant.

5.2.3. Precipitation of Essential Nutrients

Elements such as aluminum, beryllium, and titanium readily precipitate phosphate and render it unavailable to the plant. In this way phosphorus deficiency can result.

5.2.4. Catalytic Decomposition of Essential Nutrients and Metabolites

Elements such as lanthanum have a strong catalytic action on the decomposition of metabolites such as ATP [79].

5.2.5. Combination with the Cell Membrane and Reduction of its Permeability

Elements such as copper, gold, lead, and mercury are able to reduce the permeability of the cell membrane [642] and prevent the free passage of potassium, sodium, and organic molecules.

5.2.6. Replacement of Structurally Important Elements in the Cell

Some elements replace others in the cell but have an inert physiological role. Examples of this are the replacement of sodium by lithium and of chlorine by bromine [653,654].

It is obvious that a further discussion of toxicity is beyond the scope of this work, but it is hoped that the above discussion will be of assistance in providing workers in this field with a few basic ideas concerning the relationship of the elements in plants and soils. For a further discussion of toxicity of the elements to plants, the reader is referred to Hewitt and Nicholas [352] and Bowen [79].

It will be noted from Table 5.2 that there are so many different morphological changes produced in plants by mineralization in the substrate that not even an expert could be expected to realize the full significance of each variation in vegetation. However, each of the major types of change will now be considered.

5.3. MORPHOLOGICAL AND MUTATIONAL CHANGES

5.3.1. Abnormality of Form

Abnormality of form, excluding dwarfism or gigantism, is often a sign of the presence of boron or radioactive elements in the substrate. The work of Buyalov and Shvyryayeva [143] on the effect of boron on vegetation in the Soviet Union serves as a model of the successful use of morphological and mutational changes in plants for mineral exploration. These authors noted various variations in the vegetation depending on the boron content of the soil. At intermediate boron concentrations, individual plants such as *Artemisia lercheana* and *Kochia prostrata* were larger and had fresher vegetation than did plants growing over soils with normal boron levels. This effect was particularly striking in summer when much of the other vegetation was desiccated. At high boron levels in the soil, toxicity symptoms began to appear in the vegetation. This took the form of dwarfism and deformation in species such as *Eurotia ceratoides, Anabasis salsa,* and *Salicorna herbacea*. In some cases a prostrate form became apparent and there was increased branching of the stems. At very high concentrations of boron, the only surviving species were the indicator plants *Limonium suffruticosum* and *Salsola nitraria* (see Table 4.1). Similar findings were reported by Kantor [4101, who studied deep-rooted species in the Soviet Union and found that positive responses could be obtained for boron for depths of up to 30 m.

Unusual and unpredictable changes of form are produced by radioactivity. The first result of a mild dose of radioactivity is a stimulatory effect on vegetation. After the nuclear explosion at Hiroshima, exceptional yields of various crops were obtained in the following season [160]. There were also cases of grotesque mutations in plants. Fortunately, natural radiation is never as high as that encountered at Hiroshima and these levels are seldom enough to produce stimulatory effects in vegetation. However, there is ample evidence that prolonged, low-level radiation can produce morphological changes in plants. Shacklette [743] has described variations in the fruit of the bog bilberry (*Vaccinium uliginosum*) growing in an area of natural radioactivity in Northern Manitoba. Six main fruit variants were found and are illustrated in Figure 5.1. The colony of unusual variants was situated directly above a large deposit of pitchblende which outcropped on each side of it.

In experiments with plants grown artificially in carnotite, Cannon [157] reported the usual stimulatory effects in various species. Unusual plant forms included enlargement of the basal root stem of *Grindelia* sp. in which

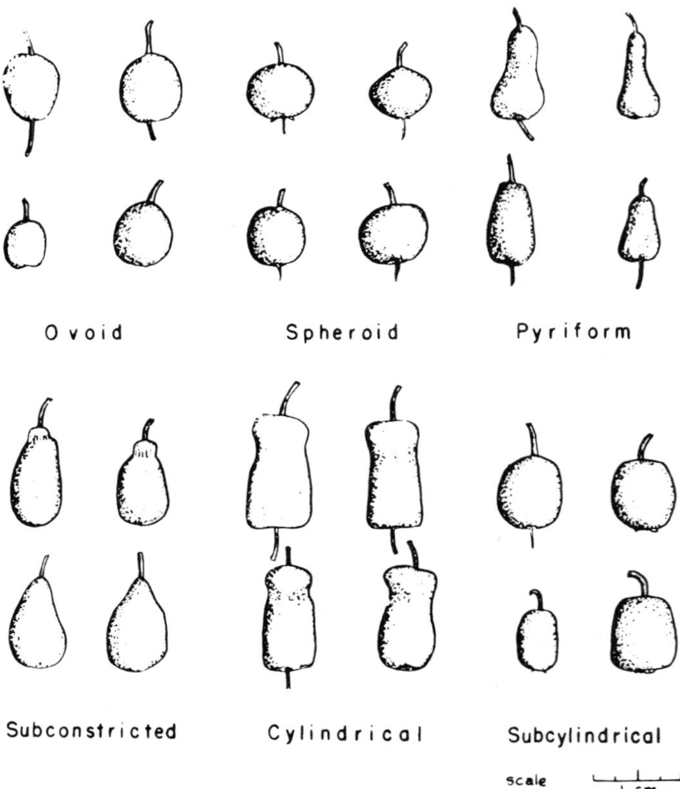

FIGURE 5.1. Fruit variation in *Vaccinium uliginosum* under the influence of natural radioactivity in the Churchill area, Manitoba. The upper right-hand drawing of each group represents the normal form; the other three represent the mutations. *Source:* Shacklette [743].

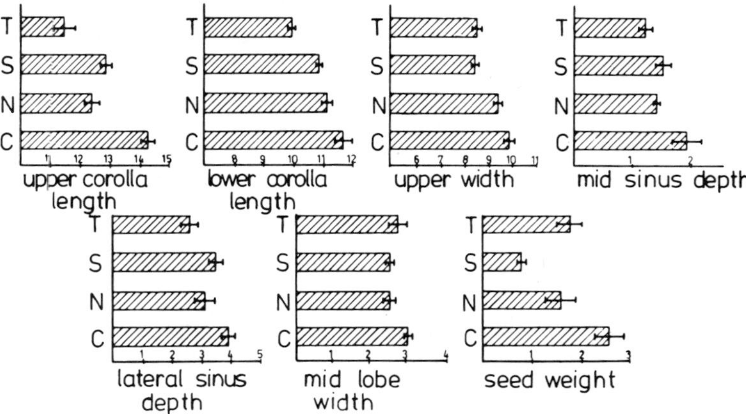

FIGURE 5.2. Morphological variation in seed weight and corolla dimensions of four populations of *Becium homblei*. The histograms record the means for each character with the relative standard deviation shown as an error bar. T = Tipperary Claims (~30 μg/g copper); S = Silverside Mine (>1000 μg/g copper); N = Norah Claims (>1000 μg/g copper), C = Chinamora (~8 μg/g copper). *Source:* Howard-Williams [371].

the basal rosette of leaves was raised 30 cm from the ground. *Stanleya pinnata* growing in the same plots produced stalks of imperfect flowers having no petals or stamens and carrying greatly enlarged green sepals.

Significant differences in seed weight and corolla dimensions were observed for four different populations of the copper flower *Becium homblei* [371]. These are illustrated in Figure 5.2. The general effect of increased copper levels was a reduction of seed weight and corolla size.

5.3.2. Chlorosis of Leaves

One of the most common field guides to mineralization is the presence of chlorosis in the vegetation. This characteristic yellowing is nearly always an indication of iron deficiency in the plant. It is caused very rarely by low levels of iron in the substrate. Rather it is a result of the reduction of iron uptake caused by antagonistic effects of other elements present in the soil in inordinately high amounts. Chlorosis is also a symptom of manganese deficiency, although in this case, the effect is one of chlorotic patches on the leaves rather than a complete overall yellowing. In a study of the moss *Pohlia nutans* growing over a copper bog in Canada, Boyle [84] observed up to 2.4% copper (dry weight) in this plant. The amount of copper appeared to be roughly proportional to the degree of chlorosis of the leaves.

As can be seen from Table 5.1, chlorosis is an indication of excessive concentrations of chromium, cobalt, copper, manganese, nickel, or zinc in the soil as all these elements are antagonistic to iron uptake by plants. The observation of chlorosis in leaves is therefore of little use to the prospector in deciding which element is present in excessive amounts in the substrate.

TABLE 5.2. Mutational Changes in the Color of Flowers of *Epilobium angustifolium* (from Port Radium, Canada) Under the Influence of Radioactivity

# Petals	Sepals	Filament	Anther	Pollen	Capsule
Abnormal Forms					
1 Pale rose pink	Cerise	Pure white	Dusky pink	Pale blue	Red above, paler beneath
2 Magenta paler than typical	Reddish purple	Pure white	Yellow	Very pale blue	Paler than usual
3 Very pale rosy pink	Clear rose red	White, pink at base	Purplish	Greenish-blue	Dusky pink
4 Very pale pink	Pale to clear rose pink	White, rose at base	Purplish	Bluish-green	Spreading, shorter than normal
5 Intense magneta; deep fuchsia	Darker than petals, purplish	Pure white	Purple	Very pale, slightly bluish	Erect, pale pink
6 Most intense magenta of entire population	Very slightly darker than petals	Pure white	Reddish	Pale blue	Erect, medium length
7 Pale magenta	Rosy-purple	Pure white	Yellow	Blue	Erect
8 Pale magenta	Rosy-purple	Pure white	Yellow	Blue	Erect
Normal Forms					
9 Light pink	Clear red	White to pale pink at base	Pale pink	Yellowish-green	Ascending, red above, pale green beneath
10 Magenta	Dark magenta	Very pale pink turning white	Greenish	Light blue-green	Erect, reddish above, greenish beneath

Source: Shacklette [744].

5.3.3. Color Changes in Flowers

Perhaps the most obvious of all plant mutations is that of change of color of the flowers. Provided that a field trip can be carried out during the flowering period, very effective results can be achieved in a short period of time. Color changes in flowers are usually either the result of radioactivity or of the presence an excess of certain elements in the soil. Shacklette [744] has observed extensive flower variation of *Epilobium angustifolium* growing in Canada over uranium deposits adjacent to those described in Section 5.3.1. These variations are shown in Table 5.2, which compares data for eight specimens of radioactive plants and two normal individuals. The table gives some idea of the factors that can be recorded by a trained geobotanist, but it is obvious that work of this nature is really the domain of the expert rather than the amateur.

In looking at color variations, due attention must be paid to the normal variability of this factor in background areas, since geobotanical interpretation must depend on an assessment of greater-than-average frequency of occurrence in the area under study. If precise values cannot be ascribed to the factor being considered, it is usually sufficient to assign terms such as rare, infrequent, common, and abundant to its frequency of occurrence.

Metal ions as well as radioactivity can effect the color of flowers. The gardener's trick of adding iron or aluminum to red hydrangeas to turn them blue is of course well known. The theory behind such color changes is interesting and may have some bearing on mineral exploration.

The majority of plant colors are produced by a surprisingly small number of pigments. Apart from **carotenoids,** which are important in yellow and orange flowers, it is mainly the **anthocyanins** that are responsible for the color range from orange to deep blue. Bayer et al. [50] have suggested that in the absence of certain metals, the anthocyanins form red oxonium salts which become blue when they are complexed with excessive amounts of iron, aluminum, or other elements. This is precisely the mechanism involved in the color change of hydrangeas referred to above.

Besides iron and aluminum, other elements such as chromium, tin, titanium, and uranium can form stable complexes with anthocyanins. It is therefore possible that excessive amounts of some of these elements could produce a blue tint in flowers that are normally red or pink and that this could be a useful field guide in mineral prospecting. I have personally seen a blue-red form of flowers of the manuka (*Leptospermum scoparium*) growing over a soil containing 6% chromium in an ultrabasic area in New Zealand.

In studies on the nickel hyperaccumulator *Hybanthus floribundus* in Western Australia, Severne [738] found that color variations in this species (white to deep blue) were to some extent related to the nickel content of the soil. White forms were restricted to soils containing less than 700 μg/g nickel whereas the deep blue form covered the whole range from 400 to 1650 μg/g.

There are several other references in the literature to color changes produced in flowers by mineralization in the substrate. Examples are color variations in *Eschscholtzia* [51], and in *Papaver commutatum* [566]. This latter mutation, said to be due to copper-molybdenum mineralization is illustrated in Figure 5.3. Cannon [162] reported that the flowers of *Pera-*

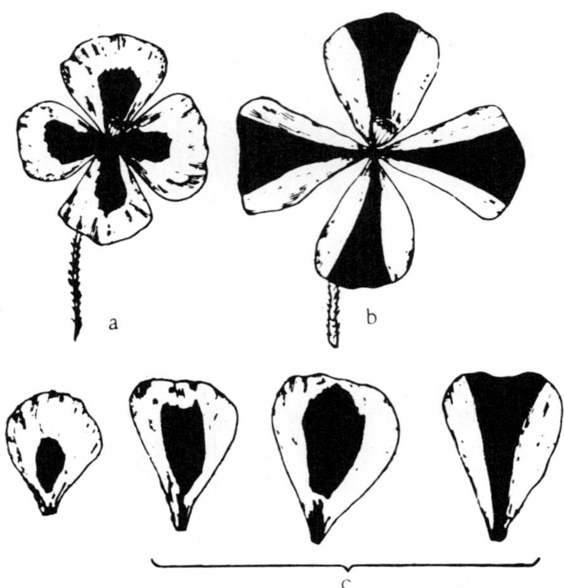

FIGURE 5.3. Change in the color of petals of the flowers of *Papaver commutatum* under the influence of copper-molybdenum mineralization. (a) normal flower, (b) modified flower, (c) degree of mutability of the corolla. *Source:* Malyuga et al. [566]. Copyright Plenum Press 1964. Reprinted by kind permission.

phyllum ramossissima (squaw apple), growing over a molybdenum deposit in Nevada, were white instead of pink.

5.3.4.   Dwarfism and Stunting

Stunting of growth, or dwarfism, is one of the most common toxicity manifestations in vegetation. Moreover, it is very difficult to assign dwarfism to any particular cause since it is likely to be due to deficiency or excess of any one of a number of metals. However, the presence of ultrabasic rocks in the substrate invariably is indicated by dwarfism of serpentine floras. One of the most marked examples of dwarfism known to me is that of the New Zealand shrub *Pittosporum rigidum* which under normal conditions grows to 5 m, but in serpentine areas becomes a cushion plant only a few centimeters high [62].

In some cases the degree of stunting can be a semiquantitative indication of the degree of mineralization. For example, Jacobsen [384] showed a progressive decrease in the height of nonspecialized plants as the copper content of the substrate increased. Stunted forms of *Tephrosia longipes* and *Combretum zeyheri* with enlarged fruits were observed over copper deposits in Zimbabwe [929].

5.3.5. Gigantism

Gigantism is considerably less common than dwarfism and is often associated with the presence of bitumen or boron. Care must be taken not to confuse gigantism with the stimulatory effects produced from an abundance of plant nutrients or even from the presence of radioactivity.

There are frequent references in the literature to the beneficial effects of petroleum products on vegetation. These effects are produced partly by conditioning mechanisms in which the structure of the soil is improved so that the temperature and moisture capacity are increased. Other effects include an increase in the availability of phosphate and in the intensity of nitrification [688]. In certain cases, however, petroleum products can cause abnormalities in vegetation resulting typically in gigantism. One of the first records of this type of effect was by Shchapova [753] who found forms of *Zostera nana* five times their normal size growing in a bituminous area on the shores of the Caspian Sea. A thorough investigation of the effects of bitumen on vegetation was carried out by Vostokova et al. [867] in West Kazakhstan and on the coast of the Caspian Sea. These workers observed gigantism in 29 species. Figure 5.4 shows typical data for *Atriplex cana* for a bituminous soil and for a background area as control. It is clear that the two areas can be differentiated from height and diameter data. The above species were those showing uniform gigantism in all dimensions. Other plants showed increased height with no lateral increase in dimensions and others showed the opposite effect. In both cases the species could be used to characterize soils with a high bitumen content. The above work has demonstrated that the geobotanical method can be used even in the search for petroleum, and is not restricted to inorganic constituents of the substrate.

5.3.6. Rhythmic Disturbances in Plants

One effect of mineralization on plants is a disturbance of the rhythm of flowering and foliation. Sometimes flowering occurs early or late, in other cases there is a second flowering, usually in the fall. Examples of this have been reported by Buyalov and Shvyryayeva [143] for plants growing over boron deposits in the Soviet Union. Vostokova et al. [867] have reported a similar effect for vegetation growing in soil contaminated with bitumen. Second blooming was noted for 12 species. Late blooming was also noted in *Stanleya pinnata* exposed to radioactivity [157]. All the rhythmic disturbances reported above have one feature in common. They were all caused by minerals whose main effect was one of stimulation rather than depression. It is probable therefore that in field work, abnormal flowering times may be linked to the presence of such stimulants as bitumen, boron, or radioactivity.

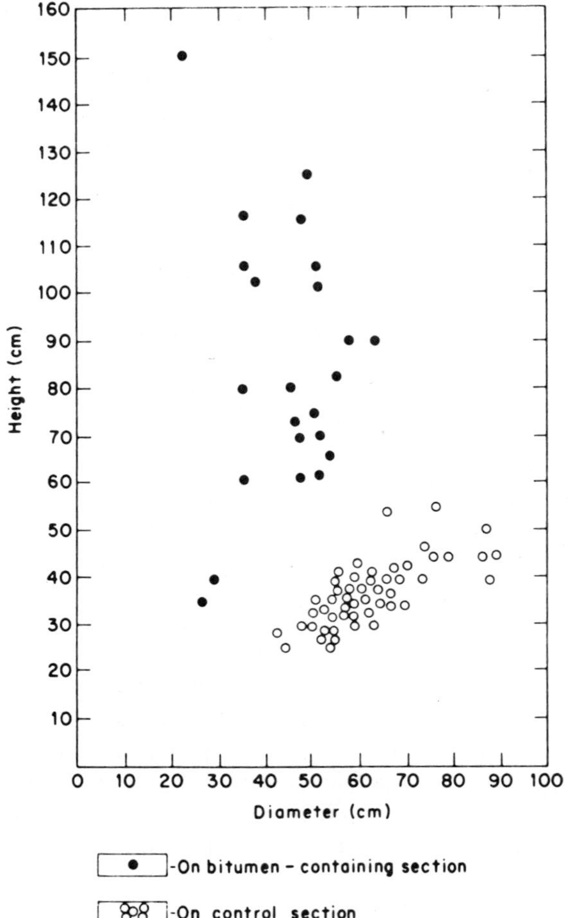

FIGURE 5.4. Graphical representation of size ratios of *Atriplex cana* growing over bituminous and nonbituminous ground. *Source:* Vostokova et al. [867]. Used by permission of the American Geological Institutes.

An intriguing possibility for mineral exploration has been advanced by Canney et al. [149]. They postulated that deciduous vegetation under geochemical stress differs not only in its infrared reflectance [98,150,216] (see Chapter 6), but may show autumn coloring at an earlier date and may also be more subject to insect attack (see Chapter 11) because of its weakened state.

The reason for early development of autumn colors is not fully understood. It is suggested [149] that it may be a function of the chlorophyll content, which is dominant until leaf fall when chlorophyll production ceases and the red, orange, and yellow anthocyanin pigments become the more dominant. These colors show up more readily when the chlorophyll concentration is low, as would be the case where heavy metals depress the iron uptake by the plant and cause chlorosis (see Table 5.1).

It is well known in silviculture [578] that insects readily perceive reduction in the defense mechanisms of sick plants and preferentially attack such subjects. Indication of excessive insect attack may therefore be yet another manifestation of heavy metals in the substrate. The insect attack theory is still only a theory, but there is some evidence for an early coloration of geochemically stressed vegetation in the autumn, so that the suggestions of Canney et al. [149] are well worth a follow-up.

5.4. CONCLUSIONS

Certain general conclusions can now be made concerning morphological changes in plants as a guide to mineralization. It is clear that to identify many of these changes, the services of a skilled botanist or ecologist will be required. This is particularly true for recognition of plant abnormalities and for the proper understanding of the true significance of chlorotic effects in leaves. Other variations such as dwarfism, gigantism, and color changes in flowers may be readily observed by less skilled personnel.

It is concluded that most chlorotic symptoms will be associated with iron deficiency occasioned more likely by an antagonistic effect from other elements, rather than as a result of iron deficiency in the substrate.

Early or second flowering are probable indications of stimulants such as excess of nutrients like boron, nitrogen, phosphorus, or potassium, or may be due to the presence of bitumen or radioactivity. Lateness of flowering, however, may be a toxicity symptom linked with the vigor of the plant.

When used in conjunction with the study of indicator plants and characteristic floras, the study of morphological changes in vegetation should be an effective aid in geobotanical exploration work. Perhaps one of the most exciting developments in indicator geobotany is in multispectral remote sensing of vegetation from satellites (see Chapter 6) and it is probably in this field, that future significant advances in assessing morphological changes will evolve.

6

REMOTE SENSING OF VEGETATION

6.1. INTRODUCTION

During the past 20 years, the development of sophisticated aircraft, new techniques in remote sensing, and the birth of the space age have introduced new and exciting dimensions to the field of indicator geobotany. Greater pressure on natural resources caused by expanding populations and rising standards of living throughout the world have lead to geobotanists taking to the air in ever-increasing numbers in order to carry out an inventory of the vegetation cover and to assess its significance in the search for water, petroleum, and minerals.

The development of artificial satellites during the past two decades has opened up limitless possibilities in aerial mapping. Technological progress has also provided greatly improved techniques for remote sensing of natural resources by use of detectors sensitive to the whole of the electromagnetic spectrum from gamma radiation at the one extreme to radar at the other. Figure 6.1 shows the wavelengths commonly employed in various types of remote sensing and shows their relative absorption by the atmosphere which, of course, is the limiting factor in their use.

Remote sensing of vegetation depends on measurement of spectral reflectance patterns to measure either chlorosis due to metal stress or to recognize difference in plant communities (see Chapter 3) influenced by mineralization. Figure 6.2 shows typical reflectance patterns for various vegetation types. There is a small maximum at 550 nm in the green part of the spectrum as well as a much larger one in the near infrared.

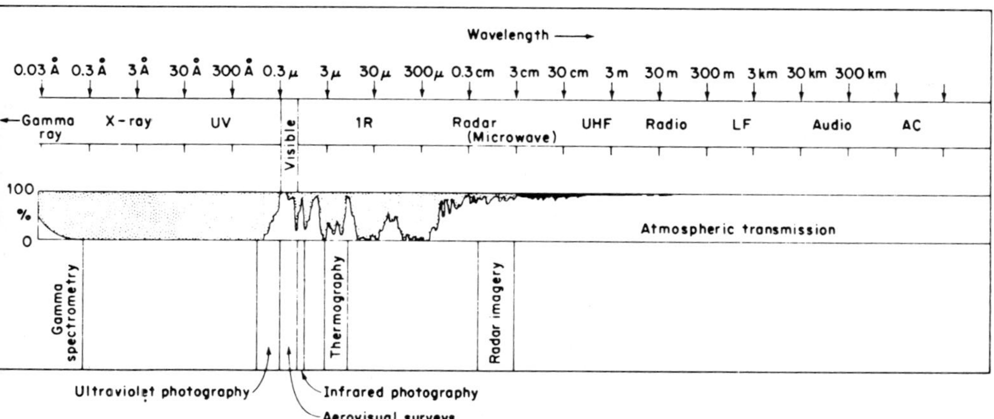

FIGURE 6.1. Schematic representation of the electromagnetic spectrum and its significance in remote sensing. *After:* Colwell [217]. Reproduced with permission from *Photogrammetric Engineering.* Copyright 1963, The American Society of Photogrammetry.

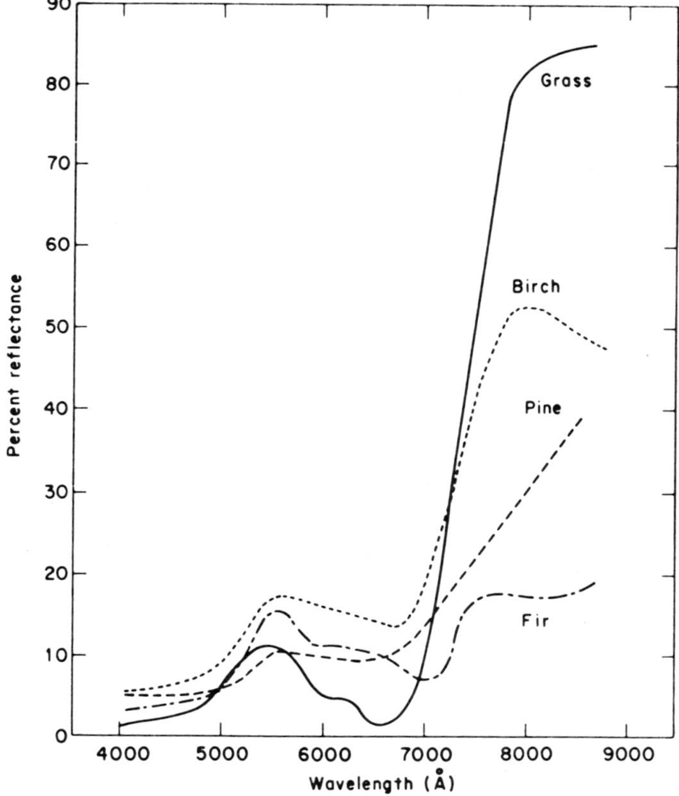

FIGURE 6.2. Spectral reflectance curves for various foliage types. *After:* Fritz [305]. Reproduced with permission from *Photogrammetric Engineering.* Copyright 1967, The American Society of Photogrammetry.

It has been determined by several workers [359,369,665,962] that metal-induced chlorosis results in lower absorption of the green part of the spectrum by chlorophyll and a concomitant increase in reflectivity at this wavelength. An excellent review by Horler et al. [359] discusses the reflectance of vegetation at various wavelengths and its application for determining geochemical stress in plants. Reflected radiation depends not only on the structure, pigment content, and water content of individual leaves which may be affected by stress, but also depends on such factors as vegetation ground coverage, leaf area, topography, and the degree of solar illumination. These variables may complicate or render impractical, stress detection methods using broad band radiation based on physiological factors such as the change in the chlorophyll content of plant tissue. It has been suggested [359] that these problems can be largely overcome by techniques utilizing the shape of reflectance spectra rather than radiance values themselves. This approach requires high-resolution spectral measurements and some type of waveform analysis.

To record reflectance spectra (R) of leaves, the above authors have used first derivative reflectance spectra ($dR/d\lambda$) using a spectrophotometer. One of the most striking features of plant reflectance spectra is the change of R at about 700 nm at the limit of chlorophyll absorption towards the red end of the spectrum. This feature is known as the "red edge." The wavelength at the red edge (λ_R) is found to change according to the chlorophyll concentration in the leaf, as is illustrated in Figure 6.3 for a healthy and a chlorotic maize leaf. This effect is predicted by Beer's Law, which can be expressed as

$$I = I_0 10^{-cl\epsilon}$$

where I is the light intensity at a given wavelength transmitted by a solution of thickness l; I_0 is the intensity of the original light; c is the concentration in moles/liter of the absorbing material; and ϵ is the molar absorptivity coefficient (often known as the extinction coefficient). A leaf, unlike a solution, is a highly scattering medium, but the transmittance of light through leaves or canopies can often be described by Beer's Law. Since the reflectivity of leaves closely approximates their transmittance, it is reasonable to expect that calculations based on Beer's Law will be applicable to leaf reflectances.

Spectral analysis has been applied to the red end of the chlorophyll absorption band [359]. Since chlorophyll is present only in plants, this technique is one method of isolating the vegetation component of a viewing area by isolating the known spectral shape of vegetation. It is reasonable to expect that the derived data would be relatively independent of factors that cause variation in reflectance values, such as canopy density. The method can be used for spectral analysis around any strong absorption band

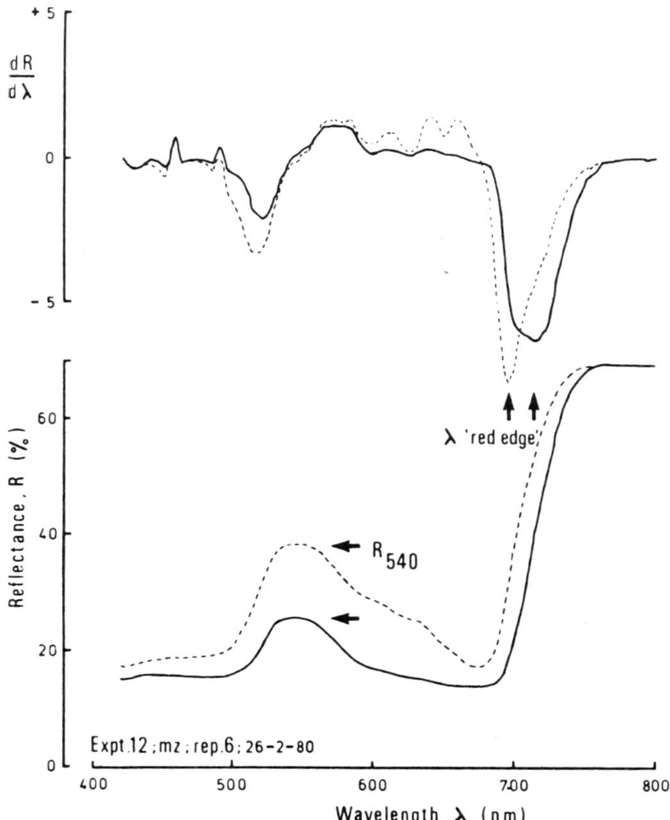

FIGURE 6.3. Reflectance and first derivative reflectance spectra of stressed (broken lines) and unstressed (continuous lines) maize leaves. *Source:* Horler et al. [359].

and could be applied to the infrared water absorption bands on which very little work so far has been carried out.

It will be noted from Figures 6.2 and 6.3, that reflectance of all types of vegetation is far greater in the near infrared than in the green part of the spectrum. This infrared reflectance is a measure *inter alia,* of the surface area of the canopy. However, chlorotic plants usually have a smaller leaf area so that the ratio of 550/800 nm is usually increased in most, though not all, cases. This is illustrated in Figure 6.4, which always shows an increase of the reflectance of stressed plants at 550 nm, though in two cases there is an increase at 800 nm and in two other cases, a decrease [665]. The principle, however, of comparing reflectance ratios at two different wavelengths is now well established and will be discussed later in this chapter.

The various types of remote sensing now will be discussed in order of increasing wavelength. They do not include gamma scintillometry since radioactivity in plants is usually far less than that of the background of the

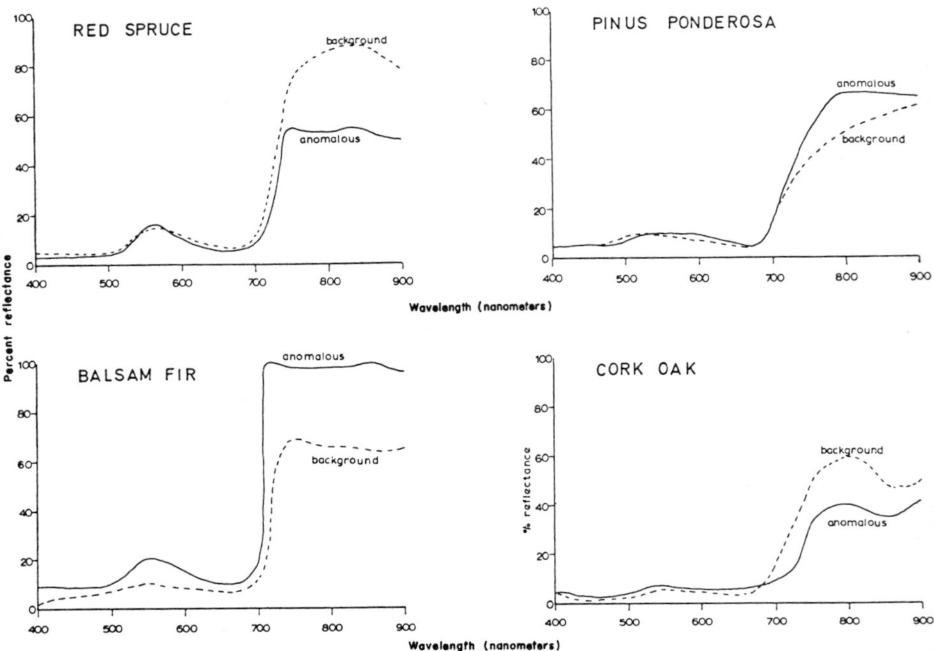

FIGURE 6.4.　Reflectance spectra for various plant species showing the effects of geochemical stress. *Source:* Press [665].

terrain, even where plants are growing over radioactive ores with no surface expression of mineralization. In some cases indeed, vegetation can absorb some of the radiation derived from the substrate.

The physiological factors that influence the reflectance of vegetation depend entirely on the part of the spectrum that is examined. These factors are summarized in Table 6.1.

TABLE 6.1.　**The Main Physiological Factors Influencing Remotely Sensed Radiation from Plants**

Measurement	Controlling Physiological Factor
1. Chlorophyll fluorescence (200–400 nm)	Chlorophyll content and organization Inhibition of photosynthetic electron transport
2. Visible reflectance (400–800 nm)	Pigments, mainly chlorophyll
3. Near infrared reflectance (800–1200 nm)	Leaf and canopy structure
4. Infrared reflectance (1200–2500 nm)	Water content
5. Thermal emission (>2500 nm)	Plant water status, evapotranspiration

6.2.   REMOTE SENSING IN THE ULTRAVIOLET AND VISIBLE RANGE

Aerial photography in the ultraviolet and visible parts of the spectrum has been carried out for over 100 years. The earliest work of this kind involved hot air balloons, kites, and even homing pigeons with miniature cameras attached to their legs. During both world wars the techniques were used extensively in military reconnaissance, but it was not until after World War II that the procedure began to be applied to any great extent to vegetation surveys. Considerable work in this field has now been carried out in the United States, the Soviet Union, and elsewhere. Most developed countries now have a complete range of aerial photographs for the whole of their territory.

The interpretation of aerial photographs in terms of geology is known as **photogeology** [679] and is finding increasing use among geologists and geomorphologists. In the Soviet Union a clear distinction is made between the use of aerial photographs for geobotany [849] and aerovisual surveys of an area with the aid of photographs carried in aircraft [417].

The association between the geology and vegetation of an area as discussed in Chapter 3 has been used for the preparation of geobotanical maps from aerial photographs. The Soviet Union has excellent geobotanical maps for a large part of its territory [760]. Similar maps are now available for countries such as Australia and the United States. The usefulness of aerial photographs (particularly in color) is shown from the following information from Viktorov et al. [849].

The color of vegetation in arid areas in summer depends to a great extent on the availability of ground water. Color photographs of plant communities taken during the summer and winter months can be compared. These photographs should show the presence of subterranean water resources because vegetation growing above ground water will remain green throughout the year. In several areas containing bitumen (see Chapter 5), plants are stimulated into gigantism and luxuriant growth with a strong tendency to second blooming and second vegetation. Areas showing luxuriant vegetation in the fall are therefore worth a follow-up.

In the steppe areas of West Kazakhstan, leguminous plants appear in the vegetation cover wherever phosphate beds are located close to the surface and can be distinguished readily by aerial color photography. The same authors have also observed that extensive areas of lichens and mosses occur on sandy deserts, such as the Kara-Kum, whenever gypsum deposits lie on or close to the surface. Absence of vegetation in aerial photographs should also be given close attention since this may be a measure of the presence of ultramafic rocks.

Although aerial photographs are extremely useful in the interpretation of plant cover, they do require visual supplementation carried out from aircraft at different altitudes. Techniques of aerovisual surveying have been reviewed by various authors [417,849,866].

It is not well appreciated that aerial photography as a means of detecting mineralization is not a new technique at all. In the early 1930s, some 132,000 sq. km of Zambia and Zaire were mapped with panchromatic film to discover "dambos" (vegetation clearings caused by excess copper and/or cobalt in the soil)[583]. As a result of this survey, three of the eight major orefields of the Zambian Copperbelt were discovered.

Although it has been suggested above that images produced from a ratio of spectral reflectances in the green and infrared parts of the spectrum may be advantageous, it is also possible to ratio reflectances in the visible part of the spectrum. For example, Birnie and Francica [68] used the ratio 565 nm/465 nm for aerial photography in Washington state. They found that a ratio greater than 1.7 for a forest canopy dominated by Douglas fir with lesser amounts of western larch indicated a pyrite halo from a porphyry copper deposit. An 87% probability of success was achieved working over areas of known mineralization. Further discussion of wavelength ratioing is given below in Section 6.6.

Very little work has been done in the ultraviolet end of the spectrum because reflectance of vegetation is low in this region. However, some plant species do exhibit solar-induced fluorescence. This type of radiation has been discussed by Horler et al. [359]. It is feasible to record chlorophyll fluorescence using either "active" or "passive" remote sensing systems. Measurement of passive (i.e., solar-stimulated) fluorescence has been reported using a Fraunhofer Line Discriminator (FLD) operating in aircraft and on the ground. It is also planned to use an FLD in the Space Shuttle. Horler et al. [359] report that plants growing over copper-molybdenum mineralization in the United States had a lower chlorophyll fluorescence intensity than the surrounding vegetation. This is consistent with reports of heavy-metal effects on the fluorescence of algae and higher plants. The effect of water stress on plants is to increase the fluorescence intensity. However, it has been difficult to show conclusively the effects of metal stress on the fluorescence of vegetation when a passive system is used. Using an active (dual) system, it was possible to demonstrate striking differences in pea plants grown in cadmium-rich nutrients compared to those grown in normal substrates. It is possible that an active system using laser excitation might be operated from aircraft. It should be emphasized that this form of remote sensing is only in its infancy and that so far it has only been shown to be theoretically feasible. Very much more research will be required before it becomes a routine tool in mineral exploration.

6.3. REMOTE SENSING IN THE NEAR INFRARED

As early as 1939 [383] it was shown that various types of vegetation could be distinguished easily with infrared black-and-white film. A typical film of this kind will have an emulsion that is sensitive to the whole of the visible part of the spectrum and to the near-infrared up to 900 nm. Maximum sensitivity will be in the range 770–840 nm. It is customary to use a red filter with this type of film in order to remove the blue end of the spectrum. From my own experience in New Zealand and Australia [98], better results can be achieved with infrared color film; this will now be considered further.

Infrared color film (CIR) comprises three image layers which are sensitive to green, red, and infrared instead of blue, green, and red as with normal color film. A yellow filter (e.g., Wratten 12) is used to withhold blue light toward which the layers are sensitive. Film of this kind has several disadvantages over conventional film. The first of these is that the exposure latitude is very limited (½ stop) and unless exposures are exactly correct, disappointing results will ensue. Processing of the film will not normally be undertaken by most commercial interests and the services of a specialist will have to be sought. Other disadvantages are that the infrared-sensitive layer is somewhat unstable, has a short life, and must be kept cool at all times. Focusing is also a problem for close work because the position for optimum focus differs among the layers so that, in effect, a compromise has to be obtained.

In spite of the above difficulties, good results can be obtained by relatively unskilled operators using conventional cameras operated from a light plane or helicopter. It is advisable to set the camera at a suitable speed (e.g., $\frac{1}{250}$ sec) and give trial exposures at all f-stops including the intermediate half stops. This is very wasteful of film but film is very much cheaper than is aircraft time. When experience has been gained, the photographer will get away with only one or two exposures for each scene.

The infrared reflectance of vegetation becomes increasingly less detectable at higher altitudes because of absorption of the radiation by water vapor in the atmosphere and also due to Rayleigh scattering which is especially pronounced at the blue end of the spectrum. Although the Wratten 12 filter normally used with CIR film cuts out wavelengths below 520 nm, where scattering is particularly serious, there is still sufficient remaining interference from this source to render infrared photography a problem at high altitudes. For example, at 12,000 m, as much as 90% of the infrared radiation is lost.

Pease and Bowden [646] discussed this attenuation problem and have shown that auxiliary filters with additional minus-blue filtration can greatly enhance the infrared signal from vegetation. For moderate enhancement, Kodak filters 82B or CC30B are recommended, and for drastic enhancement use may be made of filters 80B, 80C, or CC50B. The use of these auxiliary filters does however present additional problems. For aerial work, a shutter speed of at least $\frac{1}{250}$ sec is recommended. The filter factor for CC50B is 1.5 stops and this involves working near the limits of the aperture range of most cameras. In practice, it is advisable to compromise by reducing the shutter speed to $\frac{1}{125}$ sec and suffer some loss of definition. Although the infrared signal is enhanced when auxiliary filters are used, there is less differentiation of vegetation in the image compared with photographs taken at lower altitudes without the use of special filters.

The most obvious potential of remote sensing in the infrared part of the spectrum is in the preparation of geological maps from consideration of the distribution of those plant species that are strongly affected by the nature of the substrate.

Until the 1970s, much of the work carried out with infrared film was essentially monoband. Perhaps one of the most comprehensive of these surveys was by Paarma et al. [633] who combined extensive aerial coverage with on-foot determination of the so-called ground truth. Successful results were obtained even with with altitudes of up to 9000 m. Others workers [794, 796] have used infrared film in combination with black-and-white panchromatic film for geological and geobotanical interpretation.

The modern trend is to use multiband scanners for aerial photography and to use wavelength band ratios rather than single bands. Such multiband (as many as 11) scanners, however, usually do include a near-infrared channel: as, for example, work carried out over the well-studied Sokli carbonatite formation in northeast Finland [794]. Infrared imagery will be discussed further in Section 6.6.

6.4.   THERMOGRAPHIC IMAGERY

Thermography is sometimes confused with near-infrared imagery but, in fact, depends on an entirely different system which operates in the range >1200 nm. No photographic film is capable of operating in this wavelength range. Instead, the thermal signal is detected by some form of optical mechanical scanner [217] in which the energy is converted into an electrical signal which is used to generate visible light to produce an image. This image may be photographed in black and white or maybe recorded on tape for further computer processing.

Until recently, thermography usually has not been used for vegetation studies because its main use has been to measure soil moisture conditions [665] or anomalous thermal patterns from ore deposits [148,210]. The only biogenic material previously recognizable by thermography was peat [795] which is often colder on the surface than are the surroundings except during the day when the surface dries out under solar radiation.

Horler et al. [359] have recently shown that thermography may indeed have wider applications to vegetation studies. There are several physiological mechanisms whereby metal stress may affect the thermal emission from plants, of which the most important may be the effect of heavy metals in inhibiting stomatal opening. The above authors have found good correlation between the stomatal resistance of leaves, the metal concentration in which plants were grown and the radiant canopy temperature. Increases of canopy temperatures of up to 1°C associated with visible copper and cadmium toxicity were observed. However, it was found that thermal responses to metal treatment were obscured by moderate water stress which can cause canopy temperatures to increase by up to 3.5°C at pre-wilting stages.

6.5. RADAR IMAGERY

Remote sensing with radar has interesting applications in thickly forested terrain since it is capable of appearing to remove the vegetation completely and of exposing underlying geological and geomorphological features. Radar waves of long wavelength will readily penetrate vegetation and several feet of soil. Shorter wavelengths can be used to differentiate different species of vegetation.

In radar imagery, a high-frequency signal is generated from the aircraft and strikes the ground beneath. The reflected signal is analyzed by a special type of receiver which, as with thermography, converts the energy into a visual form from which a conventional photograph is obtained.

Airborne facilities for radar imagery are now freely available. Changes in frequency and phase of polarization of the reflected signal can be detected readily and can be used for identification of the terrain and its vegetation. Wavelengths of 1–10 cm are characteristically used for this work.

6.6. MULTIBAND SATELLITE IMAGERY

6.6.1. Introduction

It will be clear from the above discussion that no single unichannel sensor will be capable of solving the entire problem of geobotanical surveying. There is the further constraint that several overflights at different times of the year might be advisable to detect subtle changes in vegetation due to metal stress from ore deposits. Both of these problems have been overcome by the development of multiband satellite imagery, albeit with some loss of definition because of the great altitudes involved.

The first available satellite imagery was from the Gemini orbiters of the late 1960s [206]. Later, however, the four-channel LANDSAT-I was launched and was designed specifically for aerial imagery of the earth. This was followed by LANDSAT-II launched in 1975. The satellites contain a multiband scanner (MSS) with visible bands at 500–600 nm, 600–700 nm, 700–800 nm, and an infrared band at 800–1100 nm. These are known as MSS4, MSS5, MSS6, and MSS7, respectively.

6.6.2. Techniques of Satellite Imagery

Each image of the earth from the LANDSAT satellites is recorded on an on-board tape recorder and later transmitted to a ground receiving station in the United States. NASA processing of the high-density computer tapes yields either monochrome photographic images produced on an electronic beam recorder, or digital data stored on a set of computer-compatible tapes (CCT).

Each of the four LANDSAT images for a given scene covers 185 sq. km and comprises 2340 × 3264 (= 7,637,760) pixels (picture elements). Each pixel corresponds to an area of 79 × 79 m² (about 1 acre) and is established by the scanner assigning a number between 0–255 (thus giving 256 single *signatures*), which is proportional to the reflectance intensity of that part of the scene.

The fact that imagery data are available on CCT gives limitless opportunity for computer handling of the information. Two types of maps can be produced: unsupervised maps which classify pixels into groups of similar signatures based only on statistical considerations, and supervised maps which subdivide pixels of similar signatures into further groups depending on ground truth determined before or after imagery. Maps can be uniband or they can be obtained by ratioing any pair of the four LANDSAT bands to produce up to six different images [38,59,60,537]. As far as vegetation is concerned, the best pair of channels to choose is MSS5/MSS7 since the denser the vegetation, the more it absorbs MSS5, whereas the healthier it is, the more it reflects MSS7. This ratio therefore gives a good estimate of the density and physiological state of the vegetation cover [60]. Ratioing also eliminates differences of brightness due to uneven solar illumination of the scene.

6.6.3. A Case History from Canada

The following example of use of ratioed LANDSAT data is from Bélanger [59,60] and concerns the Thetford Mines area of Quebec Province, an important asbestos and chromium mining area in a large ultrabasic complex with a characteristic serpentine flora. The area of about 7000 sq. km was overflown by LANDSAT-II in June 1975 and November 1978. Figure 6.5 shows histograms of pixel signatures for the MSS7 (infrared) band [59] from which it will be noted that there is a great difference in the reflectance levels for the two seasons. The differences are mainly due to senescence of deciduous trees and shrubs which reflect infrared to a much greater degree in the early summer.

The above histograms are obviously of little use for mineral prospecting but they serve to display the seasonal variation in reflectance levels of the pixels, so that the correct choice can be made of ratios of a suitable pair of bands which will give the best discrimination of vegetational differences due to substrate variations.

The MSS7/MSS5 ratio was used for the interpretation of the Thetford scene. This ratio gave the so-called *biomass index* which is shown to the right of Plate 6.1. The biomass index plot is obtained by means of an "Applicon" color plotter. The plotter scans the LANDSAT raw data, ratios them, and sprays ink jets at computer-regulated densities from three nozzles

PLATE 6.1.   Right: Biomass index map for the Thetford Mines area of Quebec. Colors are artificially generated by an ''Applicon'' plotter and are based on ratios of reflectance levels (pixel signatures) for LANDSAT bands 7/5. There are nine basic color shades with further subdivisions. See text for description of significance of various colors. *Source:* Bélanger [59]. Top left: SURTRACE equipment in use in the southwestern United States. Photograph by courtesy of A. R. Barringer. Bottom left: The ore-searching dog Hessu looking for uranium in Finland. Photograph by courtesy of P. Mattson. *Note:* This plate is featured in color on the dust jacket of this book.

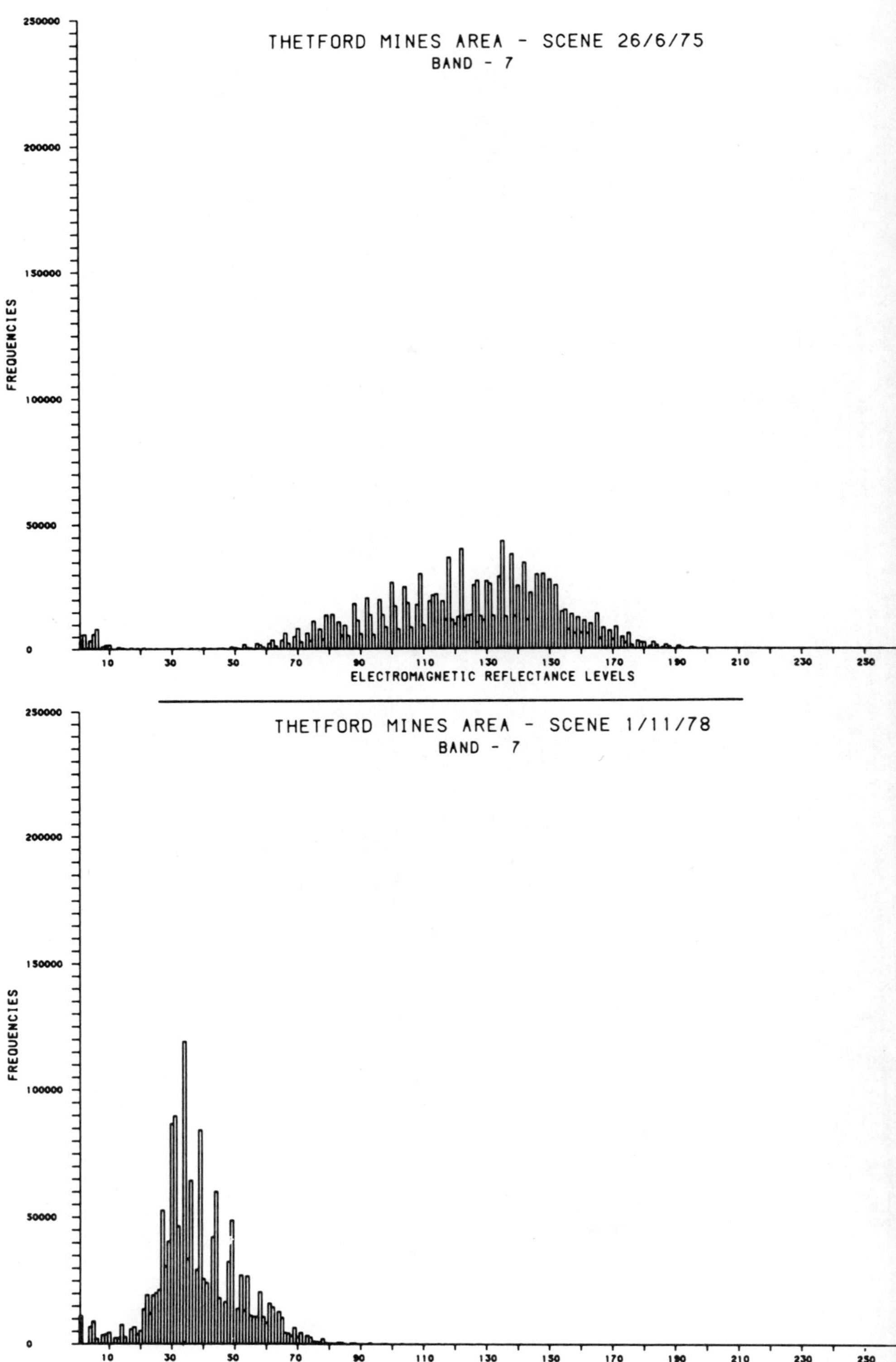

THETFORD MINES AREA - SCENE 26/6/75
BAND - 7
250000
200000
150000
100000
50000
0
FREQUENCIES
10 30 50 70 90 110 130 150 170 190 210 230 250
ELECTROMAGNETIC REFLECTANCE LEVELS
THETFORD MINES AREA - SCENE 1/11/78
BAND - 7
250000
200000
150000
100000
50000
0
FREQUENCIES
10 30 50 70 90 110 130 150 170 190 210 230 250
ELECTROMAGNETIC REFLECTANCE LEVELS

corresponding to the colors yellow, blue, and red. The maps produced in this way consist of a large number of ink dots which correspond to the individual pixels and comprise 25 shades of 9 main colors. The red shades correspond to regions of low biomass indices (water and regions bare of vegetation) while the blue shades correspond to high indices (dense, healthy vegetation). Even when vegetation is healthy, the various types do not have the same biomass index. Hence it is possible to differentiate between stands of conifers, deciduous trees, bushes, and herbs. Referring once again to the main color classifications in Plate 6.1 (reproduced in color on the dust jacket) the numbers correspond to: 1 — water; 2 — mine dumps; 3 — suburbs and villages; 4 — roads and agricultural crops; 5 — agricultural crops; 6 — conifers; 7 — shrubs and open woodland; 8 — mixed woodlands; 9 — leafy forest.

The biomass index is useful for indicator geobotany by revealing anomalies in the vegetation. For example, if the biomass index map is superimposed on a topographic map, it can be see that hills generally coincide with shades of blue or high indices. However, the hill inside the square (Plate 6.1) coincides with a low biomass index and ground truth revealed that it corresponds with a serpentine outcrop whose chemical composition is unfavorable for plant growth. Other regions of red coloration not concerning water are also primarily serpentine outcrops, particularly the area of the Thetford Mines. Further research is currently underway in this area to study certain anomalies in the spectral signatures of plant species found within the ultrabasic dispersal train in order to assist in prospecting efforts in other areas where mineral deposits are likely to be found.

In an earlier part of the survey of the Thetford Mines area, supervised scenes from LANDSAT-I were examined for ground truth [61]. The experiments involved testing all four channels but without use of ratioing. Table 6.2 shows the correlation between vegetation communities and the spectral

TABLE 6.2.   Reflectance Levels (Pixel Signatures) on LANDSAT Bands 4, 5, 6, and 7 for Five Vegetation Associations in the Thetford Mines Area, Quebec

Community Type	Reflectance Level			
	Band 4	Band 5	Band 6	Band 7
Scrub community	35	25	83	6
Coniferous forest	35–39	25–29	82–83	87–92
Mixed deciduous forest				
poplar	42	26 31	125–131	145–158
birch	36–39	25–27	106–147	125–172
Deciduous forest	40	28	139	154
Agricultural land	42	33	39	120

Source: Bélanger et al. [61].

FIGURE 6.5.   Histograms of band MSS7 from a LANDSAT scene taken over the Thetford Mines area, Quebec, during two different seasons. *Source:* Bélanger [59].

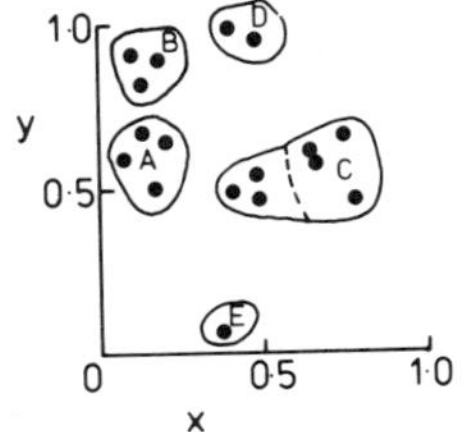

FIGURE 6.6. Ordination of 17 plant communities in the Thetford Mines area, Quebec. Each cluster represents a similar plant community: A—Scrub, B—mixed conifers, C—mixed deciduous trees, D—coniferous forest, E—deciduous forest. Values for the communities are statistical numbers based on discriminant analysis of LANDSAT reflectance data. *Source:* Bélanger et al. [61].

signatures of the pixels (i.e., 0–255). Bands 4 and 5 did not clearly differentiate any of the plant communities except for agricultural land, which tended to have slightly higher reflectance levels. Bands 6 and 7 clearly differentiated conifers from deciduous forests. The coniferous forests had significantly lower reflectances than did deciduous forests. Similarly, scrub communities had lower reflectance levels than did deciduous forests. There seemed to be a tendency for higher reflectances in deciduous forests as they became more dominated by a single species. For example, in a birch forest with more than 90% *Betula papyrifera,* the reflectances were 147 (Band 6) and 172 (Band 7) while in a forest dominated by birch, poplar, and balsam fir, the levels were 106 and 125, respectively.

Seventeen plant communities were ordinated statistically (see Chapter 19) as shown in Figure 6.6. The groupings were based on the frequency of dominant tree species. The procedure [91] assigns statistical numbers (eigenvalues) to each community and positions them within a set of axes such that communities of similar species composition form clusters. The proximity of any two communities in the diagram is proportional to the degree of similarity between the communities involved. The plant associations ranged from pure coniferous forest (D) of *Picea mariana, P. glauca, Abies balsamea,* and *Tsuga canadensis;* to a completely deciduous forest (E) of sugar maple (*Acer saccharum*). Other associations included a mixed deciduous forest (C) dominated by *Betula papyrifera, Populus tremuloides,* and *P. balsamifera;* a mixed coniferous forest (B) of *Picea* spp., *Abies balsamea,* and scattered *Betula* spp.; a scrubby association (A) of *Alnus* spp., *Salix* spp. and various spruces, balsam fir, poplar, and birch. It is clear from this figure that all five communities are completely separated with this statistical treatment. The work of Bélanger and his associates in this area shows what can be done by a combination of satellite imagery, computer technology, use of sophisticated statistical procedures, and intelligent determination of ground truth.

6.6.4. Further Examples of Use of LANDSAT Imagery in Geobotany

There have been a number of other examples of use of LANDSAT imagery, often integrated with other forms of remote sensing. For example, Cole [210] used thermography and multiband aerial photography (true color, infrared false color, and black and white) for interpretation of ore horizons in the Mount Isa-Cloncurry district of Queensland. Interpretation of these photographs together with LANDSAT imagery readily indicated anomalous plant communities in various ore occurrences.

The main problem with LANDSAT imagery is that the resolution is not as good as that which can be obtained by low-flying aircraft. Nevertheless, Bølviken et al. [76] were able to use the MSS 7/5 ratio to delineate a copper-lead-poisoned area at Karasjok in Norway. Other similar areas could not be detected because of inadequate resolution.

Lyon [537] has reviewed the use of LANDSAT imagery for geobotanical mineral prospecting in regions of varying vegetation cover. The localities and results are listed in Table 6.3. The reader is referred to the above references for further information on these case histories.

An interesting case listed by Ballew [38] involved correlating LANDSAT multispectral digital data with surface geochemical data consisting of 126 samples from the Washington Hill mercury mining district. The samples (mainly soils) had been analyzed for mercury, gold, silver, lead, copper, and bismuth. The four MSS channel values and their six ratio pairs (using a linear and a logarithmic scale) were correlated statistically with metal values, pixel by pixel. The pixel signatures were sometimes due to the vegetation cover and sometimes due to bare rock. Multiple regression analysis (see Chapter 19) was used to determine the best channel functions for

TABLE 6.3. Localities and Results of Some Case Histories of Geobotanical Anomalies Detected by LANDSAT Imagery

Nature of Cover	Location	Results of Investigation
Very low to zero	Yerington, Nevada	Oxidized and nonoxidized ore (sulphide) may be differentiated in the pit and sulfide rock discovered in the tailings pond.
	Goldfield, Nevada	Alteration zones and gossans surrounding gold-alunite mineralization identified.
Moderate	Pine Nut Mts., Nevada	A molybdenum-bearing skarn with a biogeochemical anomaly in the piñon pine and juniper, independently located by color-ratio images.
Heavy	Karajok, Norway	A known copper biogeochemical anomaly [76] was relocated by LANDSAT data.
	Tifalmin, New Guinea	Vegetation over known copper-bearing intrusives was studied, though results inconclusive.

Source: Lyon [537].

predicting the metal content in the area. Metal anomalies corresponding to alteration patterns surrounding highly bleached areas, known mines, and known prospects were identified readily. Many of these anomalies were associated with ponderosa pines (*Pinus ponderosa*) which are often found on soils of low pH and low phosphate content [67]. The work by Ballew [38] is another outstanding example of what can be achieved by LANDSAT imagery.

6.6.5. Future Developments in Satellite Imagery

Future probable developments in satellite imagery have been reviewed by Ellis et al. [272]. The third LANDSAT satellite, LANDSAT-C, was launched at the end of 1977. This craft has an identical Multispectral Scanner to that on LANDSAT-II with the addition of a fifth thermal spectral band of reduced resolution. The other instrument, the Return Beam Vidicon (RBV), has a single panchromatic band of 40 m resolution. The thermal infrared band may provide the first synoptic view of the world's geothermal regions. The RBV imagery should help users to assess the gains to be made in, for example, urban monitoring and cartography by using this higher resolution.

LANDSAT-C is believed to be the last of the Earth Resources series in which data from the major instruments are recorded by an on-board recorder. LANDSAT-D with a thematic mapper of 30 m resolution and with six spectral bands will be launched shortly. Outside of the United States, data will be available only to users with access to a local receiving station. The French SPOT (Système Probatoire d'Observation de la Terre) satellite due to be launched in 1984 will have an even better resolution (20 m) than LANDSAT-D.

It seems that the most exciting current developments in geobotany are in the field of remote sensing. New satellites with improved resolution and new computers will greatly increase the flow of more and better information but, in the long run, the most successful workers in the field will be those who establish a good basis of ground truth and make intelligent use of appropriate statistical procedures.

7

AN ASSESSMENT OF GEOBOTANICAL EXPLORATION METHODS

7.1. INTRODUCTION

The effectiveness of geobotanical methods of prospecting is difficult to assess in order to place it in perspective with other methods. This is because so much depends on the skill of the field worker. For example, it is not difficult to get relatively untrained personnel to sample soils, but geobotanical surveying requires more skill, and although a botanical knowledge is useful, it is not absolutely essential.

A significant factor hindering a proper evaluation of geobotanical methods of exploration is that much of this work has not been carried out by exploration companies, but instead has been undertaken by organizations such as geological surveys or university groups. A common pattern of this type of work is that the research organization develops the procedures by working in a region of known mineralization and then provides the data to the exploration company, which then should presumably apply the method in unknown areas where minerals are to be sought. In fact, this vital second step is seldom carried out (although there have been important exceptions to this). Although scientific knowledge has been enriched by the survey, it becomes a piece of pure, rather than applied, research unless the mining company itself is willing to apply the technique or at least persuade the original workers to continue until the method has been applied in unknown areas. Does the reluctance of exploration companies to apply what they have paid for stem from an innate conservatism, or does it arise from the fact that some geologists, secure and confident in their own speciality, become faltering and hesitant once the boundary into another scientific discipline has been crossed?

Almost without exception, successful use of the geobotanical method has been the result of thorough integrated studies combining the initial survey with a subsequent exploration program and incorporating other procedures, such as biogeochemical, geochemical, and geophysical methods of exploration.

Another factor that can affect the success of an operation involves the selection of a suitable area to carry it out. Clearly a region completely devoid of vegetation would never be considered for geobotanical prospecting. On the other hand, there could well be areas that, although carrying extensive vegetation, would be no more promising, although this might not be obvious initially. Examples of this would be the presence of artificially introduced species in the area of the orientation survey but not found elsewhere. Other unpromising regions would be where the vegetation cover was sparse and where the anomalies were small in area. In such cases the minimum size of quadrat needed to define the vegetation assemblage might be appreciably larger than the target, so geobotanical work would probably be meaningless in that area.

7.2. SOME EXAMPLES OF SUCCESSFUL USE OF GEOBOTANY

7.2.1. The Sokli Carbonatite Complex

It is probably in the growing field of remote sensing (see Chapter 6) that geobotany will have its greatest impact in the future. One of the great success stories in this connection is the aerial mapping of the Sokli carbonatite complex in Finland [794,634]. Aerial surveys (visible and near-infrared bands) were combined with a very thorough evaluation of ground truth and enabled the complex to be delineated with great accuracy. Indeed before geobotanical aerial surveys, the massif was originally thought to have an area of only 20 sq. km; later investigations [634] have now increased this to 150 sq. km.

The spectacular geobotanical anomaly at Sokli is due to a luxuriant growth of vegetation over the phosphate-rich carbonatite, which contrasts with the nutrient-poor skeletal soils derived from the surrounding Archaean basement rocks.

7.2.2. Kimberlites

In a recent review, Cole [212] has shown that aerial remote sensing can be used to pinpoint kimberlite pipes because of their usually higher phosphate and potassium contents compared with the surrounding rocks. She has cited examples of kimberlites in eastern Botswana where the Orapa pipe (1.5 km across) produces a distinctive circular anomaly. Unfortunately many kimberlite tubes have a very small surface expression so that satellite imagery may only be able to identify the larger ones. A fertile field for further studies which may well have been undertaken already, would be a search for kimberlites in northwestern Western Australia which is likely to become an important new producer of diamonds in the immediate future.

In an earlier paper, Buks [138] used aerial photography to study kimberlite pipes in the Yakut ASSR of the Soviet Union. It was observed that these kimberlites contained an inordinately high concentration of the larch (*Larix dahurica*) compared with the surrounding forest. The vegetation was more luxuriant on the kimberlites where the diameters of the larch trunks were, on the average, twice as great as elsewhere. The same pattern was observed for *Alnus* spp. (alder).

In retrospect, it now seems that there are indeed plant indicators of diamonds in spite of the earlier somewhat fanciful claims (see Chapter 4) of Spix and Martius [774], in the early nineteenth century, and of Freise [303] a century later.

7.2.3. Copper Deposits

It is perhaps in the case of copper deposits that the most success has been achieved by the geobotanical method. Some of these cases have already been discussed (Chapter 4), but further mention should be made of the famous "copper flower" of Central Africa. *Becium* (originally *Ocimum*) *homblei* was used successfully by the geologist G. Woodward who, at the time (1949), was employed by Roan Selection Trust in Zambia. The following extract [23] describes the early work.

The more information was collected, the more convinced some of the investigators became that the "copper flower" theory was fact. So faithfully did the flower follow the line of deposits, the flower charts were almost identical in outline to the underlying copper deposits. Where the ore body narrowed, so did the distribution of the flowers and where the ore widened as at Roan Extension, Muliashi, the floral spread not only followed suit but appeared in greater density. Over known deposits, the abruptness with which it ceased to grow at almost the precise edge of mineralization, showed that here indeed was a sensitive copper indicator of remarkable selectivity. . . . By 1954, observations had been made at 30 widespread copper occurrences from near Lusaka to the Congo border. In all but two cases the faithful floral signposts were there. In both exceptions the soils over the mineralized formations were deeply leached and had a low copper content. . . . It is now nine years since Copperbelt geologists' researches began and in that time they have more than once had the satisfaction of seeing the occurrence of *Ocimum* (i.e., *Becium*) *homblei* in places where copper was not previously known, lead to soil sampling, pitting, and finally drilling with successful result, . . . pits were being sunk on a long-known ore body about three quarters of a mile from its then known limits, and extensive growth of the copper flower had been previously recorded on a floral map. . . . Drill rigs came in and among the sage (i.e., *Becium*) clumps that had brought them there, the diamond bits went down to intersect large new reserves.

As it turned out, *B. homblei* was ultimately proved to have a widespread occurrence in Central Africa [370] (see Chapter 4) even over nonmineralized ground. Nevertheless this does not detract from its original effectiveness as a local indicator in the Zambian Copper Belt. This reader is referred to Chapter 4 for other examples of successful use of the geobotanical method.

7.3. ECONOMICS OF GEOBOTANY

It is not easy to give a general evaluation of the economics of geobotanical methods of prospecting since these techniques are themselves so diverse. In the case of studies carried out on foot, the execution is quite simple and the only cost lies in mapping the occurrences of plant species. For this part of the work, costs must surely be lower than for any other geochemical technique since no expensive equipment is needed and there are no costs of analysis to be considered. A further advantage is that results are available almost immediately. These considerations, however, do not take into account the cost of a preliminary orientation survey which might be needed to establish the identity of indicator plants or plant communities. Too much emphasis should not be placed on such costs because once a universal indicator has been discovered, its use should be applicable in a large number of other areas.

In these days of hyperinflation, there is a natural reluctance to cite costs in absolute units of dollars and cents since rising costs will have made the figures meaningless even before a book is printed. However, at the time of writing (1983), the cost of a LANDSAT scene covering 34,000 sq. km is about $500, if digital data are required for the four channels. The same area covered by conventional aircraft would cost about $250,000 for an image scale of 1:50,000, or $7.35 per sq. km. Against this the LANDSAT unit cost of $1.45 is negligible. This does not, of course, take into account the cost of processing the digital data and subsequent plotting of maps, but even when these are taken into account, the cost is still negligible. There is also the poorer resolution of LANDSAT to be taken into consideration but later generations of satellites are planned with much better resolution than the earlier LANDSAT craft. There can be little doubt that satellite imagery represents the best cost-effective technique for carrying out surveys of large areas.

7.4. ADVANTAGES AND DISADVANTAGES OF GEOBOTANY

A number of advantages and disadvantages of geobotanical methods of prospecting are listed below.

Advantages

1. Once an orientation survey has been carried out, further costs are extremely low.
2. Different geological formations, as well as mineralization within them, can be detected by geobotany.
3. Aerial methods can be applied with consequent saving of time and effort.
4. Indicator plants can sometimes indicate mineralization at depth in cases where there is no surface expression of this mineralization.

Disadvantages

1. A high degree of skill is required.
2. Data may not be universally applicable.
3. Sometimes the method can be applied only at certain seasons; for example when plants are in flower.
4. The method can only be applied where favorable vegetation exists.
5. There is little coordination of this type of work at present.

The above listing of advantages and disadvantages of course is far from complete, but the main points have been covered. A consideration of paramount importance is that geobotanical methods are just a few of many techniques and should supplement rather than replace other procedures. There are at present far too few people engaged in and trained in this type of work. This deficiency should be remedied as soon as possible so that every progressive exploration company will have a pool of trained staff members available for carrying out such studies.

7.5. THE FUTURE

It is probable that the most important future developments in geobotany will be in improved satellite imagery combined with more sophisticated methods for evaluation of ground truth. Better resolution remote sensors, combined with more sophisticated statistical and computational procedures, should do much to improve the effectiveness of the technique and to remove to a greater extent, the degree of subjective personal judgment which at present must be applied to the method.

It might at first seem to be difficult to predict what future improvements might be made in the field of study of indicator plants and indicator plant communities, since such operations tend to be visual and do not lend themselves to automation. However, if efficient instruments could be designed for *in situ* measurement of elemental concentrations in rocks, soils, and minerals, this will be immensely useful in orientation surveys. At present the geobotanist has to rely on chemical analysis of the substrate for a thorough study of indicator plants. A step in this direction has been made by the development of portable X-ray fluorescence units [681] which in theory make such measurements a possibility. The problem so far with these units is their relative lack of sensitivity, but if significant improvements in their performance can be made in the future, they may well prove to be useful for the geobotanist. The only *in situ* analytical instrument at present that is particularly useful for the geobotanist is the portable scintillometer for measuring radioactivity, an instrument of inestimable value in this type of work.

A final factor should be considered now. In time, and certainly in the foreseeable future, a very large part of the earth's surface will have been surveyed by various geochemical and geophysical prospecting techniques. Geologists will have left until the last what is to them the least familiar of potential methods. When that time comes, geobotanical methods of prospecting will be the last of the potential techniques to be fully exploited in the search for minerals, which will become progressively more elusive with each passing year.

2

GEOZOOLOGY IN MINERAL EXPLORATION

8

INTRODUCTION TO GEOZOOLOGY

At first sight the idea of using animals as field assistants might seem to be ludicrous in the extreme. Yet it has been found that such dissimilar animals as dogs and termites can be of use in prospecting for minerals. A survey of the literature shows that not only these but other animals have been used for this purpose either directly or indirectly.

The problem immediately arises as to a suitable nomenclature for what is largely an uncollated new science. To be consistent with botanical methods of exploration, techniques involving direct analysis of animal material (e.g., trout livers, see Chapter 10) should be termed **biogeochemical,** whereas those involving a visual study of animals or animal disease or behavior should be termed **geozoological.** There then remains the problem of classification of cases where animals are used directly as field assistants, as in the case of ore-detecting dogs (see Chapter 9). There is the further example of the analysis of termite mounds for inorganic matter brought to the surface by these creatures (see Chapter 11). Is this pedological prospecting?

In view of the above problems of terminology, I have decided to use the term **geozoology** to describe all methods of prospecting involving the direct or indirect use of animals, whether these methods be visual or involve chemical analysis.

The science of geozoology, though new in nomenclature has its roots in antiquity when Herodotus (see Chapter 11) reported that the natives of India studied termite mounds to detect gold mineralization. It is only recently, however, that the field has begun to have a wider application in mineral prospecting. Its potential is obviously not as great as that of botanical methods because of the basic problem of the relative mobility of most (though not all) animals. In only a few cases are animals so restricted in their habitat that they can provide favorable conditions for prospecting for minerals.

In this section of the book there will be a discussion of how various animals such as mammals, birds, fish, and insects can be used in the search for minerals. Particular attention will be paid to the use of two of these (dogs and termites) as voluntary and involuntary field assistants. It is hoped that the ideas presented will assist in bringing to the attention of research workers a very neglected field of mineral exploration.

9

LAND MAMMALS AS INDICATORS OF MINERALIZATION

9.1. INTRODUCTION

There are several ways in which land mammals may be used directly or indirectly to indicate mineralization. The first of these is by clinical symptoms of poisoning by heavy metals, and the second is by symptoms caused by an excess of one element producing a deficiency of another in the animal. The third method is by direct use of animals as field assistants. Each method will now be considered.

9.2. LAND MAMMALS AFFECTED BY TOXIC METALS

Variations in the health, form, and vitality of plants and animals under the influence of various chemical factors in the environment from which they obtain their nutrients have been known for many years. An excess or deficiency of a particular element or group of elements usually is linked with some form of endemic disease or condition which readily distinguishes the district from others in the neighborhood. Classical examples of this are the effects of iodine deficiency on humans and animals and the effect of serpentine soils on vegetation (see Chapter 3). Vinogradov [852] has assigned the name **biogeochemical provinces** to such areas, shown in the following quote:

We apply the name biogeochemical provinces to regions on earth which differ from adjacent regions with respect to their content and chemical elements and which as a result experience a different biological reaction by local flora and fauna. In extreme cases as a result of marked deficiency or excess of the content of any chemical element or elements in the province, plants and animals will suffer biogeochemical endemias. By regions of the earth we mean rocks, soils, and water basins which are populated by organisms and which have limits in time and space. We include in this definition parts of the sea, atmosphere, or sea and atmosphere together. With respect to the content of one or more chemical elements in a particular region, that is, in rocks, water, and the organisms of this region, we mean either the normal level or an inadequate or excess content of a particular chemical element or several elements (or their compounds). It should be noted that it is important to know not only the absolute content of an element in a particular region, in its rocks and soils, but also the contents of any particular form of this element. For example Fe^{2+} or Fe^{3+}; Mn^{2+} or Mn^{4+}; Se^{4+} or Se^{6+}. Thus we often encounter not only an absolute shortage or excess, but also a relative deficiency or excess depending on the degree of availability of any particular chemical element to particular types of flora and fauna. For instance, cobalt and many other chemical elements in areas where there is an alkaline soil reaction, are quite unavailable to plants. On the other hand, under these conditions, molybdenum is easily taken up as a result of easily soluble molybdates while in an acidic medium, this element is less available to plants due to its leaching into compounds which are not easily soluble. The biological reaction of flora and fauna of a particular region arising under the influence of an excess or deficiency of any chemical element (or elements) is the most important and basic criterion for the delimitation of a biogeochemical province.

Biogeochemical provinces are said to be **zonal** (strongly influenced by climate and soil type) or **intrazonal.** Zonal provinces usually involve deficiencies in elements such as calcium, cobalt, and iodine due to the nature of zonal soils in the area. Intrazonal provinces are often influenced by local enrichment of elements due to the existence of ore bodies and their associated dispersion halos.

In domestic animals, nutrition and health are influenced by the composition of the herbage they consume from a variety of soils, not only by gross deficiency or toxic excesses of particular elements but also by variations in the relative proportions of these metals. The administration of supplementary diets that are not necessarily of local origin can often confuse the picture by modifying the influence of the biogeochemical province.

The link between human health and geology is even more complex because humans, unlike animals, are very mobile and their diets (or even drinking water) are not necessarily derived locally. The study of the geographical nature of human disease is known as **epidemiology.** For the reasons stated above, the study of human disease patterns is not a good way of attempting to locate mineralization in a biogeochemical province, unless perhaps in the case of primitive communities isolated from twentieth-century civilization.

Epidemiology is a very controversial field, and almost any findings can be subject to widespread questioning and criticism. One of the many human diseases said possibly to be caused by heavy metals in the environment is **multiple sclerosis** (MS). It has been alleged that high levels of heavy metals such as copper, lead, and zinc are to be found in areas of high incidence of the disease, compared to areas of low incidence. The same elements are said to produce the disease known as **Balkan neuropathy** [357] and high levels of arsenic are said to produce **blackfoot disease** in Taiwan [357]. It is well known that high levels of arsenic in drinking water supplies near Halifax, Nova Scotia, cause sickness and even death and are derived from nearby gold mine workings where arsenic is geochemically associated with the gold.

Domestic animals are much more suitable for geozoological prospecting for minerals since they are usually confined to a small area and derive most of their nutrition therefrom. The relationship between animal diseases and mineralization was highlighted by a monumental study carried out by J. S. Webb and his associates [802, 913, 915, 916] at Imperial College, London. In 1965, the group carried out a geochemical reconnaissance survey of stream sediments in England and Wales. Some 50,000 samples (one per sq. mile) were analyzed for about 20 elements and the data were transferred to the pages of a synoptic geochemical atlas with an individual sheet for each separate element and with elemental concentrations indicated by solid circles of varying diameter. The survey highlighted many areas of trace element excesses or deficiencies and, in retrospect, was able to correlate these with previously recorded instances of diseases caused by excess or deficiency of heavy metals.

One of the most important of the above discoveries is shown in Figure 9.1. A geochemical map of molybdenum in Derbyshire [916] showed significant anomalies in the streams which drain a contact between limestone and younger sandstones and shales. The same figure shows the distribution of the occurrence of the disease hypocuprosis (deficiency of copper in the blood) among cattle. The copper deficiency in this case is certainly caused by the antagonistic effects of high molybdenum levels in the environment. The degree of correlation in this case is remarkable. A similar finding was made at Limerick, Ireland [913], where high molybdenum in sediments correlated well with **hypocuprosis** in cattle.

A more direct indication of mineralization (also in Derbyshire) is afforded by the residual toxicity of ancient lead mines dating back as far as Roman times. For example, Chisnall and Markland [191] investigated farmland where several cattle and ponies had died from suspected lead poisoning adjacent to Roman lead workings, and found up to 200 μg/g of lead (dry weight) in the grass of the pastures of the area.

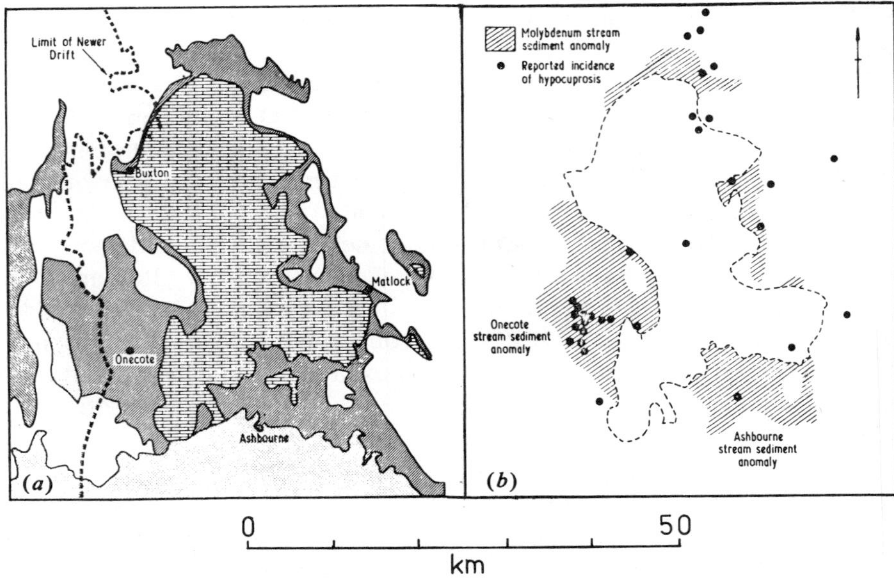

FIGURE 9.1. (*a*) Distribution of molybdenum in stream sediments of Derbyshire, England. (*b*) Incidence of hypocuprosis in cattle and sheep in Derbyshire. The central region bounded by a broken line is a limestone outcrop. *Source.* Webb et al. [916]. Reprinted by permission from *Nature.* Copyright 1968, Macmillan Journals Ltd.

One of the best known areas of copper/lead/zinc mineralization in the United Kingdom is in North Wales, where the alluvial soils support grasslands that have occasionally caused lead poisoning in cattle [802] and toxic levels of zinc in cereals [802].

Easily the best known of ore discoveries originating from observation of toxic effects in animals, was made on the Colorado Plateau. It has already been referred to in Chapters 3 and 4. For many years if had been recognized that certain plants (particularly those of the genus *Astragalus*) were extremely poisonous to stock [53]. This problem is so widespread in the Western United States that very large areas cannot be grazed at all. For a long time the cause of the toxicity was unknown until finally selenium was implicated as the culprit. The *Astragalus* plants accumulate large amounts of selenium, but since the seleniferous ground often contains associated uranium, Cannon [154] was able to use the distribution of the *Astragalus* plants to indicate uranium mineralization at depth. The final uncovering of uranium was therefore the end member in a chain of four factors: sick animals; plant indicators of selenium; association of selenium and uranium; discovery of uranium.

The above study was of great importance because it was not retrospective. It was carried out as a logical follow-up of observed diseases in animals. The other studies were retrospective but there is no reason to suppose that intelligent assessment of the original clinical condition should not have led to a mineral discovery. After all, Agricola [7] had pointed out as long ago as 1556, that minerals produce stunting and other morphological symptoms in plants and animals. The reason that few, if any, exploration personnel use the occurrence of animal diseases in prospecting is probably because of the increasing specialization of science. There is little contact across the psychological barrier that separates scientists of one discipline from another.

As a corollary to the above statements, it should be mentioned that spectacular mineral discoveries are not likely to result from a study of animal diseases alone. Such studies, however, may reveal biogeochemical provinces, where use of other prospecting methods may reveal the presence or ore deposits. Unlike plants, however, animals will never be identified by remote sensing methods, so that geozoology is never likely to be as effective as geobotany in mineral exploration.

9.3. THE USE OF DOGS TO SCENT OUT MINERAL DEPOSITS

The use of dogs as field assistants in the search for ore deposits seems at first sight to be a rather fanciful proposition. However, dogs have been trained to scent out sulfide-rich boulders and ore deposits in four countries. According to Brock [95], the pioneering work began in Finland with the investigations of Dr. Aarno Kahma of the Geological Survey of Finland, who conceived the idea in 1962. It was based on the assumption that a trained dog, with its very keen sense of smell, should be able to detect a weathered sulfide boulder from a distance of several meters and possibly even below the overburden.

Together with an assistant, Kahma selected an Alsatian dog and trained it for 2 years before putting it to a practical field test in 1964. Preliminary field tests were successful and, in 1965, the dog "Lari" (shown in Plate 9.1) was pitted against a human float-boulder prospector on an unprospected field of some 3 sq. km. The field was gridded and the finds marked on a map. The results were very impressive. Lari found 1330 sulfide-bearing boulders, some of which were below an overburden of 10–20 cm. The prospector found only 270 surface boulders. In that same summer, Lari also located pyrite and chalcopyrite floats on another prospecting site, and as a result of this discovery, a copper orebody of economic significance was later encountered by diamond drilling. A reward of $2000 was paid to the trainer by the Finnish Government and a smaller sum to the originator of the idea. Lari was rewarded with four frankfurters. The Geological Survey of Finland has a number of prospecting dogs in use and uses these animals in routine prospecting. Kahma's work was not reported until much later [269, 408], but the idea of using dogs quickly spread to Sweden [623, 624], the Soviet Union [630], and Canada [95]. Plate 6.1 (bottom left) shows the ore dog "Hessu" searching for uranium in Finland.

PLATE 9.1.   Lari, the world's first ore-searching dog. Photograph reproduced by kind permission of P. Mattson.

Experience of the use of dogs in Sweden has been summarized by Nilsson [624], from whose work the following extract is reproduced by kind permission of the Institution of Mining and Metallurgy, London.

. . . "Today the usefulness of the dogs cannot be questioned since they are able without difficulty to locate boulders containing relatively small amounts of pyrite, pyrrhotite, chalcopyrite or arsenopyrite which are covered by 10–20 cm of soil. Several cases have been reported of dogs having located boulders buried at a depth of 1m in a boulder pile. One report from the Soviet Union stated that dogs have located orebodies covered by alluvial sediments 12 m thick [630].

At the present time (i.e., 1973) the Geological Survey has four prospecting dogs. . . . So the dogs can be used effectively, the Geological Survey in turn gives the trainers a course in the techniques of boulder searching and integrates them with the Survey's prospecting teams.

Several breeds of dogs are thought to be acceptable for Training Center has from the beginning chosen German Schäfers (Alsatians) since they possess the strength and energy for boulder tracing. The Labrador retriever is another dog which because of its fine sense of smell is thought to be suitable. Labradors however have not been tested in the field, so a proper judgment cannot be made at this time. Choosing the right dog is probably more a question of picking a specific dog with the right characteristics, than of choosing a certain breed of dog. For this reason, at the Dog Training Center each dog must pass suitability tests before it is put into a training program. These tests involve: sense of smell, defence, and endurance. Positive test results in such areas as sense of smell must be produced first. . . .

The training of prospecting dogs begins when the animals are 1–2 years old, the programme itself taking about four months. This period is followed by approximately two months of practical field training. Although the dogs normally perform satisfactorily on completion of this training, they are usually not at their best until they have been at work for a full field season. Each year thereafter, the dogs are given a three-week refresher course. . . .

It is advantageous if the first part of a dog's basic training can take place under winter conditions. Snow covering the ground and low temperatures reduce the number of outside odors that can confuse the dogs. In addition, they seem to react better to the weak smell of sulphur compounds in sulphide-bearing boulders under these conditions.

The training is intended to develop a desire on the part of the dogs to locate sulphide-bearing rocks and to defend them. Advantage is taken of the dog's fighting spirit. Initially the dogs are taught to retrieve various sizes and types of sulphide-mineralized rocks thrown out by the trainer. Some of the rocks contain no sulphides. If the specific rock does contain sulphides, the trainer tries to get the dog to defend it by pretending that he is going to take it away. . . . When the dog's defence of sulphide boulders has reached the point of aggressiveness, it is then trained to search for mineralized boulders on its own initiative and to mark the spot by barking. . . .

In the field the dog is led to his searching by the trainer. The trainer chooses a suitable path over the area to be worked, taking into consideration the direction of ice movement and wind conditions. The dog is allowed to search freely along this path. Theoretically the dogs can be used in all types of terrain for the whole year round. The Geological Survey however has not needed the dogs during the winter and therefore has no field experience from this season. Russian experiments [630] have shown that snow up to 0.5 m thick does not present any difficulties.

A dog's sense of smell functions best under humid conditions. The dogs however have not shown any noticeable difficulties in marking boulders during the dry summer period. . . . Dogs do not seem to be able to distinguish between different sulphide minerals. This means that areas containing an abundance of uneconomic sulphides (e.g. pyrite) are less suitable . . .

It is estimated that a prospecting dog will perform effectively for about seven years. During this time a healthy well-trained dog can put in a normal working day, i.e. eight hours a day.''

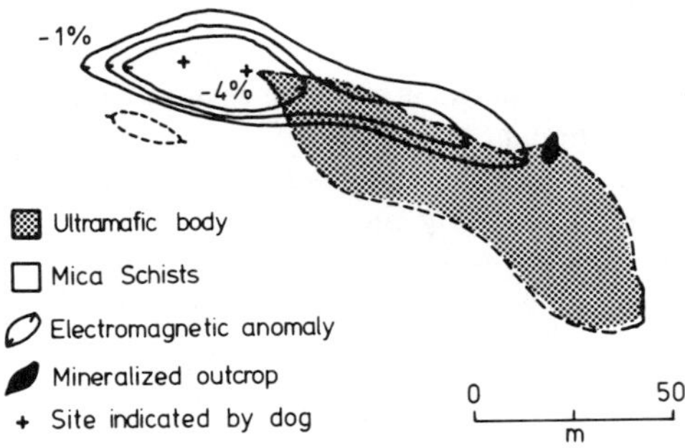

FIGURE 9.2. Site where a dog detected sulfide mineralization beneath a 2-m overburden at Pielavesi, Finland. *Source:* Ekdahl [269].

A case history from Finland is illustrated in Figure 9.2 which represents a site where weakly mineralized outcrop was known. A prospecting dog showed heightened interest at the sites marked with crosses on the map. When this area was surveyed by geophysics, an electromagnetic anomaly was detected. Trenching of the anomaly revealed the presence of pyrrhotite-chalcopyrite dissemination in a highly weathered horizon overlain by a 2-m thick sand and clay overburden.

In Canada, experience with dogs has been rather less favorable than in Fennoscandia [95]. Nevertheless some success was achieved with two Alsatian dogs named Buddy and Jai. Training was different from that used in Sweden and was based on the animals' love of play rather than on their aggressive instinct. The dogs were trained to retrieve rocks that the trainer threw out. Retrieval of a sulfide-bearing rock earned a reward, whereas nonmineralized retrievals involved a reprimand. At a later stage the fetching reward was replaced with verbal praise. It was noted that the interest of the dogs waned after long periods without successful finds. After about 2 months, the dogs were able to find medium- to high-grade rocks which had been buried in a test plot under 3 cm of soil.

In field operations in the Yukon, Saskatchewan, Quebec, and New York State, it seemed that the dogs could definitely smell and retrieve sulfide floats and boulders. High temperatures, high humidity, and thick vegetation seemed to hinder the dogs. Although the degree of success obtained with the Canadian dogs was much less than with the Fennoscandian animals, this may have been because the Canadian trainers had had less experience of the art, or indeed because the first two dogs may not have been very suitable.

The obvious ability of dogs to detect sulfide mineralization may have applications that overlap into geobotany. It has been my experience in Western Australia that most native plant species have a characteristic odor which the field worker can readily recognize and which can serve as a guide to identification. For example, I can readily recognize the nickel indicator and hyperaccumulator, *Hybanthus floribundus* [209, 740] (see Chapter 4), from its odor alone. If this is the case, a well-trained dog should be able to perform very much better than a human and be able to assist the geobotanist in identification of indicator plants.

The limiting factor for future use of prospecting dogs will undoubtedly be the cost and time expenditure that is required for training. For this reason they are not likely to be used on a widespread basis throughout the world, and will probably be confined to cold-temperature and sub-Arctic regions where there is an extended winter period without distracting extraneous odors.

10

BIRDS AND FISH AS INDICATORS OF MINERALIZATION

10.1. BIRDS AS INDICATORS OF MINERALIZATION

There are numerous references in the pre-1900 literature to accumulation of noble metals such as gold in birds [85]. Almost all of the earlier reports are suspect, mainly because of lack of suitable methods of analysis. During the past 20 years, however, there have been a number of reliable data on the subject. For example, Razin and Rozhkov [680] found up to 30 ng/g (ppb) gold in five species of Siberian birds. The data are of no practical value in prospecting because of the vast distances covered by the birds during their migratory travels. Even local birds can be found over a wide range of territory.

There is a most interesting case of a bird that is alleged to have been used successfully in prospecting for gold. It has already been referred to briefly in Chapter 4, and was first mentioned by Pagliuchi [635] (later cited by Boyle [85]). The following quotation is taken from a section entitled "A Feather Guide to Gold."

> . . . I constantly heard the sharp cry of a bird which is seldom seen as its habitat is in the dense foliage of tall trees. The bird is gray in color and about the size of a robin. It is called "el minero" (the miner). The natives informed me that this bird is always found in the vicinity of gold mines or wherever gold is abundant. Thereafter, as I rode through the country, wherever I hear the cry of "el minero" either I saw a quartz ledge extending across the trail

or was able to find one by looking about a little. Apparently the natives were correct. Why the affinity which seemed to exist between the bird and quartz ledges (i.e., reefs)? My investigations and attempts to solve the mystery left me still puzzled. Finally a miner told me that if I wanted gold I must always look for the mora tree. He pointed out to me some trees which grew next to the quartz ledge that I was examining at the moment. Then the solution of the puzzle dawned upon me. The mora tree grows and thrives only on siliceous ground, and "el minero" feeds on the berries of this tree; hence whenever his cry is heard, one is assured of finding a quartz vein in the vicinity. Of course the quartz may or may not be auriferous, but when prospecting in a country known to be auriferous, it must be admitted that the bird is a helpful quide to the prospector.

In spite of the above optimistic assessment of the usefulness of "el minero" it may well be that this is yet another example of mythology (see Chapter 4). From correspondence with D.W. Snow of the British Museum, it appears that "el minero" is in fact a common South American bird known as the screaming piha (*Lipaugus vociferans*). This bird is also known as the "gold bird." In Bolivia it is known as the "seringuero" (rubber bird) and in Guyana as the "greenheart bird" because of its respective association with rubber trees and greenheart (a useful native timber tree). Its loud and penetrating call is one of the most characteristic sounds of the forest. The fruit of the mora tree (*Mora excelsa*) has a large woody pod quite unsuitable for fruit eaters such as *Lipaugus*. The male birds gather together in traditional places in the forest (known as leks) in order to call the females and to carry out mating. These places tend to be in more open parts of the forest (as, e.g., in nutrient-deficient areas of quartz reefs). This tenuous link may be the only real evidence for supposing that "el minero" does in fact indicate quartz reefs at all.

10.2. FISH AS INDICATORS OF MINERALIZATION

Fish may be used more successfully than birds in mineral exploration. The first reference to this possibility is in a paper by Warren et al. [901]. These workers analyzed the livers of 96 rainbow trout (*Salmo gairdneri*) and cutthroat trout (*S. clarkii*) from various localities in British Columbia. All the livers with >50 μg/g zinc (wet weight) were from fish living in waters known to be associated with economic zinc mineralization. Of 17 localities where livers showed >60 μg/g copper, four were known to contain significant mineralization and another seven merited further investigation on the basis of their known geology. The same authors also analyzed trout livers from a highly mineralized stream in West Devon, England, and found highly anomalous copper levels (100–340 μg/g) in three of the samples.

A later paper by Ward [877] reported molybdenum levels in tissues of rainbow trout and kokanee salmon (*Oncorhjyncus nerka*) from waters varying greatly in molybdenum content. There was a good correlation between the two variables, though increases in the fish were only slight for increasing amounts of molybdenum in the waters. The livers of the rainbow trout contained a mean of 2233 μg/g (wet weight) compared with 100 μg/g in *O. nerka*.

Heavy metals (cadmium, copper, iron, mercury, manganese, and zinc) were determined in organs of 111 specimens of *S. gairdneri* and *S. trutta* (brown trout) in order to study emanation of these elements from geothermal regions of New Zealand [111]. The results showed a good correlation between elemental levels and the degree of geothermal activity in the collection locality. Livers contained up to 506 μg/g (wet weight) of copper and up to 60 μg/g zinc. They also contained up to 1.3 μg/g mercury, which is well above the normally accepted level of 0.5 μg/g established by most health authorities.

Bollinberg [74] was one of the first to suggest that seaweed, shellfish, and pelagic fish could be used for mineral prospecting. His work was carried out in the vicinity of relatively undisturbed zinc/lead ores of the Quamarujuk Fiord in west Greenland. Lead and zinc concentrations in the seaweeds *Fucus distichus* and *F. vesiculolus* and the mussel *Mytilus edulis,* increased at stations nearest to the ore deposit. The lead content of the seaweed increased from 2.3 μg/g (wet weight) to 200 μg/g, whereas zinc increased from 3.7 to 488 μg/g. Similar findings were obtained for other chalcophile elements such as copper (3.1 to 8.9 μg/g) and cadmium (0.6 to 8.0 μg/g). A similar pattern was observed for soft portions of *Mytilus edulis:* 1.1 to 5.4 μg/g copper, 2.2 to 22 μg/g lead, 5.2 to 36 μg/g zinc, and <0.5 to 0.6 μg/g cadmium. Up to 142 μg/g zinc was found in the liver of the wolffish (*Anarrhicas minor*). In a later survey, Bollingberg and Johansen [75] studied the lead content of organs of the wolffish in the same area, before and after exploitation of the lead/zinc ores. They found a five-fold increase of the lead content of livers, which had been high even before mining began.

It is obvious that the analysis of fish livers can only be effective if the species are confined to a fairly small area. For example, migratory salmon could hardly be used, and it is difficult to imagine that any oceanic pelagic fish might be used for this purpose. The idea of using shellfish, however, is a good one. For example, Butterworth et al. [142] found anomalous levels of cadmium and zinc up to 140 km downcoast away from the Avonmouth smelter in England. This was admittedly a question of pollution from a smelting complex, but because of the very high concentration of sea-water trace elements found in shellfish (of the order of one million times for some elements), the shellfish should prove to be very sensitive indicators of potential mineralization and have the further advantage of being easy to obtain and of having a relatively fixed habitat.

CHAPTER

11

INSECTS AS INDICATORS OF MINERALIZATION

11.1. INTRODUCTION

It already has been suggested (Chapter 5) by Canney et al. [149] that insects could be used to detect metal-stressed trees which presumably have a lowered resistance to insect attack. The main flaw with this argument is that many other factors apart from metal stress may predispose insects to attack a given plant, and it is possible that geochemical stress may feature in only a small proportion of such attacks.

If insects were to be used directly for mineral exploration, there would be no advantage compared with soil or vegetation sampling which would be easier and more reliable. It is not surprising therefore that the literature concerning elemental uptake by insects is very sparse. Babička et al. [35] reported gold in cockchafers from Oslany in Czechoslovakia. However, as some of the earlier work reported from this area is suspect (see Chapter 4), it may well be that these data are also unreliable.

An interesting survey of the metal content of insects was carried out by Wild [936] on the termites *Odontotermes transvaalensis* and *Trinervitermes dispar* located in termitaria (termite mounds) on serpentinites of the Great Dyke of Zimbabwe. Workers of both species contained high levels of nickel and chromium of up to 2000 and 1500 μg/g (dry weight), respectively. The soldiers contained much less (up to 300 μg/g for both elements), and the queens even less (20 μg/g and 30 μg/g, respectively). The explanation for this is that the workers feed directly on the nickel and chromium-rich vegetation, whereas the soldiers and queens are fed second-hand by a liquid

101

diet administered by the workers, who presumably have decontaminated the saliva by absorption of heavy metals into their own bodies. A similar high level of nickel and chromium (2000 and 2500 μg/g dry weight, respectively) was observed in a tenebroid beetle (*Catamerus* sp.) from the same area.

In the above examples it was hardly a worthwhile proposition to analyze termites at random for the nickel and chromium content as a means of detecting serpentinite, because such areas are so well defined by the visual appearance of the vegetation alone (see Chapter 3). However, although insects cannot be used directly for mineral exploration, there have been reports of two novel indirect methods, which will now be described.

11.2.   BEES AS PROSPECTORS

It was first suggested by Makarochkin and Udenich [542] that the composition of honey may indicate the elemental content of the soil in which the nectar-producing plants grew originally. They reported concentrations for Si, Al, Mg, P, Mn, Fe, Ti, Mo, and Cu in the ash of honey from the Lake Ilmen area of the Soviet Union. Elevated levels for molybdenum (200 μg/g) and copper (100 μg/g) were presumed to be due to elevated concentrations of this element in the local soils.

In a later paper [541], analysis of honeys from six widely separated districts of the Soviet Union indicated geographical differences in trace element contents of the soils. In a further extension of this work, in a region of the southern Urals [543] it was observed that the molybdenum, copper, and titanium content of honey ash was significantly higher than in the local rocks. The same authors discussed the possibility of using this for prospecting purposes.

There appears to be no reason why honey should not be used as a sampling type in mineral prospecting. The bees themselves have seldom been analyzed, though Razin and Rozhkov [680] reported about 20 ng/g (ppb) gold in a bee (*Vespidae* sp.). Bees usually do not forage more than a kilometer or so away from the hive and the honey that they produce is therefore indirectly indicative of the immediate environment. The technique at this stage is admittedly somewhat esoteric and will need to be proved by extensive experience and scientific testing until it has widespread acceptance by the scientific community.

11.3.   TERMITES AS PROSPECTORS

Of all the geozoological methods mentioned in this book, the use of termites as field assistants is one of the best proven and potentially most useful. In 1965, W. F. West, manager of the Leopard Mine, Zimbabwe, suggested that termite mounds be sampled for mineral prospecting [921]. The following quotation is from this original paper.

> Termite colonies abound in this country and they cannot exist without water. Owing to the intermittent rainfall, they must depend on underground water supplies to keep them alive for six to nine months each year, and the water must come from water-carrying fissures. On the Leopard Mine their water-carrying passages extend down to the 60 m level in the mine which is the present water level. The termites mine their way down the softer sections of the surface rocks to obtain the water, these softer sections obviously being the dried-up portions of the water-carrying fissures and, in the case of the Leopard Mine, the ore-bearing fissure. In mining their way downwards to water, they remove the necessary soil and bring it to the surface to be deposited in the form of a heap known as an ant heap. Two things are obvious: (a) that the termites have located a fissure, and (b) that they have brought a form of sample of the material contained in that fissure, to the surface. Some of the termite heaps on the Leopard and Leopardess Mines yield more than 3 g of gold per tonne of material to sampling, and these are located on fissures containing payable orebodies that are at present being profitably mined. The plotting and sampling of termite heaps, irrespective of the nature of the overburden such as Kalahari sands, is a useful and accurate method of prospecting with this advantage, the onerous part of the work has already been done by the termites at no cost to the prospector.

Mr. West was not the first to suggest that termitaria could be used for mineral prospecting. Even in 450 B.C., Herodotus [349] wrote:

> . . . these are the most warlike of the Indian (i.e., from present-day India) tribes, and it is they who go out to fetch the gold—for it is in this part of India that the desert lies. There is found in this desert a kind of ant of great size— bigger than a fox though not as big as a dog. These creatures as they burrow underground throw up the sand in heaps just as our own ants throw up the earth, and they are very like ours in shape. The sand has rich content of gold, and it is this that the Indians are after when they make their expeditions into the desert.

One suspects that Herodotus may have exaggerated the size of the termites, but there seems no good reason to doubt that the early Indians may have used termitaria for prospecting purposes. A typical termite mound is shown in Plate 11.1.

The anthill theory of West was at first greeted with ridicule by some mining circles [24] but ultimately he excited the interest of the Minister of Mines [25] who observed that the technique had been used to discover a gold anomaly concealed by an overburden of Kalahari sand, a common covering of much of south-central and southern Africa which effectively conceals outcrops and renders the prospecting work much more onerous.

Follow-up work carried out by Watson [912] did not substantiate West's claims and it was asserted that the termites may have carried the mound

PLATE 11.1.   Typical termite mound of *Macrotermes natalensis* from southern Kasai (Zaïre). Photograph by R. P. Bouillon, supplied by the Musée Royal de l'Afrique Central, Tervuren, Belgium.

material from a depth of less than 3 m, rather than from the water table at 227 m. The same worker [911] reached similar conclusions for a zinc anomaly covered by Kalahari sand in western Zimbabwe.

Independent work by West [922] did appear to vindicate his earlier thesis that termites indicate gold at depth. An area of 10 sq. km in the Silobela area some 80 km west of Que Que, Zimbabwe, was selected for a test. This area was covered with an overburden of 1–3 m of soil derived largely from nonmineralized Kalahari sand. The area included three gold mines that had been observed by the ancient (fifteenth-century) gold miners and had been worked in modern times. Samples were taken from 45 cm below the highest point of each mound and were analyzed for gold. It was evident that there were anomalously high gold levels in the groupings of ant hills. These areas coincided almost exactly with the payable sections of three mines. Another traverse in a different area showed that termitaria were indicating gold, zinc, lead, molybdenum, and silver mineralization. Two anomalies (zinc-

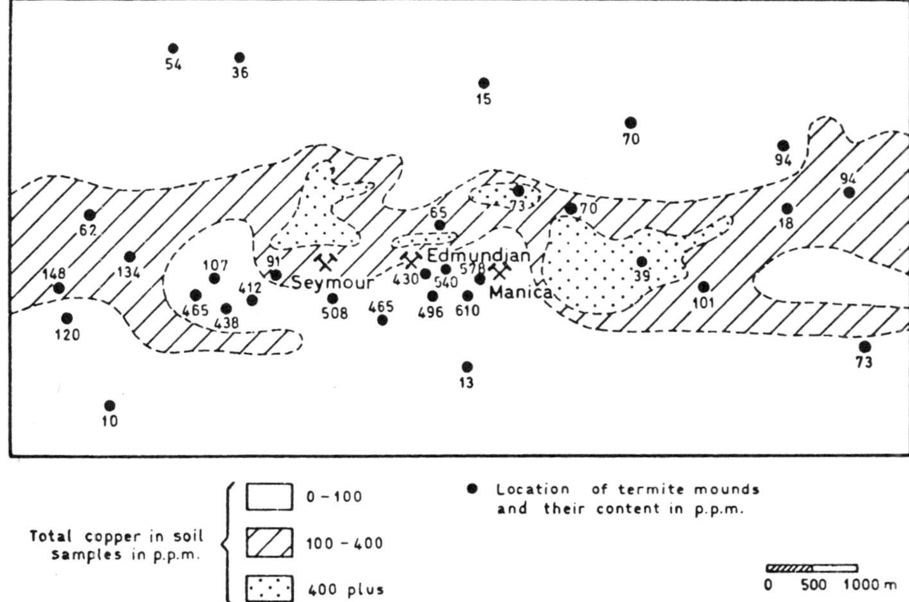

FIGURE 11.1. Distribution of copper in termitaria and soils in Mozambique. The position of known ore occurrence is indicated by crossed hammers. The concentration of copper in the soil are poorly correlated with the location of the ores. *Source:* d'Orey [247]. Copyright 1970, The Institution of Mining and Metallurgy, London.

silver and gold) were of particular interest. The Mines Department sank a shaft into the former anomaly and exposed a zinc-silver reef at 10 m below the surface. The gold anomaly assayed 8–25 μg/g of gold.

Further confirmation of the effectiveness of termitaria in prospecting for minerals was furnished by the work of d'Orey [247] over a small copper deposit at Vila Manica in Mozambique close to the Zimbabwean border. The ore deposits were developed in instrusive serpentinite covered by an extensive transported soil which was more than 15 m deep in places. The mounds belonged to the family Macrotermitidae whose members are known to excavate to a considerable depth. Anomalous concentrations of nickel and copper were found above the known ore deposits hidden by deep transported soil. This is illustrated in Figure 11.1.

A final word seems to be the ultimate confirmation of the efficacy of termites as field assistants [402]. It is in the form of a letter to me from the editor of *Chamber of Mines Journal* (Zimbabwe):

Just before the past conflict brought prospecting to a close, a gold mine did come into operation based on information supplied by anthills. That mine is still in operation and is suitably named "Termite Mine."

3
BIOGEOCHEMISTRY IN MINERAL EXPLORATION

12

AN INTRODUCTION TO BIOGEOCHEMICAL PROSPECTING

Biogeochemical methods of prospecting depend on the chemical analysis of elements in vegetation. In such investigations it is assumed that the biological absorption coefficient (BAC; see Chapter 15), that is, the concentration of an element in plant material divided by the concentration of the same element in the soil, remains relatively constant. Therefore, anomalies in the soil or bedrock will be determined via the plant material.

Because biogeochemical methods involve chemical analysis of plant material (usually at trace element levels), the technique is much younger than the associated field of geobotany. It dates back only to 1938, when Tkalich [812] discovered that an arsenopyrite ore deposit in Siberia could be delineated by the iron content of the vegetation growing upon it.

Brundin [134] initiated similar studies at about the same time and tested the new biogeochemical method for vanadium in Sweden and for tungsten in Cornwall, England. The pioneering work of Brundin and Tkalich was followed by other work (mainly in Fennoscandia) [675, 676, 855–860, 862] during World War II and immediately afterwards. Lack of communication at this time prevented another pioneer, H. V. Warren of British Columbia, from being aware of the Scandinavian work when he commenced his studies in 1944. While engaged in removal of 50,000 tonnes of overburden from an area of possible mineralization, he reported [887]: "It was impossible to overlook the way in which the roots of trees and various lesser plants penetrated into the very places which we were reaching with so much effort.''

In 1945, Warren and his co-workers began field operations which showed that the copper and zinc content of some trees and lesser plants could reflect to some extent the presence of these elements in the underlying soil formations. From 1948 on, the pace of development of the method as used in British Columbia accelerated considerably. Thus, by 1949, Warren was able to report that:

"The reaction to biogeochemistry in British Columbia had changed from one of overall disbelief to benevolent skepticism."

In the period 1947 to 1966, Warren and his associates published some 27 papers on the biogeochemical method. Although now semiretired, Warren is still active and continues to publish papers in this and related fields.

The classical work of Warren and his associates has done much to lay the foundations for biogeochemical prospecting as a respectable exploration tool. This is despite the fact that this work was undertaken before the advent of modern multielement analytical procedures such as atomic-absorption spectrometry (see Chapter 18) and sophisticated statistical procedures rendered possible by the advent of the computer (see Chapter 19).

The biogeochemical method was refined and widely used in unprospected areas by the U.S. Geological Survey in the 1950s and 1960s. One of the first to apply the then-new method in the United States was Harbaugh [340] who investigated base metals in soils and vegetation in the Tri-State region of Missouri, Kansas, and Oklahoma.

Work carried out by the U.S. Geological Survey on the Colorado Plateau, as part of a program of exploration for uranium, was undertaken mainly by H. L. Cannon and her co-workers [154, 157, 160]. The geobotanical work has been reported elsewhere (Chapters 3 and 4) but the biogeochemical studies also made a significant contribution to the discovery of several uranium anomalies and economic ore deposits. The biogeochemical work at the above institution has been continued by other workers such as H. T. Shacklette, M. A. Chaffee, J. A. Erdman, and many others, and has contributed to the deserved prominence of the U.S. Geological Survey in the field of biogeochemical prospecting.

Following the pioneering work of Tkalich [812], Russia has remained the main center of biogeochemical prospecting in the world. After the interruption caused by World War II, research into biogeochemical work was resumed in 1948 on a large scale, although a biogeochemist had been included in most exploration surveys since 1945. The leading Soviet biogeochemist today is probably A. L. Kovalevsky at Ulan Ude in Siberia. Much of his pioneering work (particularly the concept of barrier-free biological sample types; see Chapter 17) is summarized in his recent book which has now been translated into the English language [477].

The successful utilization of biogeochemical methods of prospecting involves a knowledge of many disciplines, including botany, chemistry, ecology, geology, phytochemistry, plant physiology, soil science, and statistics. It is for this reason, more than any other, that the development of the method has been slower than that of many other prospecting methods. In Part 3 of this book an attempt has been made to bring together elements of the above disciplines insofar as they are significant for biogeochemical prospecting. It is hoped that this work will be instrumental to some small degree in bringing together the field worker and the various specialists. Lack of communication between different scientific disciplines is a universal problem not confined to biogeochemistry in mineral exploration. However, until it is overcome, progress will be slower than it should be.

Topics discussed in Part 3 of this book include the formation and classification of soils, and the factors that influence uptake of elements by plants. A practical field guide to biogeochemical prospecting is given, and there is a large section devoted to the very important topic of statistical interpretation of biogeochemical data. Both of these topics are given a longer and more thorough treatment than in my earlier work [97]. However, the section on chemical analysis has been both brought up to date and abridged, since it is impossible in a single chapter to give full justice to the theory and mode of operation of every important analytical procedure, though the basic elements have been presented, nevertheless.

A chapter is devoted to the use of herbarium material for biogeochemical prospecting; there is also some discussion of aerial sampling as well as analysis of vegetation: a new and exciting field that promises to revolution- ize biogeochemical methods of exploration.

In the treatment of biogeochemistry, it is not claimed that the method is a universal panacea for the search for minerals, and indeed under some conditions, it may not be wise to expect too much from the method. How- ever, if used in the right place by suitably trained personnel well versed in its potential and application, it will prove to be a valuable auxiliary or even primary method in the continuing search for minerals throughout the world.

13

SOILS AND THEIR FORMATION

13.1. INTRODUCTION

Volcanic action and the disintegration and decomposition of large masses of solid rock have left a blanket of unconsolidated material on the surface of the earth. This layer of mineral material is known as the **regolith** and contains varying amounts of organic material from the decomposition of living organisms growing upon it.

Soil forms the upper layer of the regolith and is separated into horizons of varying depths differing morphologically and chemically from the parent material.

A basic knowledge of soils and soil formation processes is an essential prerequisite for the proper understanding of biogeochemical prospecting techniques. This knowledge is particularly necessary since vegetation and the soil are so intimately connected and interdependent.

The science of **pedology** has developed only comparatively recently and is still relatively young compared with other natural sciences. The neglect of pedology and soil science in general has historical reasons. Before the nineteenth century, the soil was regarded as being as lowly as the humble peasant or serf who tilled it and was considered by many scientists as being less worthy of study than "more respectable" subjects such as chemistry and physics. This neglect was encouraged by a second factor, this was the fact that, in western Europe at least, so much of the readily accessible land was already under the plow that little undisturbed soil remained. It was not surprising therefore that the development of soil science began in Russia, where there was still so much virgin soil.

The great Russian pedologist Dokuchaev (1846–1903) usually is credited as being the founder of soil science. In his famous book *The Russian Chernozem,* Dokuchaev [246] noted that **chernozems** (heavy black soils) were always associated with grassland vegetation and he considered them to be intimately related. This theory was in marked contrast to those of classical geologists who had always considered that chernozems belonged to a particular geological formation. Dokuchaev also studied the **podzols** of forested areas and noticed the various horizons found in these soils. He laid the foundations of the **zonal** theory of soils, which established that soil formation is related strongly to climate.

Another great Russian soil scientist, Glinka [1867–1927], a pupil of Dokuchaev, was responsible for the organization of the subject of soil science, and offered the world's first course in soil science at the University of St. Petersburg in 1911. Part of the relative popularity of biogeochemical methods of prospecting in the Soviet Union springs from the preeminence of that country in soil science as well as from the existence of soil maps for a large part of the country, which in many cases were prepared over 50 years ago.

Soil science is now a well-established subject in most countries and is receiving ever-increasing attention as a useful discipline for increasing the efficiency of agriculture to feed the world's growing population. This recognition of the great value of soil science does not always apply to workers in other fields. Hawkes and Webb [347] have commented that:

"It is perhaps surprising that geologists in general have little understanding of the soils which mantle so much of the earth's surface since any information which can be gained from soils concerning the bedrock geology must be of value."

With the increasing use of soil sampling in geochemical exploration, a proper understanding of the nature of soils and the factors involved in their formation is becoming ever more important.

The purpose of this chapter is to give a short introduction to the subject of soils and soil formation with particular emphasis on the biological principles involved in **profile** development. In the Soviet Union, soil sampling, particularly of the humic layer, is considered to be a part of biogeochemical prospecting and in view of the interdependence of soils and vegetation, this definition probably is justified. In this present work, however, a clear distinction between the two is made, particularly as soil sampling has been more than adequately covered by Hawkes and Webb [347]. Soil sampling therefore will be considered only as it is an adjunct to prospecting methods involving plant analysis; that is, for determining BAC values (see Chapters 14 and 15) or for comparing the relative performance of the two sample types in detecting mineralization.

13.2. SOIL HORIZONS AND SOIL PROFILES

A properly formed soil consists of a series of layers known as **horizons,** which are subdivided further into **subhorizons.** The top horizon (A) is usually the most humic, and there is a steady decline in the organic content down the soil profile. Material is leached from the A horizon, which is said to be **eluvial** because of removal of soluble salts and colloidal material. This material is deposited into the **illuvial** B horizon. These two horizons comprise the **solum,** which overlies the C horizon. The C horizon is derived from weathering of the original rock (parent material) and is formed by a process known as **polygenesis.** Eluvial material usually consists of clay minerals and soluble salts of iron and calcium. Figure 13.1 shows a typical soil profile.

Subhorizons have subscripted designations such as A_0, A_1, B_1, B_2, and

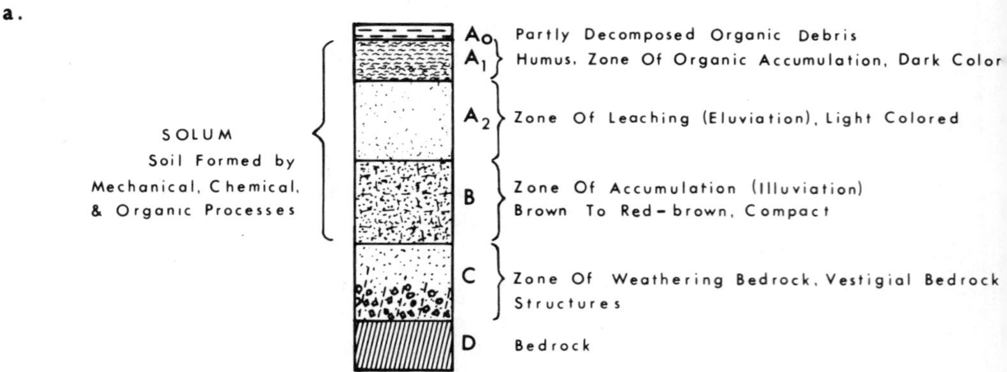

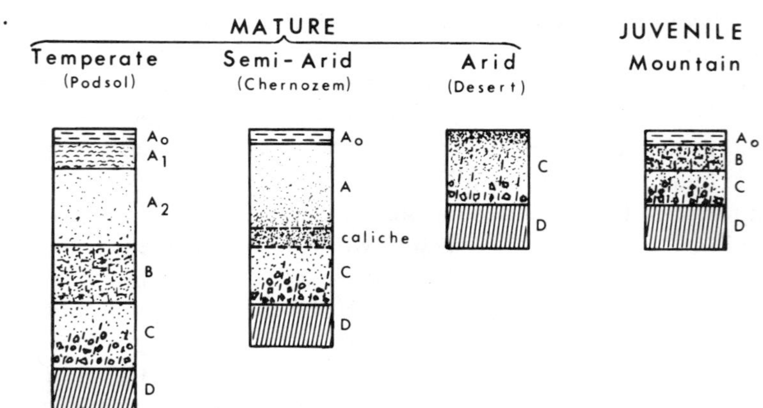

FIGURE 13.1.   Diagrammatic representation of soil profiles: (*a*) typical soil profile, (*b*) climatic variations. *Source:* Andrews-Jones [22]. Copyright 1968, Colorado School of Mines Press.

so on, and classification can be a very complicated matter. Exploration personnel, however, are seldom pedologists and the simple (if naive) approach using only three main horizons (i.e., A, B, and C) is usually followed in geochemical exploration. This is because there is seldom time to study the soil sufficiently well beforehand in order to justify the recognition of more than three horizons.

13.3. FACTORS AFFECTING SOIL FORMATION

13.3.1. Climate

Climate is perhaps the main factor affecting soil formation. Such soils are said to be **zonal** in character, whereas those that are affected by factors other than climate are said to be **intrazonal.** A third group of young soils, influenced by the time factor, are said to be **azonal.** Figure 13.2a shows, in

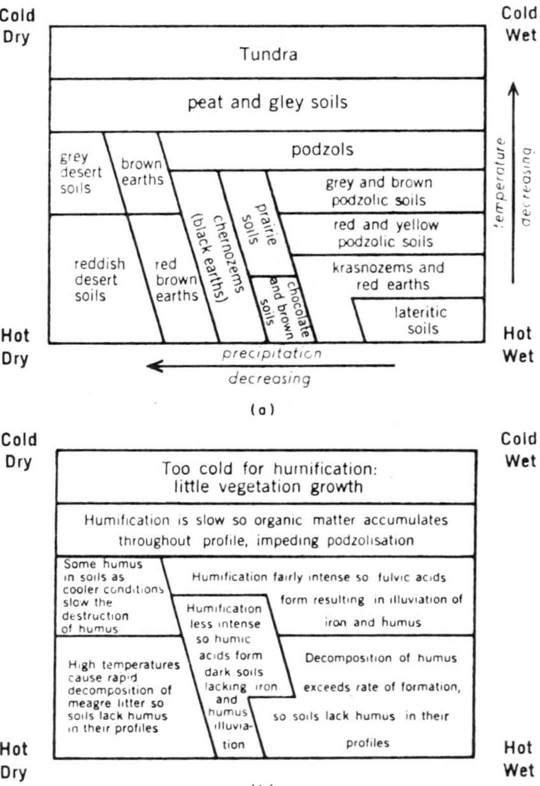

FIGURE 13.2. (*a*) The effect of climate on soil zonation, (*b*) the role of organic matter in soil formation in response to climate. Boundaries of (*a*) and (*b*) are coincident. *Source:* Corbett [222]. Copyright 1969, Longman Cheshire (Pty) Ltd.

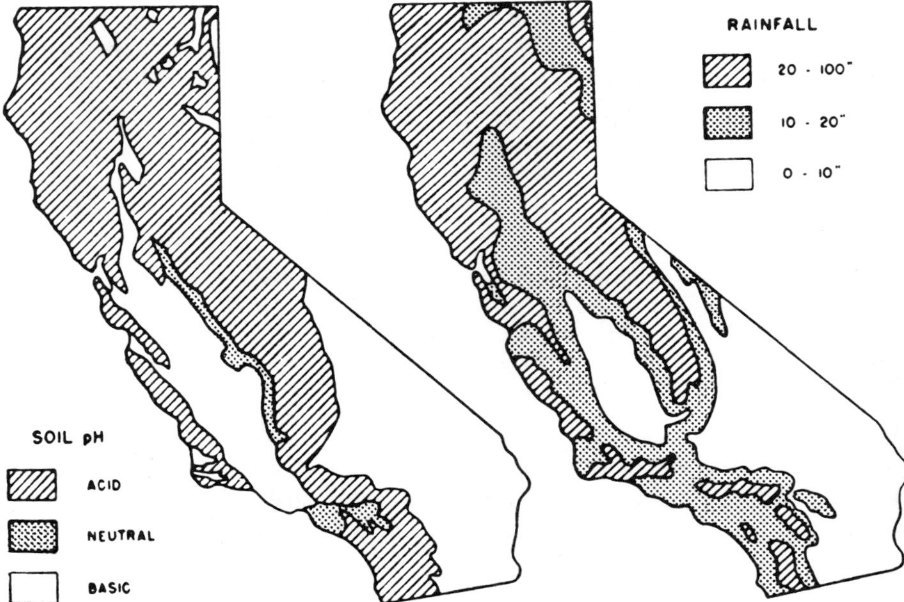

FIGURE 13.3. Map of California showing the close relationship between soil pH and rainfall. *Source:* Jenny [400].

simplified form, the main effects of climate on soil development. Under hot, wet conditions, humic material is oxidized quickly so that laterites and krasnozems lack organic matter, and their color reflects the oxides of their main constituents (usually Fe_2O_3). Peat and peat bogs tend to form in colder areas where oxidation is slowed. Under hot and dry conditions, desert soils again reflect the color of their inorganic oxides because of the lack of humic material.

The pH of soils is largely a function of the humus content and therefore indirectly of rainfall. This close association is illustrated in Figure 13.3. Variation in pH has an important influence on ore formation and of course on the uptake of trace elements by plants. Figure 13.4 shows that in well-leached laterites of high pH, the silica can be removed, leaving behind deposits of bauxite. Similarly, soils of low pH can often be deficient in aluminum and have high residual silica.

13.3.2. The Effect of the Parent Material

As previously mentioned, the parent material is not the dominant factor in soil formation. This is true at least for the physical nature of the soil, though not for its chemical composition. Indeed, if the soil did not reflect the chemical composition of the bedrock, it would never have been possible to use soils for prospecting purposes [347].

The more juvenile the soil, the more closely it will reflect bedrock.

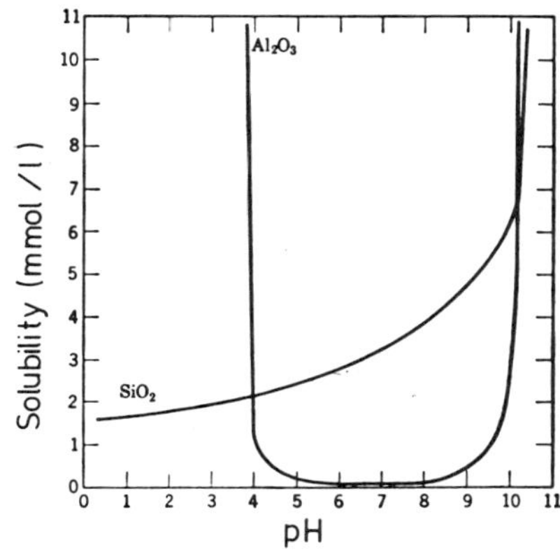

FIGURE 13.4. The solubility of silica and aluminum as a function of pH. *Source:* Mason [576]. Reprinted by Permission of John Wiley & Sons, Inc.

Juvenile soils tend to overlay rocks that are not weathered readily. The rate of weathering of rock minerals is in the sequence (most resistant first) of quartz, muscovite, potash felspar, biotite, alkalic plagioclase, alkali-calcic plagioclase, hornblende, calcic-alkalic plagioclase, augite, calcic-plagioclase, and olivine. This sequence is known as the Goldich Series [319] and is the reverse of the well-known Bowen Reaction Series [81], which arranges silicate rocks in the order of their crystallization from cooling magma; that is, olivine is the first (highest melting point) mineral to crytallize from the magma and is at the same time the most easily weathered of all.

13.3.3. The Effect of Relief

Relief produces a localized microclimate and causes significant differences in profile development. Steeper slopes produce shallower soils and less profile differentiation. Mobile elements are leached easily from the upper parts of the terrain and are reprecipitated downslope in regions of gentler relief. Lime is often deposited in these regions and gives the soil a higher pH downslope. Poorly drained soils in areas of low relief tend to have a higher organic content than in well-drained parts of the landscape. It is here that peat bogs are formed.

13.3.4. The Effect of Time

The time taken for a well-developed soil profile to appear is largely a function of climate. Under hot and wet conditions a soil profile will develop much faster than in a dry climate. Quantitative measurements of rates of soil formation have been rendered possible by studying new geological formations formed after natural disasters or by observing soil development on moraines of retreating glaciers.

The volcanic explosion of the island of Krakatoa in 1883 afforded an excellent opportunity to study the rate of soil formation on freshly weathered material. After 45 years, a layer 36 cm deep was observed to be fairly well differentiated from the parent material in conditions that were admittedly favorable because of the tropical climate. By contrast, Chandler [180] studied the rate of podzolization of a moraine from a retreating glacier in Alaska and found that 500–1000 years were needed for a well-developed soil profile to appear. The appearance of the volcanic island of Surtsey off Iceland about 10 years ago will provide scientists with a good opportunity for studying soil development in a cold climate, though significant differentiation hardly could be expected during the lifetime of those now living.

13.3.5. The Effect of Biological Activity

The first plants to colonize the surface of weathered rocks are usually mosses and lichens. The decay of these plants increases the organic content of the substrate which simultaneously has an increased water-holding capacity. Higher plants can then colonize the site. The concomitant development of vegetation and soil is known as **seral progression** and terminates with the so-called **climax vegetation.** Seral progression takes place very slowly in cold climates but much faster in tropical areas. Once again the Krakatoa explosion afforded an excellent opportunity for scientists to study seral progression. A few years after the explosion, grasses had become re-established on the volcanic ash. By 1900, a savanna-type drought-resistant shrub layer had appeared. Later, forest trees recolonized the soil profile. Today, the vegetation is similar to that existing in 1883, although somewhat poorer in species.

Removal of the climax vegetation does not necessarily mean that it will be re-established later. The soil supporting the tropical rain forest of many tropical regions is so fragile and infertile that removal of the forest can lead to erosion and a residual soil that is so impoverished that it will not support the former climax vegetation.

The accumulation of humus in soils is largely a function of climate, as mentioned above. The distribution of organic matter in the soil is controlled by the rainfall and permeability of the soil. Thick chernozems found under semiarid grasslands have a uniform distribution of humus through the A and B horizons, whereas heavily leached podzols have discrete zones of humic material.

Humus contains a large number of colloidal organic compounds which are grouped under the headings of **humic** and **fulvic** acids. These two compounds differ greatly in their mobility and effect on the soil. The fulvic acids assist in the mobilization of various elements, particularly iron and aluminum. Marked illuviation will occur when these acids are present. Humic acids by contrast are far more stable and tend to impede illuviation. The ratio of humic to fulvic acids varies appreciably among different soil types. In chernozems, the increased amount of humic acids results in less leaching of metals in the profile of the soil and therefore less elemental differentiation.

TABLE 13.1.   Production of Organic Matter by Vegetation

Type of Vegetation	Annual Production (tonnes/ha)
Alpine meadows	0.10–0.20
Short-grass prairie	0.30
Tall-grass prairie	0.40–0.90
Mixed tall-grass prairie	1.10
Pine forest	0.70
Beech forest	0.73
White pine needles	1.04
Beech-birch-aspen forest	1.42
Tropical primaeval forest	5.60
Tropical savannas	6.65
Monsoon forest	11.10
Tropical legumes	12.20
Tropical rain forest	22–45

Source: Jenny [400].

In podzols, by contrast, the increased amounts of fulvic acids encourage illuviation.

The mass of organic material produced per annum is very much a function of the climate and type of vegetation. As might be expected, tundra and desert associations are the least productive, whereas tropical rain forests produce the greatest amounts. Table 13.1 lists the production of organic matter for different types of vegetation in different climates.

The pH of humic material is, to some extent, a function of the type of vegetation overlying the soil. Grasses tend to have a higher calcium content than do woodland species, with the result that calcium humates with a basic reaction are produced to a greater extent.

Figure 13.2*b* shows the role of organic matter in soil formation in response to climate. The soil boundaries are the same as in Figure 13.2*a*. Figure 13.5

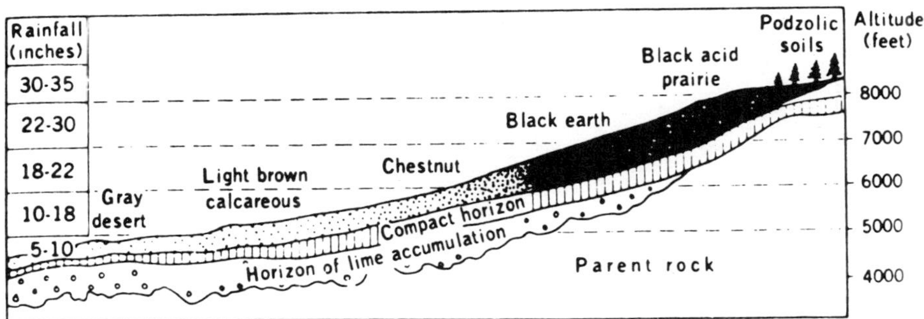

N.B. Scale of soil profiles is greatly exaggerated

FIGURE 13.5.   Gradation of soil types from desert to humid mountain top at Big Horns, Wyoming. *Source:* Hawkes and Webb [347], and Thorp [803].

shows the gradation of soil types under the influence of climate and relief and summarizes pictorially some of the above discussion of soil formation.

13.4. THE DISTRIBUTION AND MOBILIZATION OF MINOR ELEMENTS IN SOILS

13.4.1. Introduction

Minor elements are not distributed uniformly in the soil profile. The concentrations and distributions are influenced, of course, by the nature of the parent material, but factors such as relative mobilities and climate are also important.

Figure 13.6 shows the normal abundance of minor elements in soils. The more extreme ranges are indicated by broken lines.

In most soils the concentrations of minor elements in the various horizons of the profiles are far from constant. Table 13.2 shows the distribution of a number of elements within profiles of five common zonal soils collected during a nationwide soil survey in the Soviet Union. All profiles, except those of grey desert soils, show an accumulation of all five elements in the upper humic horizon. In all cases, with the same exception, there is a

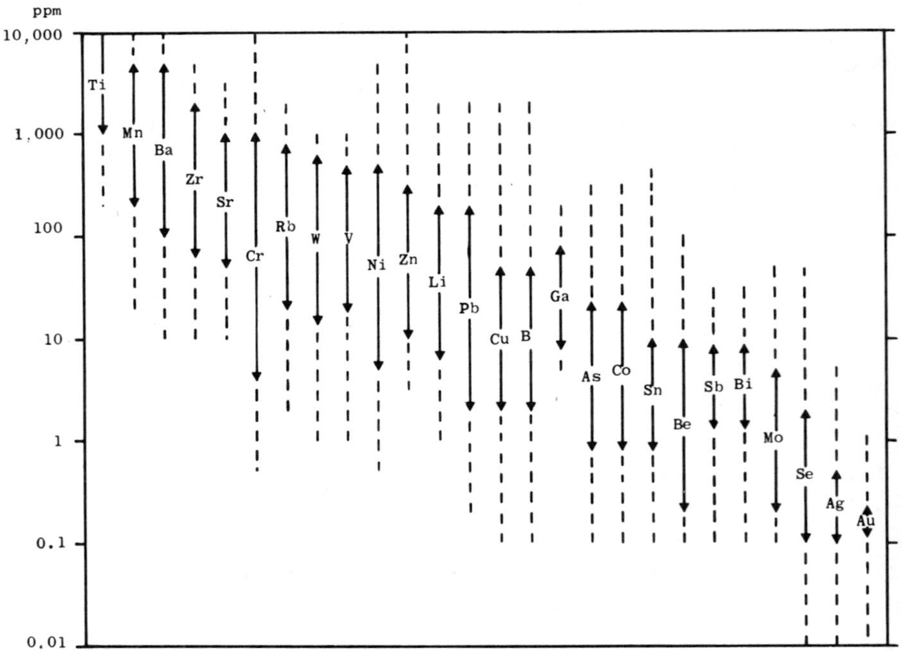

FIGURE 13.6.   Range of normal abundances of minor elements in soils. Broken lines indicate extreme values. *Source:* Andrews-Jones [22]. Copyright 1968, Colorado School of Mines Press.

TABLE 13.2.   Elemental Concentrations (μg/g) in Horizons of Some Zonal Soils of the Soviet Union

Soil	Horizon	Depth (cm)	Co	Cr	Cu	Ni	Zn
Gleyed tundra	A	0–15	4.9	40	23	25	76
	A	15–25	1.5	—	11	10	—
	B	25–50	2.2	5	13	12	74
Podzol	A	0–18	14	180	8.3	24	33
	A	18–34	9.3	65	5.5	22	25
	B	34–60	6.6	290	5.4	16	25
	B	60–110	8.3	220	6.8	22	34
Grey forest	A	0–5	8.1	570	12	27	—
	A	20–25	6.4	140	10	17	28
	B	40–45	6.7	31	9.4	22	26
	B	75–80	7.8	440	8.4	2.2	30
	C	100–105	7.2	27	7.6	21	47
Chernozem	A	0.5	6.7	400	17	39	90
	A	24–32	5.6	630	12	33	90
	B	80–82	7.5	220	19	35	63
	C	128–144	12	200	16	48	—
Grey desert	A	0.5	3.4	570	5.1	10	39
	B	65–70	6.1	570	8.4	28	41
	C	160–170	8.8	570	8.4	28	41

Source: Malyuga [563]. Copyright 1964 Plenum Press. Reprinted by kind permission.

relative depletion of the elements in the A_2 and B_1 horizons due to leaching. The desert soils show depletion of the A horizon probably due to lack of humus and clay minerals in this part of the profile.

The uneven distribution of trace elements in the various soil profiles highlights the serious problem faced by the exploration geochemist in soil surveys. If the same horizon is not sampled consistently throughout the project, false anomalies may easily appear in the final data. For example, referring to Table 13.2, a soil survey for chromium in gleyed tundra soils using the B horizon as the sampling medium (5 μg/g chromium) could turn up anomalous values (40 μg/g) whenever the A_0 horizon was sampled in error. This sampling problem is one that is overcome by the use of vegetation, because a large plant can effectively sample all the soil horizons and produce a mean value.

Particle size is also important in soil sampling. For example, Williams [941] has shown that 98% of the cassiterite in soils and sediments from Stewart Island, New Zealand, is concentrated in the $+20$ mesh fraction of the samples. A more spectacular example is found in Western Australia. One of the world's great nickel provinces is located in the Archaean greenstones of the Kalgoorlie area. Nickel mineralization was only discovered in the late 1960s because soil analyses had concentrated on the fine -100

mesh fraction where the nickel concentration is quite low. As with the cassiterite example cited above, most of the nickel is found in the coarse fraction of the soil.

13.4.2. Mobilization of Minor Elements in the Primary Environment

Hypogene mobility occurs at depth under conditions of high temperature and pressure. The major constituents of the earth's crust form a sequence of minerals dependent on the prevailing temperature and pressure [81]. The minor elements usually occupy spaces in the lattices of these minerals according to the rules of **diadochic substitution** [320]. The residual fluids deposited as pegmatites or hydrothermal veins are usually extremely rich in trace elements. The same sequence of mobility is found in metamorphism when the last-formed minerals are the first to become liquid again as the temperature rises. The mobilization and transport of the elements in the primary environment is known as **primary dispersion.** The reader is referred to Hawkes and Webb [347] for a fuller discussion of this subject.

13.4.3. Mobilization of Minor Elements in the Secondary Environment

Supergene mobility in secondary disperson within the surface environment is of great importance in elemental differentiation within soils and takes place under conditions of low temperature and pressure. Mobilization is influenced strongly by Eh, pH, and the stability of the minerals that have to be decomposed. The factors that ultimately cause the breakdown of the minerals are mechanical, physical, chemical, and biological. Garrels [311] has calculated the theoretical mobilities of trace elements in drainage waters from consideration of the physical chemistry of the associated mineral and ionic species. Equilibrium between mobile and immobile phases in the natural environment is seldom encountered, however, due to interferences from a wide variety of factors such as adsorption on to clay minerals and humic material.

Perel'man [648] has calculated a factor known as the **coefficient of aqueous migration** (K) which is a measure of the mobility of a particular element and is defined by the following expression.

$$K = \frac{100M}{aN}$$

where M is the concentration of the element in drainage waters (mg/liter), N is the concentration of the element in rocks (%), and a is the mineral residue (%) contained in the water. This treatment takes into account the Eh and pH of the water. Table 13.3 shows the relative mobilities of the elements in the supergene environment. This table shows the pH dependence of the mobility of some elements, particularly the group cobalt, copper, mercury, silver, and uranium. The group molybdenum, selenium,

TABLE 13.3. Relative Mobilities of the Elements in the Supergene Environment

Relative Mobilities	Environmental Conditions			
	Oxidizing	Acid	Neutral to Alkaline	Reducing
Very high	Cl, I, Br, S, B	Cl, I, Br, S, B	Cl, I, Br, S, B, Mo, V, U, Se, Re	Cl, I, Br
High	Mo, V, U, Se, Re, Ca, Na, Mg, F, Sr, Ra, Zn	Mo, V, U, Se, Re, Ca, Na, Mg, F, Sr, Ra, Zn, Cu, Co, Ni, Hg, Ag, Au	Ca, Na, Mg, F, Sr, Ra	Ca, Na, Mg, F, Sr, Ra
Medium	Cu, Co, Ni, Hg, Ag, Au, As, Cd	As, Cd	As, Cd	
Low	Si, P, K, Pb, Rb, Ba, Be, Bi, Sb, Ge, Cs, Tl, Li	Si, P, K, Pb, Li, Rb, Ba, Be, Bi, Sb, Ge, Cs, Tl, Fe, Mn	Si, P, K, Pb, Li, Rb, Ba, Be, Bi, Sb, Ge, Cs, Tl, Fe, Mn	Si, P, K, Fe, Mn
Very low to immobile	Fe, Mn, Al, Ti, Sn, Te, W, Nb, Ta, Pt, Ce, Zr, Th, rare earths	Al, Ti, Sn, Te, W, Nb, Ta, Pt, Cr, Zr, rare earths	Al, Ti, Sn, Te, W, Nb, Ta, Pt, Cr, Zr, Th, rare earths, Zn, Cu, Co, Ni, Hg, Ag, Au	Al, Ti, Sn, Te, W, Nb, Ta, Pt, Cr, Zr, Th, rare earths, S, B, Mo, V, U, Se, Re, Zn, Cu, Co, Cu, Ni, Hg, Ag, Au, As, Cd, Pb, Li, Rb, Ba, Be, Bi, Cs, Ge, Cs, Tl

Source: Andrews-Jones [22]. Copyright 1968, Colorado School of Mines Press.

uranium, and vanadium mobilizes readily under oxidizing conditions, since in all cases the higher oxidation states of these elements are much more mobile. This is particularly true for the UO_2^{2+} and U^{6+} ions.

Mobilization of trace elements in soils is controlled by four main factors: chemical weathering, biological weathering, humus and clay minerals, and the biogeochemical cycle. Each will now be discussed.

13.4.4. Factors Affecting the Mobilization of Minor Elements in Soils

Chemical Weathering. The stability of ions in soils is heavily dependent on the ionic potential (Z/r) where Z is the ionic charge and r is the ionic

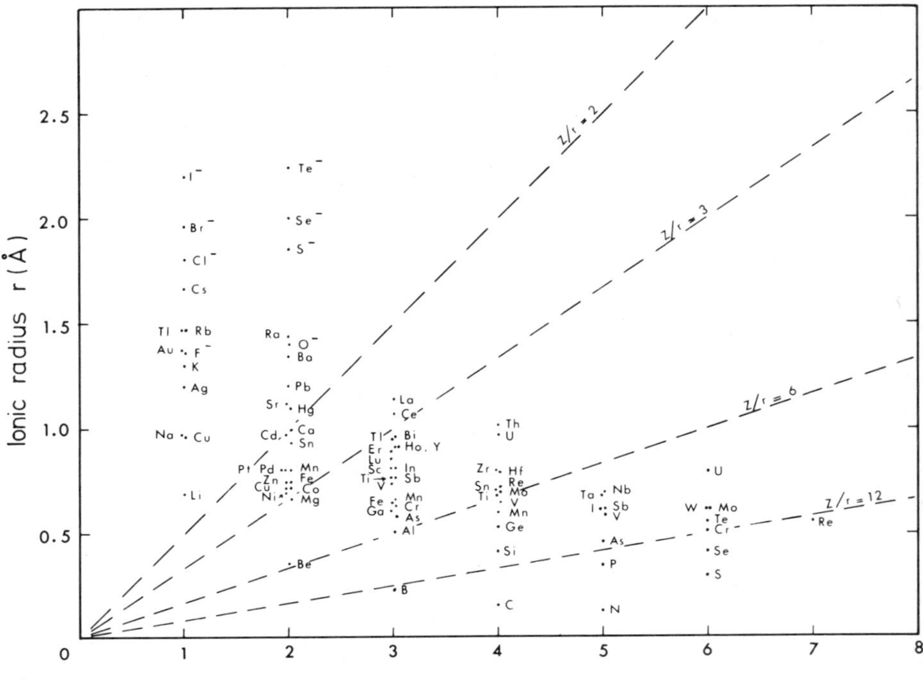

FIGURE 13.7.   The ionic potentials (ionic charge/radius) of the elements. *Source:* Brooks [97].

radius. Ions with low ionic potentials (<2) tend to remain in solution, whereas those with intermediate potentials (8–10) tend to precipitate as hydroxides. When the ionic potential exceeds 12, soluble oxyacids such as nitrates and sulphates tend to form. The ionic potentials of the chemical elements are shown in Figure 13.7. The concept of the ionic potential is extremely important in geochemistry and has a bearing on the uptake of trace elements by plants (see Chapter 14).

Biological Weathering.   Although a wide range of organisms has an ability to weather soil minerals, it is usually the lower organisms, such as lichens and bacteria [644], that have the most effect. Many plants attack minerals by carbon dioxide generated at their roots systems, whereas lichens produce mucous substances that attack the minerals mechanically. For example, Malyuga [563] reports that even the immensely stable mineral garnet almandine is broken down by the lichen *Rhizocarpon geographicum*.

The nitrifying bacterium *Bacillus megaterium* readily breaks down calcite, dolomite, and phosphate. Anaerobes which produce lactic and butyric acids can attack calcite, aragonite, magnesite, dolomite, and siderite. Indeed some bacteria are said to produce iron ores [57] by a reaction of the type

$$4FeCO_3 + O_2 + 6H_2O = 4Fe(OH)_3 + 4CO_2 + 167 \text{ kJ}$$

Evolved heat is used as a source of energy by the organisms. Other bacteria such as *Bacillus amylobacter* and *B. extorguens* strongly attack felspars with consequent solubilization of potassium and silicon.

Although bacteria form a very small proportion by weight of the soil mass, they can effectively handle large quantities of material due to their very high metabolic rates which are stimulated by conditions of high temperature and humidity.

Microorganisms play a major part in the cycles of carbon, nitrogen, sulfur, and phosphorus. Under reducing conditions, sulfides are formed and precipitate elements such as copper, iron, molybdenum, and zinc. The solubility of iron and manganese is particularly affected by soil organisms. Ferrous iron is reduced to the ferrous stated under favorable conditions. In contrast, reducing bacteria produce the opposite effect and give the soil profile the characteristic grey, streaky appearance of **gleyed** material. Manganese salts are oxidized similarly to insoluble manganese dioxide. Oxidation or reduction of other elements such as arsenic, molybdenum, selenium, tellurium, uranium, and vanadium can also be affected by microorganisms. For a fuller account of the role of microorganisms in biogeochemistry, the reader is referred to Zajic [964].

The Effect of Humus and Clay Minerals on the Mobilization of Minor Elements. Clays act in the same way as synthetic ion-exchange resins, in that they exchange ions at their surfaces and usually have a relatively high cation-exchange capacity, which in some cases even exceeds that of the synthetic resins. Clay particles known as **micelles** usually have a negative charge because of an excess of oxygen due to the dissociation of hydroxyl groups at the surface, or by substitution of atoms of higher positive charge by those of lower charge. The exchange capacity of clays is an inverse function of particle size. Though less important than cation exchange capacity, clays can also exchange anions by use of hydroxyl groupings on their surface.

Humus has an exchange capacity far in excess of that of clays. Whereas clays will have a capacity of up to 100 meq/g, a figure five times higher can be found in humic material. Humus may well form strong chelate complexes with metal ions, particularly at high pH values.

The Biogeochemical Cycle. The distribution of minor elements in soils is strongly influenced by the biogeochemical cycle illustrated in Figure 13.8. Trace elements are solubilized at the root system of the plants and are transferred through the aerial parts, where they are finally deposited as metal complexes in leaves or twigs. At senescence, the leaves fall and decay. The more soluble components, such as carbonates of the alkali metals and alkaline earths, sulfates, phosphates, and humic complexes of iron and manganese, are leached through the soil profile by rainwater. Sparingly soluble or insoluble compounds and complexes of other metals, such as

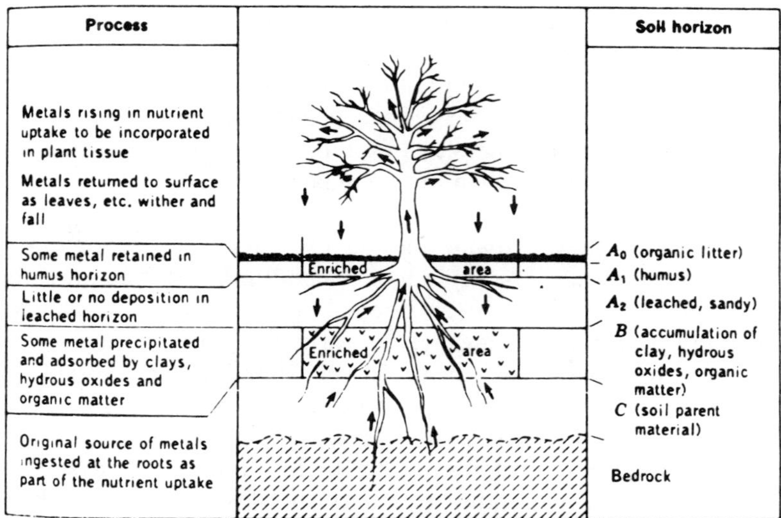

FIGURE 13.8.   The biogeochemical cycle. *Source:* Hawkes and Webb [347].

hydroxides and protein or humic complexes, are retained in the humic horizons of the soil. The cumulative effect of this process is to produce a very significant enrichment of certain elements in the topmost layer of the soil. Humic layers become particularly enriched in elements such as arsenic, beryllium, cadmium, cobalt, germanium, gold, lead, manganese, nickel, scandium, silver, tin, uranium, and zinc. The biogenic enrichment of such elements is known as the **Goldschmidt Enrichment Principle** [320] and has been deduced from a study of coal ash whose parent material was judged to be similar in chemical composition to humus.

The degree of enrichment of trace elements by plants follows the so-called **Irving-Williams Rules** [380]. These rules state quite simply that the stability of metal-organic complexes is independent of the nature of the ligand and that for divalent cations the order of stability (most stable first) is in the sequence: Pt, Pd, Hg, UO_2, Be, Cu, Ni, Co, Pb, Zn, Cd, Fe, Mn, Ca, Sr, Ba. For monovalent cations, the sequence is Ag; Tl, Li, Na, K, Rb, Cs. The corresponding values for trivalent ions are Fe, Ga, Al, Sc, In, Y, Pr, Ce, and La.

14

ACCUMULATION OF ELEMENTS BY PLANTS

14.1. INTRODUCTION

A reasonably reproducible uptake of soil elements by plants of the same species is an irreducible minimum requirement for successful use of the biogeochemical method. In this chapter, this topic will be examined in some detail because of its importance.

Nearly all of the chemical elements have been detected in plant material and it is probably true to say that for each element there will be at least one species capable of concentrating it to a remarkable degree. A good example of this is the accumulation of nickel in the sap of *Sebertia acuminata* (sève bleue = blue sap) in New Caledonia [392]. The sap contains up to 25% (dry weight) of this element.

Preferential enrichment of trace elements in plants is known as the **Goldschmidt Enrichment Principle** [320], though in fairness, it must be stated that the Russian biogeochemist V.I. Vernadsky [847] first put forward similar ideas about a decade earlier.

14.2. ELEMENTAL ABUNDANCES IN PLANTS

The relative uptake of a given element in a plant species is called the **Biological Absorption Coefficient (BAC;** see Chapter 15) and in the Russian literature is sometimes referred to as the **clarke** (after the American geologist F.W. Clarke). If BAC values are expressed on a dry-weight basis, levels for

TABLE 14.1.   Mean Biological Absorption Coefficients (BAC) for Vegetation

Biogenic Elements		Intermediate Elements		Nonbiogenic Elements	
B*	1.70	Mo*	0.04	Na*	0.01
S*	0.96	Mg*	0.034	Rb	0.007
Zn*	0.90	Ni	0.03	RE	0.003
P*	0.88	Co	0.02	Cr	0.003
Mn*	0.40	U	0.02	Li	0.0015
Ag	0.25	Fe*	0.012	Si	0.0006
Ca*	0.14			V	0.0006
Sr	0.13			Ti	0.0003
Cu*	0.13			Al	0.0003
K*	0.12				
Ba	0.12				
Se	0.10				

KEY: *—Essential element RE—Rare earths
Note: Concentrations in plant material expressed on dry weight basis.
Source: Hutchinson [374] and Cannon [156]

most elements are below unity. This is shown in Table 14.1. The table clearly shows that elements that fulfill a physiological function in plant nutrition and metabolism (shown with asterisks in the table) are usually absorbed to a greater extent than are other "ballast elements" [304].

There is a strong relationship between the BAC of an element in vegetation and its position on the ionic potential diagram (see Chapter 13) [374]. This is illustrated in Figure 14.1, from which it will be noted that essential elements such as boron, calcium, carbon, nitrogen, phosphorus, potassium,

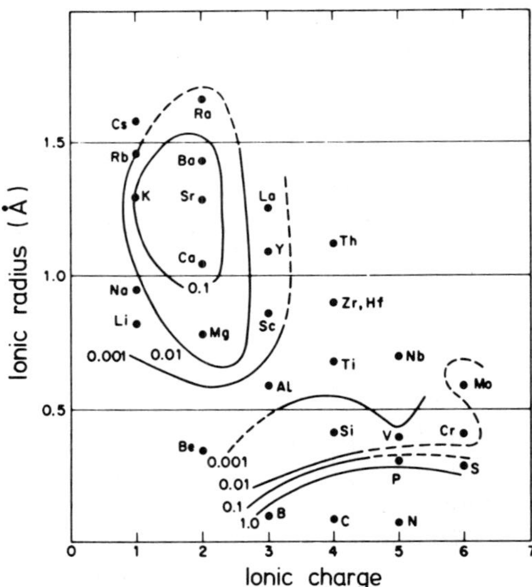

FIGURE 14.1. Biological absorption coefficients (BAC) of terrestrial plants for various elements in relation to their ionic potentials. *Source:* Hutchinson [374].

and sulfur all have BAC values greater than 0.1 and may be said to be **biogenic.** These elements have either very high or very low ionic potentials, which is not unreasonable in view of the expected correlation between the solubility of an element in water and its ease of uptake by plants.

In addition to the seven essential elements cited above, there are at least nine others: chlorine, copper, hydrogen, iron, manganese, magnesium, molybdenum, oxygen, and zinc. These 16 elements may be said to be universal nutrients, but there are other elements such as selenium, cobalt, silicon, and vanadium that may well have a physiological role in some species. This is particularly true for selenium which forms seleno-amino acids in species such as *Astragalus* [853].

14.3. MECHANISMS OF ION ABSORPTION BY PLANTS

There are three main mechanisms whereby ions may be absorbed by plants. Two of these involve uptake at root systems and the third involves foliar absorption in the aerial parts of the plant.

Uptake at root systems involves either diffusion into the plant from the soil solution or cation exchange at the surface of clay minerals. Cation exchange is the more important of the two processes and takes place by the production of carbon dioxide as a result of respiration processes. The carbon dioxide reacts with water and liberates hydrogen ions which exchange with cations held upon the clay minerals. When the cations reach the root tips, they are again replaced by hydrogen ions and the cycle is repeated [420].

Some ion absorption also occurs by simple diffusion into the cells of the root tips. A surprising feature of the accumulation of ions by roots is that their concentration in the cell fluid is often many times greater than in the soil solution. This is known to be a metabolically mediated process requiring the expenditure of cellular energy.

The ions absorbed at the roots of the plants usually are translocated in an upward direction toward the leaves. The work of Stout and Hoagland [780] seems to indicate that the **xylem** is the chief medium for this transport. There is evidence for a reverse flow of some nutrients from the leaves back to the stems or younger leaves, usually just before exfoliation. Movement of minerals from the leaves occurs via the **phloem.** There is also some evidence for periodic daily circulation of some minerals.

The forms in which the elements are translocated within the plant are quite varied. The work of my research group has shown that chromium is transported as an oxalato complex in *Leptospermum scoparium*, whereas nickel in many New Caledonian plants is transported as a citrato complex. A good example of this is the sap of *Sebertia acuminata* (see above) which is virtually pure nickel citrate [392]. Nickel in *Alyssum* species seems to be complexed partly with malic acid.

An important factor in plant nutrition is the availability of the elements present in the soil. For some elements, very little of the total content in the soil can be utilized by plants. This is due to a number of factors including drainage, pH, Eh, the nature or absence of clay minerals, antagonistic effects of other ions, and the presence of complexing agents in the soil.

A number of different tests have been devised for determining the availability of elements in soils. These procedures usually involve shaking the soil with various solutions such as ammonium acetate or dilute acetic acid and measuring the concentrations of the elements in the aqueous phase after equilibrium has been achieved. In practice the only really reliable method of measuring availability is to grow selected test species in the medium and then determine elemental uptake by chemical analysis of the plant.

14.4. VARIABILITY OF ELEMENT UPTAKE AMONG DIFFERENT SPECIES

Plants vary very greatly in their ability to accumulate elements from the soil. This is illustrated in Table 14.2, from which it will be seen that levels of selenium in plant ash vary by a factor of 4000, whereas for nickel the factor is 24. The nickel values, however, do not refer to specialized plants known as hyperaccumulators (see below and Chapter 21) which are capable of taking up much more of this element. A high degree of uptake of an

TABLE 14.2. Variable Accumulation of Nickel and Selenium by Plants Growing in Soils Containing Constant Amounts of Each Element

Element	Species	Concentrations (μg/g dry weight)	
		Plant Material	Soil
Nickel	*Cassinia vauvilliersii*	110	2500
	Myrsine divaricata	42	2500
	Gentiana corymbifera	39	2500
	Leptospermum scoparium	32	2500
	Stellaria roughii	17	2500
	Hebe odora	5	2500
Selenium	*Astragalus pectinatus*	200	2
	Stanleya pinnata	17	2
	Aplopappus fremontii	16	2
	Gutierezia sarothrae	4	2
	Zea mays	1	2
	Xanthium sp.	0.3	2
	Salsola pestifer	0.2	2
	Munroa squarrosa	0.2	2
	Helianthus annuus	0.1	2
	Malvastrum coccineum	0.05	2

Sources: Lyon et al. [536] and Williams [942].

element by a given plant species may not necessarily be advantageous. Indeed the reverse may be the case. Plants with inordinately high metal contents tend not to reflect relatively small variations in the chemical composition of the substrate [657].

14.5. HYPERACCUMULATION OF METALS BY PLANTS

There is a small number of plant species that have an ability to concentrate certain trace elements to a degree (even on a dry-weight basis) well in excess of that of the substrate. One of the first of such plants to be recognized was *Alyssum bertolonii* [593] which contained an incredible 1% of nickel in dried leaves. As a result of the analysis of several thousand herbarium specimens (see Chapter 21), Brooks et al. [115] were able to identify another 44 hyperaccumulators [110] of nickel in the same genus. Most of these were found in Turkey over a number of regions of serpentine rocks. The surprising thing about these species is the degree of localization of several of them. Some are confined to ultrabasic occurrences of only a few hectares in area.

The copper-cobalt deposits of Zaïre support an unusual metal-tolerant community of plants with very high concentrations of both elements [119]. These are listed in Table 14.3, together with spectacular accumulators of other elements.

As already mentioned, hyperaccumulation of an element by a plant is not necessarily favorable for biogeochemical prospecting. On the other hand, very high concentrations of elements by plants usually means that these levels can be detected by simple field tests. For example, for many years we have used filter paper impregnated with dimethylglyoxime (applied

TABLE 14.3. **Hyperaccumulation of Elements by Plants**

Element	Plant	Concentration (%)	References
Cobalt	*Haumaniastrum robertii*	1.02	101
Copper	*Aeolanthus biformifolius*	0.39	116,552
Chromium	*Pearsonia metallifera*	0.77	933
Lead	*Thlaspi rotundifolium* s. sp. *cepaeifolium*	0.79	Unpub. data
Molybdenum	*Epilobium angustifolium*	0.10	896
Nickel	*Alyssum masmenaeum*	2.43	115
Selenium	*Astragalus pattersoni*	4.60	157
Uranium	*Uncinia leptostachya*	0.13	927
Zinc	*Thlaspi calaminare*	4.00	682

Note: Data expressed on dry weight basis.

by soaking the papers in a 1% solution of the reagent in ethyl alcohol) in order to detect "nickel plants" not only in the field but in the herbarium as well (see Chapter 21). The leaf is crushed, the paper is moistened and folded around the sample, and after gentle pressure between thumb and index finger, the sample is examined after 1 min. A pink coloration indicates a nickel content in excess of 1000 μg/g (0.1%). Similar papers were prepared for testing for cobalt hyperaccumulators in Zaïre by use of the reagent α-nitroso-β-naphthol. This reagent, however, is carcinogenic so that the papers should be handled with gloves and should not be moistened with the tongue.

14.6. EXCLUSION MECHANISMS AND THEIR SIGNIFICANCE FOR BIOGEOCHEMICAL PROSPECTING

14.6.1. Introduction

The ability of plant species to restrict uptake of a toxic element is known as an *exclusion mechanism*. In such cases the amounts of the element in the aerial parts of the plant (twigs and leaves) remain at a constant level irrespective of the concentrations of the toxic element in the soil. This exclusion mechanism ultimately breaks down at a certain threshold concentration of the element in the substrate. Above this threshold, increased amounts of the element are accumulated over a short concentration range in the soil, until it is completely toxic to the plant. This is illustrated in Figure 14.2, which shows the uptake of four different elements by the Scandinavian "kisplanten" (pyrite plants; see also Chapter 4), *Lychnis alpina*, *L. alpina* var. *serpentinicola*, and *Silene dioica*. In the case of lead, it will be observed that there was virtually complete exclusion for *L. alpina* up to about 3000 μg/g (0.3%) lead in the soil. Above this level, there was an exponential increase in lead until the plant finally ceased to grow in soils containing more than 1% of the element. The plots also show partial exclusion for copper and zinc and nonexclusion for nickel. Even though nickel is taken up in a linear manner, a limit is ultimately reached at which the plant ceases to tolerate the substrate. From these plots, it would appear that all three species could be used successfully in biogeochemical exploration for nickel, unsuccessfully for lead, and only moderately successfully for copper and zinc.

The operation of a partial or total exclusion mechanism by plants is one of the biggest obstacles to biogeochemical prospecting. In the course of time, some elements such as nickel and uranium have acquired the reputation of being "easy" for biogeochemical prospecting, whereas elements such as copper and zinc were considered to be "difficult." For example, Marmo [573] noted the following:

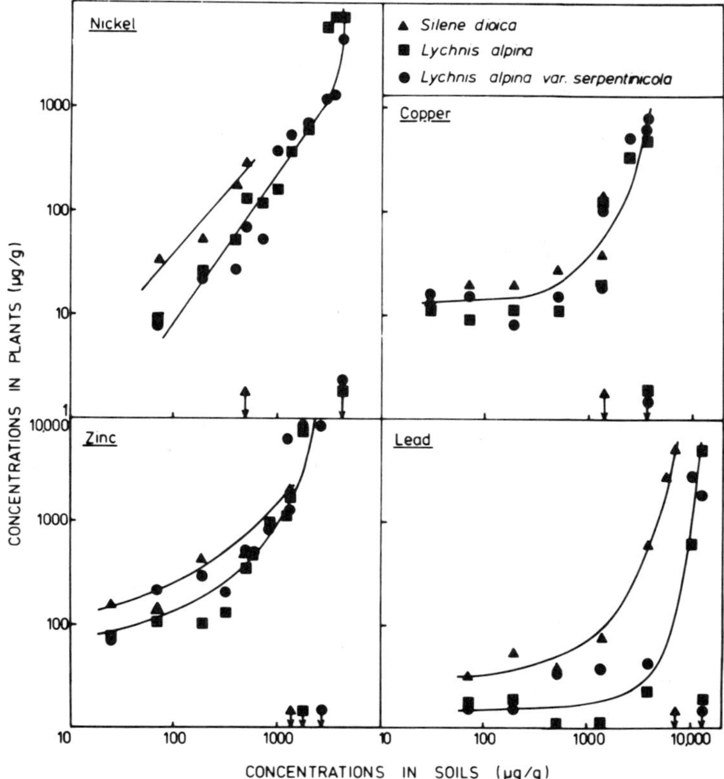

FIGURE 14.2. Curves representing the uptake of four elements by three Fennoscandian plant species. Arrows at the bottom right-hand side of the plots represent tolerance limits. *Source:* Brooks and Crooks [105].

From observations made by botanists, the concentrations of copper in plants does not increase linearly but after reaching some value separately determined for each plant, the increase of the copper content of the plant will no longer reflect the increasing content of the underlying rock as well as when comparatively small copper contents are in question.

Similarly, Boyle [83] observed a poor response to zinc in some Yukon vegetation. The poor vegetation response to copper and zinc is often concealed when, as so often happens, retrospective work is carried out over ore deposits. The concentration of ore elements is usually so great that plants are growing at the limit of their tolerance and take up elements anyway, even though they would have partially excluded them at lower concentrations in the soil. A further discussion of this problem follows in the next section.

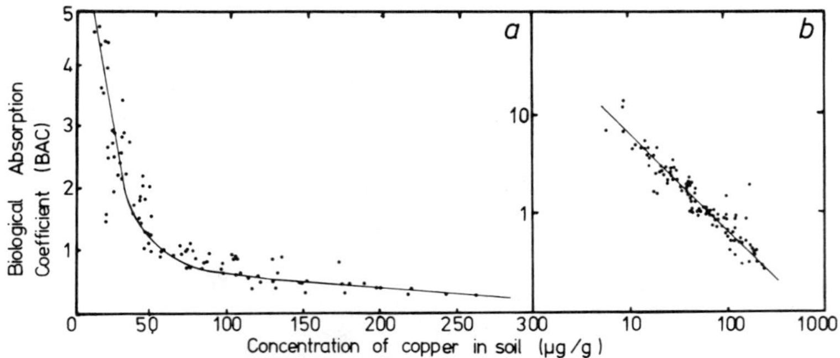

FIGURE 14.3.   Biological absorption coefficients (BAC) for copper in *Quintinia acutifolia* expressed as a function of the concentration of copper in the soil. Data expressed in both linear and logarithmic units. *After:* Timperley et al. [806].

14.6.2.   Partial Exclusion of Essential Elements

It had been obvious to my research group for many years that elements that were "difficult" for biogeochemical prospecting were, with the exception of lead, invariably essential elements for plant nutrition. If the plant/soil ratio for the concentrations of a given element was plotted as a function of its concentration in the soil (i.e., BAC vs. levels in the soil), an asymptotic curve was produced. In other words, the BAC was high at very low concentrations of the element in the soil and gradually declined to a limiting value at higher concentrations. This is illustrated in Figure 14.3. When logarithmic coordinates were used, a linear inverse relationship resulted. Linearity of this relationship implies that

$$\log y = n \log x + \log k$$

where y is the BAC, x is the concentration of the element in the soil, and k is a constant. The slope of the line (n) was very close to -1 in the plots of data obtained from field studies. Hence

$$\log y = -\log x + \log k$$

or
$$\log xy = \log k$$

and
$$xy = k$$

In other words, the BAC will decrease as the elemental concentration in the soil increases in order to preserve the constancy of k, which is clearly the concentration of the element in the plant, and corresponds to its physiological requirement level.

If similar plots are attempted for a nonessential element such as nickel (Fig. 14.4), the value of y remains constant irrespective of the concentrations of the element in the soil. It is for this reason that nickel is an "easy" element and copper is "difficult."

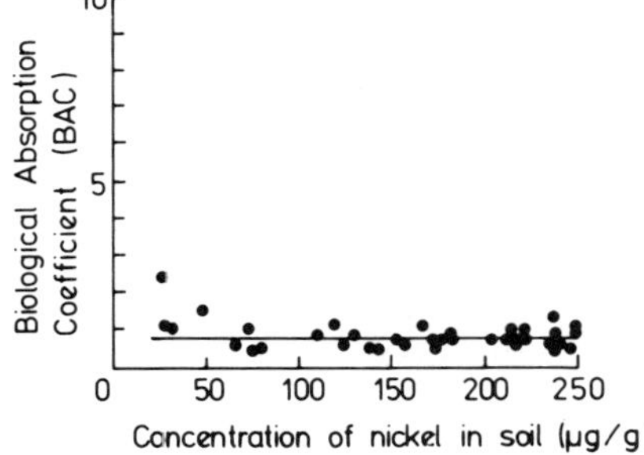

FIGURE 14.4. Plot of biological absorption coefficients (BAC) for nickel in *Nothofagus fusca. After:* Timperley et al. [806].

14.6.3. Criteria for Establishing Whether or Not an Element is Essential for Plant Nutrition

A full discussion of the concept of normal and lognormal distributions is given in Chapter 19, but it must be mentioned briefly here. Normal distributions concern data that are distributed symmetrically about the arithmetic mean in a so-called Gaussian curve. Lognormal distributions, however, are asymmetric and have a pronounced tail on the frequency diagram. However, if logarithmic coordinates are used for the x axis, a Gaussian curve is reestablished for this lognormal distribution.

It has been observed that, in human tissue at least, essential elements are normally distributed and nonessential elements are lognormal. A simple statistical test is to compare geometric and arithmetic means with the median. Agreement of the median with the geometric mean implies lognormality and agreement with the arithmetic mean implies a normal distribution. This test is not very satisfactory, as it is valid only if data are taken from one distribution. The presence of anomalous distributions in addition to background populations will nearly always give a skew to the histogram and the combined data will appear to be distributed lognormally. The situation is made worse by the fact that so many biogeochemical surveys are carried out over known ore bodies where two distributions are bound to occur if samples were taken well into background. The reader is referred to Timperley et al. [806] for a further discussion of this important topic.

It might be expected that the concentration range of essential elements in plants would be narrower than that for the nonessential elements because of regulatory mechanisms that tend to keep the amounts of essential elements constant in plants. Table 14.4 gives the relative standard deviation (*rsd*) of elemental concentrations in plants and soils from various sites in New Zealand. It will be noted that the *rsd* for copper and zinc tends to be less than for most other elements. The value of the *rsd* in plants is therefore some indication of the degree of essentiality of the element concerned. It will also be noted from the correlation coefficients (*r*) for elements in plants and soils, that values of *r* for copper and zinc tend to be less significant than for other elements, but become more significant for elements with wider ranges of concentrations (higher *rsd* values) in plant material.

TABLE 14.4.   Statistical Data for Elements in Soils (in parentheses) and Plant Ash

Sp.	Metal	No.	Median	A.M.	G.M.	rsd(%)	rsd rat.	Sig.	E.I.
A	Cu	35	100(60)	132(190)	100(162)	137(253-L)	0.54	*NS*	Yes
	Pb	35	100(148)	229(963)	123(229)	157(248-L)	0.63	*PS*	No
	Zn	35	270(150)	304(287)	282(166)	39(169-L)	0.23	*NS*	Yes
B	Cu	49	100(95)	111(148)	99(86)	75(126-L)	0.59	*S***	No
	Ni	49	90(205)	163(203)	124(172)	91(49-S)	1.85	*S***	Yes
	U	26	5(12)	5.6(20)	3(8.9)	100(121-L)	0.82	*S***	Yes
	Zn	49	250(81)	263(83)	253(81)	29(27-S)	1.05	*NS*	Yes
C	Cu	48	160(75)	160(93)	155(71)	27(73-L)	0.37	*NS*	Yes
	Zn	48	585(83)	602(88)	580(82)	29(41-S)	0.70	*NS*	Yes
	Ni	48	157(152)	193(159)	146(131)	70(53-S)	1.30	*S***	Yes
D	Cu	71	155(105)	209(149)	166(132)	61(54-S)	1.13	*NS*	Yes
	Mo	71	30(100)	87(143)	28(118)	252(65-S)	3.87	*S***	Yes
E	Cu	93	72(80)	77(120)	72(80)	47(125-L)	0.38	*S**	No
	Ni	93	87(175)	92(181)	84(157)	43(45-S)	0.96	*S***	Yes
	U	22	3(10)	7.8(97)	1.4(14)	177(271-L)	0.65	*S***	Yes
	Zn	93	200(83)	196(89)	192(86)	21(31-S)	0.68	*NS*	Yes
F	Cu	36	130(60)	179(113)	138(62)	62(192-L)	0.32	*NS*	Yes
	Pb	36	120(148)	174(643)	123(229)	78(207-L)	0.38	*NS*	No
	Zn	36	440(150)	588(22)	501(166)	71(135-L)	0.53	*S*	Yes
G	Cu	103	70(80)	77(125)	71(81)	64(119-L)	0.54	*S*	Yes
	Ni	103	105(175)	126(182)	102(156)	69(47-S)	1.47	*S***	Yes
	U	20	2(6)	3.3(11)	1.2(5.6)	130(129-L)	1.00	*S***	Yes
	Zn	103	220(83)	231(88)	221(84)	31(32-S)	0.97	*NS*	Yes

Source: after Timperley et al. [806].

KEY: Sp.—species: A, *Beilschmiedia tawa;* B, *Nothofagus fusca;* C, *N. menziesii;* D, *Olearia rani;* E, *Quintinia acutifolia;* F, *Schefflera digitata;* G, *Weinmannia racemosa*
A.M.—arithmetic mean
G.M.—geometric mean
Sig.—significance of plant-soil relationship for each element. See Section 19.4.1 for meanings of symbols.
E.I.—Essentiality indicated?
L—large range
S—small range

Although the relationship between the correlation coefficient and the degree of significance of the relationship is discussed fully in Chapter 19, some preliminary explanation should be given here. The correlation coefficient (r) is a measure of the degree of correlation between two variables. Values approaching $+1$ or -1 imply an increasingly significant direct or indirect relationship, respectively. The probability (P) of this relationship being not true is indicated by progressively smaller values of P, so that a low value of P implies a high degree of probability that a relationship does, in fact, exist.

It would seem from the above discussion and inspection of Table 14.4 that there is some justification for assuming that the magnitude of *rsd* values for elements in plants is a criterion of essentiality. This argument does not take into account the range of values of the element in the soil. Although the concentration of an essential element in plant material will be somewhat independent of the soil values, this will not apply at relatively high concentrations in the soil, because under these conditions, the plant's regulatory mechanism will break down. Indeed, it is only because of this breakdown that biogeochemical prospecting is possible for "difficult" elements such as copper and zinc.

Timperley et al. [806] have suggested that a more meaningful interpretation of essentially is to take the ratios of the *rsd* values of the elements in the plants and soils as shown in Table 14.4

If the range of values for an element in the soil is small, an equally small variation for the same element in plants should appear, whether or not the element is essential. It is only when the range of values in the soil is relatively great that large variations in the *rsd* for essential and nonessential element should be expected in the plants. For this reason, Timperley et al. [806] have classified the ratios of the *rsd* in plants and soils into two groups, depending on the concentration of the element in the soil. When the *rsd* for an element in the soil is greater than 70% this is taken arbitrarily as being a "large" range. The term "small" range is used for values less than 70%.

From Table 14.4 it will be observed that when the *rsd* for a given element in the soil is large, the plant/soil ratio for the *rsd* values (except for lead in *Schefflera digitata*) is always greater for nonessential than for essential elements. If, for the above case, a value of 0.60 is taken as a dividing line for ratios of the *rsd* values, nonessential elements are involved for values above 0.60 and essential elements for those below. A particularly significant observation is that values of *r* for the plant-soil relationship are nearly always highly significant for nonessential elements and not significant for essential elements. Therefore it would seem that it is possible to predict a favorable or unfavorable plant-soil relationship merely by considering ratios of the *rsd* values for concentrations of elements in plants and soils.

When the range of values for an element in soils is small, the *rsd* for nonessential elements in plants should decrease, and less differentiation would be expected between essential and nonessential elements. It is nevertheless still possible to distinguish statistically between the two classes of elements under these circumstances.

When there is a small concentration range of an element in the soils (*rsd* < 70%), a threshold value of 1.2 has been assigned by Timperley et al. [805] for distinguishing between essential and nonessential elements in plants as measured by the plant-soil *rsd* ratio. Using this criterion, Table 14.4 shows that it would have been possible to predict essentiality or nonessentiality in every case, except for nickel in *Quintinia acutifolia* and molybdenum in *Olearia rani*. Molybdenum is a special case because, although it is an essential element, the physiological requirement by many plants is so low that it effectively behaves like a nonessential element [805].

Overall, essentiality or nonessentiality would have been predicted cor-

TABLE 14.5. Statistical Tests to Establish the Physiological Role of Elements in Plants and to Predict the Significance of the Plant-Soil Correlations for These Elements

| | Range of Concentrations of Element in Soil in Terms of rsd | | | |
| | Large Range (*rsd* >70%) | | Small Range (*rsd* <70%) | |
Role of Element	Ratio of *rsd* (plant-soil)	Predicted sig. of *r* (plant-soil)	Ratio of *rsd* (plant-soil)	Predicted sig. of *r* (plant-soil)
Essential	< 0.60	$P > 0.001$	< 1.2	$P > 0.001$
Nonessential	> 0.60	$P < 0.001$	> 1.2	$P < 0.001$

Source: Timperley et al [806].
KEY:　　*r*—correlation coefficient
　　　　rsd—relative standard deviation
　　　　P—level of probability that a relationship does not exist

rectly in 21 out of 23 cases using the above criteria. In a further 19 out of 23 cases, it would also have been possible to predict accurately whether or not a highly significant correlation would exist between the concentrations of an element in plants and soils. A highly significant correlation is taken as being one for which the value of *P* is less than 0.01 (i.e., *S**; see Chapter 19). Table 14.5 summarizes statistical tests to prove the correct prediction of highly significant relationships. As an illustration, we may consider the case of uranium in leaf ash of *Q. acutifolia*. The *rsd* values for this element in plants and soils were respectively 177% and 271%. The range of values in the soil implied a large range for which the threshold value for *rsd* ratios is 0.60. The observed value of 0.65 indicated that uranium was nonessential for the plant and that the plant-soil relationship for this element would be highly significant at least. In fact, the value of *P* was less than 0.001 (*S***) and implied a very highly significant relationship.

14.6.4 Barrier-Free Biogeochemical Prospecting

A poor plant-soil correlation for various elements has long been recognized by many Russian biogeochemists [477, 563]. A.L. Kovalevsky was the first worker to seek a practical solution to this problem, and in an important fundamental contribution [463] set out his ideas on the use of appropriate biological sample types for biogeochemical prospecting. He recognized that many plants operate exclusion mechanisms for certain elements. These mechanisms usually occur at root systems but sometimes operate at the aerial parts of the plant. His solution to the problem was to test a very large number of what he called **biological objects,** which I have renamed **biological sample types.** These are roots, bark, twigs, needles, leaves, or even the entire aerial parts of plants. On plotting the elemental content of each biological sample type as a function of the concentration of the same element

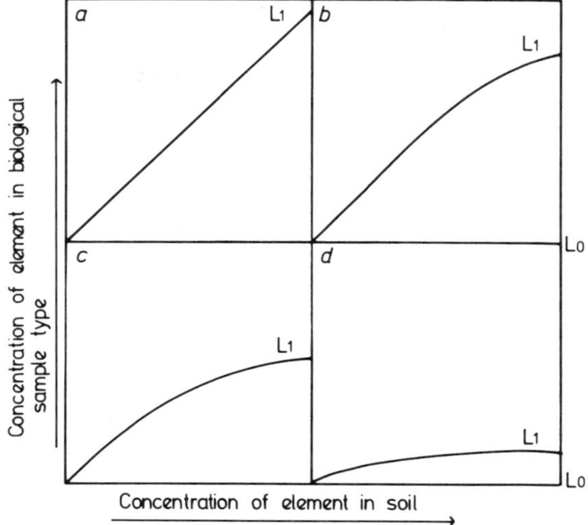

FIGURE 14.5 Idealized plots of elemental concentrations in hypothetical biological sample types and in the supporting soil: (*a*) barrier-free system ($L_1/L_0 = {>}300$); (*b*) high-barrier system ($L_1/L_0 = 30\text{-}300$); (*c*) medium-barrier system ($L_1/L_0 = 3\text{-}30$); (*d*) low-barrier system ($L_1/L_0 = 2\text{-}5$). L_0 = background level for element in sample type; L_1 = limiting elemental content of sample type.

in the soil, he found that plots were of four types, as illustrated in Figure 14.5.

In plot *a*, there is a very highly significant relationship between the two variables and the biological sample type is extremely useful for biogeochemical prospecting. It is said to be a barrier-free system. In plot *b*, there is partial exclusion at high concentrations of the element in the soil, and the sample type is less suitable (though usable) for prospecting. This is known as a high-barrier system. Plot *c* represents a medium barrier situation where the sample type is probably unsuitable for prospecting. In plot *d*, the low barrier organs are completely unsuitable under any circumstances. The four categories are defined by the ratio of the limiting concentration of the element in the plant material (L_1) compared with the concentration of the same element in samples taken from background (L_0). For example, a tree root sample with 500 µg/g lead where background samples gave 1 µg/g of this element, would be classified as barrier free, since the ratio exceeds 300. See Figure 14.5 for the ratio categories.

Kovalevsky's work has general applications, since the flora of Siberia where most of his work was carried out has many similarities with that of North America and Northern Europe. His nomenclature is sometimes a little unclear, but I have retained it in this chapter because of possible confusion when referring to his original work. For example, the term ''high barrier'' does not imply a high degree of exclusion of elements. It implies that the ratio L_1/L_0 is high. Similarly, ''low barrier'' should logically be close to ''barrier-free'' in significance, but it in fact implies a high degree of exclusion because L_1/L_0 is low.

Element	No. of Samples Tested	Percentage of Biological Sample Types Within the Category			
		Non-barrier	High Barrier	Medium Barrier	Low Barrier
Molybdenum	64	16	65	19	0
Gold	67	01	26	44	20
Lead	70	10	22	35	33
Tungsten	31	3	20	32	45
Silver	90	2	7	49	42
Beryllium	53	0	4	45	51
Uranium	65	0	3	17	80
Fluorine	60	0	3	7	90
Total	500	Av. 5	Av. 19	Av. 31	Av. 45

Source: Kovalevsky [477].
KEY: Non-barrier—no limit to BAC values
High barrier—BAC 0.50–5.0
Medium barrier—BAC 0.15–0.49
Low barrier—BAC < 0.15
Note: BAC values based on dry weight.

TABLE 14.7. Classification of Biological Sample Types Used in Prospecting for
Lead in Eastern Siberia

Category	Lead Concn. (μg/g dry wt)	BAC $\times 10^4$	Species Sampled	Organs Used
Nonbarrier	50–5000	50–5000	*Larix dahurica* (larch)	N,B,R
			L. sibirica (Siberian larch)	N,B,R
			Pinus silvestris (pine)	B,R
			P. contorta (pine)	B,R
			Vaccinium vitis-idaea (red bilberry)	A
			Polytrichum hyperboreum (moss)	A
			Peltigera aphthosa (lichen)	A
			Cladonia gracilis (lichen)	A
			C. alpestris (lichen)	A
			Cetraria islandica (lichen)	A
High barrier	5–500	15–1500	*Betula platyphylla* (birch)	L,B
			Salix sp. (willow)	L
			Rhododendron dahuricum	L
			Ledum palustre (marsh tea)	L,T
			Lonicera sp. (honeysuckle)	L
			Vaccinium uliginosum (blueberry)	L
			Larix sibirica (larch)	N
			Pinus silvestris (pine)	N,T
			P. sibirica (pine)	N,T
			P. contorta (pine)	T
			Rubus sachalinensis (raspberry)	T
			Vaccinium vitis-idaea (red bilberry)	A
			Lamium sp.	A

TABLE 14.7. (Cont.)

Category	Lead Concn. (μg/g dry wt)	BAC × 10⁴	Species Sampled	Organs Used
Medium barrier	0.50–50	1.5–500	*Populus tremula* (aspen)	L,T
			Salix sp. (willow)	L,T
			Rhododendron dahuricum	L,T
			Vaccinium uliginosum (blueberry)	L,T
			Potentilla fruticosa (cinquefoil)	L,T
			Spiraea media (meadow seet)	L
			Rosa acicularis (rose)	L
			Artemisia sp (wormwood)	L,A
			Betula platyphylla (birch)	T
			Lonicera periclymemum (honeysuckle)	T
			Epilobium angustifolium (fireweed)	A
			Bromus secalinus	A
			Lappula echinata (stickseed)	A
			Pulsatilla tenuiloba	A
			Carex silvatica (wood sedge)	A
			Laminium sp (henbane)	A
			Betula sp. (birch)	B
			Salix sp.	B
Low barrier	0.50–5	0.15–50	*Populus tremula* (aspen)	L,T
			Spiraea media (meadow sweet)	L,T
			Potentilla fruticosa (cinquefoil)	L
			Rubus sachalinensis (raspberry)	L
			Betula sp. (birch)	T,L,B
			Salix sp. (willow)	T,B
			Rhododendron dahurica	T
			Rosa acicularis (rose)	T
			Vaccinium uliginosum (blueberry)	T
			Juniperus monosperma (juniper)	T
			Larix sp. (larch)	T
			Carex sp.	A
			Atriplex sp (orache)	A
			Lappula echinata (stickweed)	A
			Sanguisorba offinalis (bloodwort)	
			Leucantheumum sp. (ox-eye daisy)	A A
			Artemisia vulgare (wormwood)	A
			A. gmelini (Gmelin's wormwood)	A
			A. frigida (cold wormwood)	A

Source: Kovalevsky [477].

KEY: A—aerial parts
  B—bark
  L—leaves
  N—needles
  R—roots
  T—twigs
  BAC—biological absorption coefficient.

141

Table 14.6 classifies some of the more common ore elements in order of effectiveness for biogeochemical prospecting, as determined from examination of some 500 biological sample types. Molybdenum appears to have the highest proportion of suitable sample types, though only 16% of the 64 organs examined were barrier free. Kovalevsky equates high L_1/L_0 values with effectiveness for prospecting. This is not necessarily the case when dealing with hyperaccumulating species (see above). It is my experience that such plants do not discriminate well between degrees of mineralization over an orebody.

It is clear that barrier-free sample types are relatively few in number (about 5% of all organs tested) and for some elements (such as beryllium) do not exist at all. Therefore it will often be necessary to use high-barrier sample types for most elements. The pioneering work of Kovalevsky is applicable elsewhere, so that because of the similarity of boreal vegetation throughout the Northern Hemisphere, data are presented for lead in Table 14.7 [463]. Space limitations do not permit a listing of all elements for which data are available, and the reader is referred to other works by this author [460, 466, 469, 474] for further details.

The discovery of barrier-free biological sample types is one of the most significant advances in biogeochemical prospecting in recent years and should do much to put the technique on a more quantitative basis.

14.7. FACTORS AFFECTING ELEMENTAL ACCUMULATION BY PLANTS

14.7.1. Introduction

The main criterion for the successful use of the biogeochemical method is that the element investigated should be taken up by the plant to an extent directly proportional to its concentration in the soil. In other words, that the BAC should be as constant as possible. It has been shown in the preceding section that identification of suitable biological sample types is a satisfactory solution to this problem.

There are probably up to 20 different variables that can affect elemental accumulation by plants. There are indeed so many of these factors that it is surprising that biogeochemistry has ever come to be used at all for mineral prospecting. The field worker should be aware of these variables, even though most of them are minor in their effects. Use of suitable procedures will eliminate or at least reduce the effects of most of these variables; they are therefore not such a serious problem as might be supposed at first sight. Some of these factors will now be considered.

14.7.2. Type of Plant Sampled

From the discussion in Section 14.6.4, it is obvious that the type of plant sampled is of the utmost importance, since the BAC values can be very

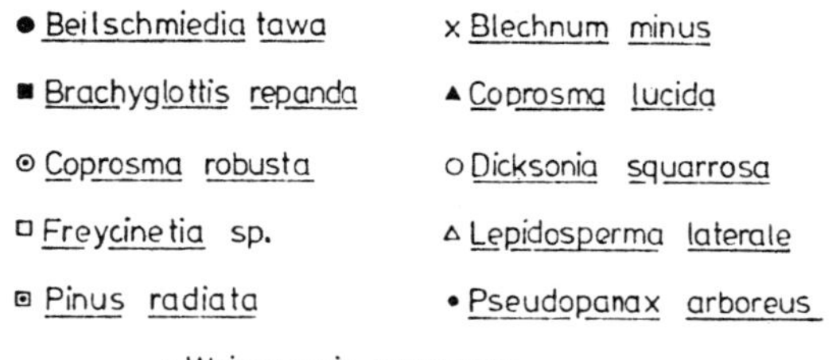

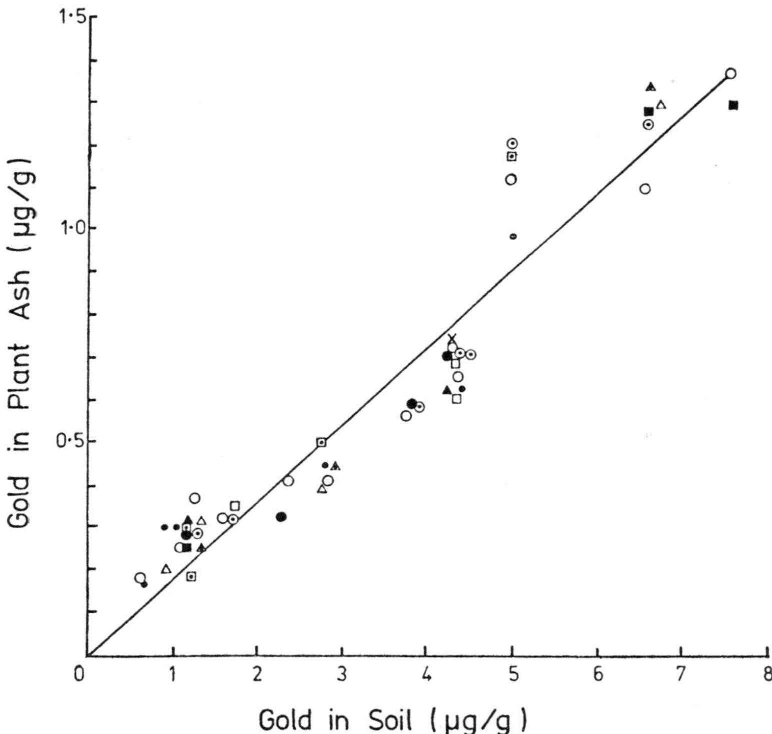

FIGURE 14.6.   Plot of gold concentrations in ashed leaves of ten New Zealand species plotted as a function of the gold content of the soil. *Source:* Ward and Brooks [878]. Reprinted by courtesy of the New Zealand Department of Scientific and Industrial Research.

variable among different species. In general, the choice of plants depends on the element being sought. For example, in the case of gold and molybdenum (Table 14.6) which are "easy" elements for biogeochemical prospecting, there is probably a much wider choice of biological sample types than in the case of beryllium, where barrier-free samples might be hard to find. An example of this is given in Figure 14.6, where it can be seen that gold uptake by some New Zealand plants seems to be independent of species.

From my own experience, the biggest contrast between metal uptake by plants growing in mineralized and background areas is afforded by shrubs and herbs rather than by large trees. Against this must be set the fact that the root systems of the former usually do not sample the same volume of soil or penetrate as deeply through the overburden.

14.7.3. Plant Organ Sampled

This topic has already been dealt with to some extent in Section 14.6.4. However, much remains to be discussed. The extent of elemental accumulation by vegetation is not necessarily an indication of its reliability for biogeochemical prospecting. There is conflicting evidence as to which organ contains the highest concentration of trace elements.

In fairly extensive studies on the above problem using native species, I have found little difference in the reliability of leaves compared with twigs in prospecting for copper, molybdenum, nickel, uranium, and zinc. An exception was that twigs of *Beilschmiedia tawa* were more reliable than leaves in prospecting for lead [618]. The lack of differentiation between leaves and twigs in New Zealand may be due partly to the fact that hardly any of our native trees are deciduous, so that a sample of leaves from any one species is likely to contain individuals of different ages. This factor seems to mitigate against seasonal effects which are very significant in deciduous vegetation.

There is ample evidence that the elemental content of the leaves of deciduous plants varies considerably between budding and leaf fall. For this reason it is advisable never to sample leaves of deciduous species. Warren and Delavault [885] recommend the use of 1–3 year-old twigs with diameters of 3–6 mm.

Kovalevsky [477] has found that plant roots are often the best plant organs to sample. There are, however, two practical difficulties with this type of sample. The first of these is that plant roots are often very hard to sample in cold climates during the depth of winter (a major advantage of the method is, in any case, the possibility of its use in winter in areas of frozen ground). The second problem is that roots are very easily contaminated by the soil and should therefore be peeled wherever possible.

14.7.4. The Age of the Plant Organ

Of all the factors that affect the uptake of trace elements by plants, the age variable has received the least attention. There is some evidence for appreciable variation in the elemental content of plants that are of the same species but are of different ages. There is also strong evidence that the elemental content of certain plant organs, particularly leaves, is influenced by seasonal factors. A classical investigation into this problem was carried out by Robinson [694] who studied the uptake of the rare earths by the hickory tree by checking the same specimen at different times of the year. He found that on June 1, the content of the leaves was 174 μg/g. A month later this had risen to 634 μg/g, and on October 1 of the same year the level had reached 981 μg/g. Robinson and Edgington [697] also observed a doubling of the boron content of hickory leaves between spring and fall. Seasonal variations in leaves have also been noted for molybdenum [45] and for copper and zinc [909]. Work in the Soviet Union has also revealed seasonal differences in the concentrations of elements in leaves. Malyuga [554] has reported that levels in the fall are two or three times greater than in the spring. Chebaevskaya [184] has shown that the content of cobalt, copper, and nickel increases toward the time of development of the reproductive organs.

It now seems to be well established that the elemental content of many deciduous leaves rises to a maximum just before exfoliation. This is probably a convenient mechanism whereby plants can get rid of unwanted ballast elements. One of the most convincing proofs of this seasonal variation has been furnished by Guha [335]. Figure 14.7 shows the variation of six elements in leaves of 18 deciduous species studied over a 12-month period. Increases in boron, iron, and manganese were noted over the growing season whereas copper, molybdenum, and zinc decreased in many cases. Some of Guha's findings contradict those of other workers, but the differences may be due to different species used.

Most of the recorded literature on the subject has indicated a tendency for elements to be concentrated in actively growing tissues such as shoots and young leaves. Lowest levels are usually in the sapwood and heartwood of tree trunks. It is interesting to record that the above concentrations are based on dried material. If data are expressed on an ash-weight basis, elemental levels in trunks and heartwood are often higher than in twigs or leaves because the ash content of the former is much lower.

14.7.5. The Health of the Plant

The ability of a plant to accumulate trace elements is partly a function of its health. There is sometimes a tendency for the elemental content to increase due to a breakdown of the regulatory mechanism of the species. Unfortunately, it is not always possible to predict whether disease will increase or decrease elemental levels. Unhealthy plants, therefore, should be avoided in a biogeochemical survey, though they may well serve as useful geobotanical indicators [149].

14.7.6. The pH of the Soil

Although there are many variables that can adversely affect the reliability of the biogeochemical method, many people think that the pH factor is the one that is most likely to affect the method. Elemental uptake by plants can be affected by the pH of the soil. How the availability of a number of elements is affected by the pH is demonstrated in Table 14.8. Elements are

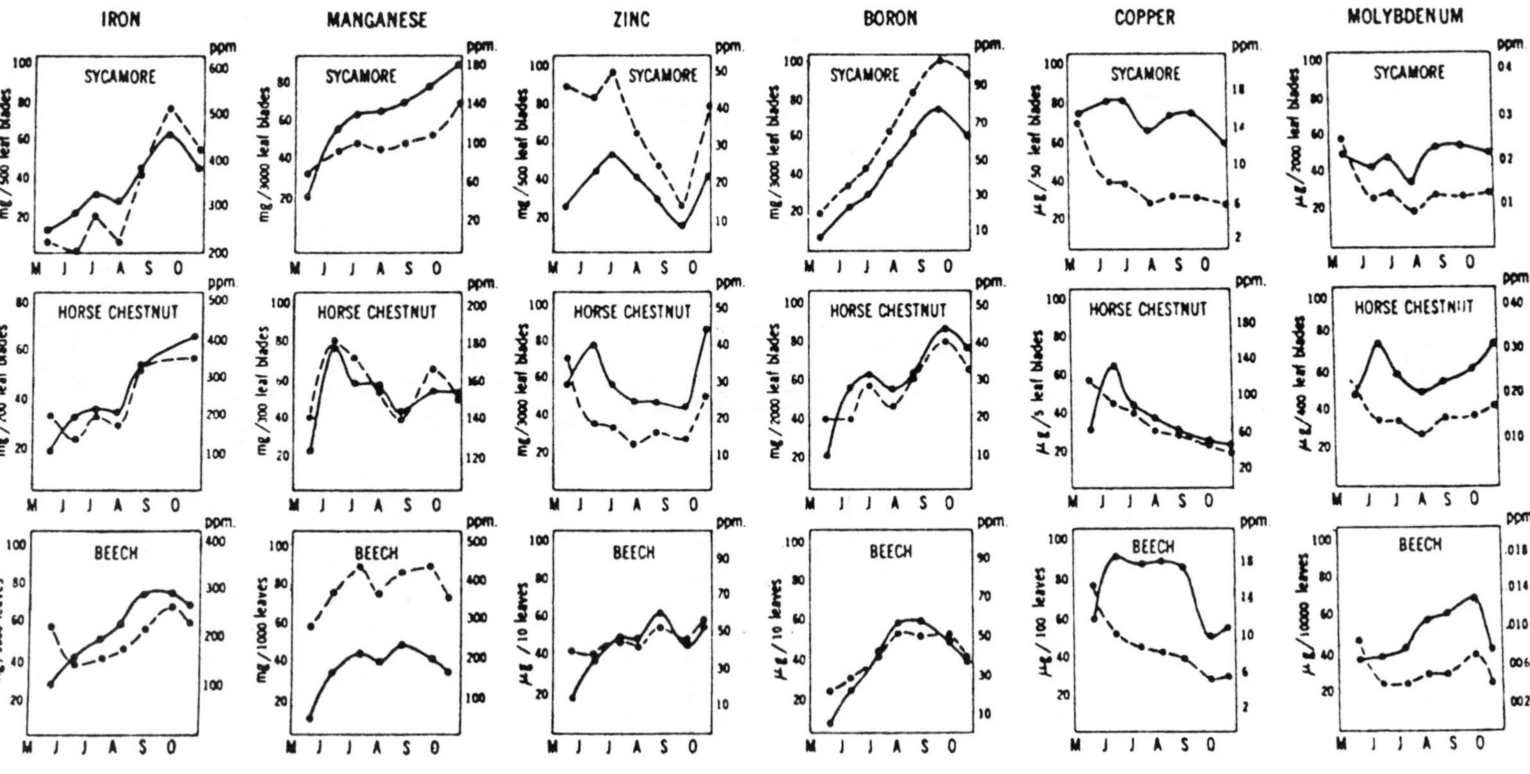

M J J A S O refer to the months of the year in which samples were collected

FIGURE 14.7. Seasonal variation of minor elements in the leaves of three species of deciduous trees. *Source:* Guha [335].

TABLE 14.8.   The Effect of pH on Elemental Uptake by Plants

Element	Effect of pH	References
Copper	100% uptake at pH 4, 33% at pH 6, less availability at increasing pH	398, 905
Manganese	Available below pH 5.5, unavailable above	12,94,308,608
Molybdenum	Available above pH 5	44,202,343
Nickel	Less uptake at increasing pH	805
Tungsten	Available above pH 5	44,202,343
Zinc	Unavailable above pH 6	12,701

either precipitated or solubilized depending on the pH, which can also affect the exchange capacity of clay minerals. The influence of pH on the uptake of ore elements by plants will now be considered in detail.

Studies by Timperley et al. [805] have shown that in biogeochemical prospecting for copper, nickel, and zinc in New Zealand, the pH of the soil was not of great importance. The BAC values for these three elements in 300 leaf samples of four different species from various parts of the country were correlated statistically with the pH of the substrate. In only one case (uptake of nickel by *Weinmannia racemosa*) was any significant relationship established. In this case an inverse relationship was noted, which showed that nickel uptake decreased as the pH was raised.

In general, changes of pH would not seem as serious as was first imagined. The most likely changes probably would occur over sulphide ore bodies, where a drastic lowering of the pH is likely to increase BAC values for most elements and thus serve as an advantage rather than a disadvantage.

14.7.7.   The Distribution of Plant Species

One of the problems with biogeochemical prospecting is that, unlike soil sampling, a sample is not necessarily available at any chosen spot. If a selected species is very common, that is not usually a problem. However, it is obviously useless to select a biological sampling type that does not combine effectiveness with a widespread distribution.

14.7.8.   The Depth of Root System

One of the main advantages of biogeochemical prospecting compared with other geochemical methods lies in its power of penetration. This penetrating power depends in turn on the depth of the root system, on the ability of the plant to translocate minerals from the roots to its aerial systems, and

TABLE 14.9   Depth of Root Penetration of Woody Plants and Shrubs in Different Climatic Zones

Zone	Species	Root Depth (m)
Tundra	*Betula nana*	0.3
	Larix sibirica	0.4
Forest podzol	*Picea excelsa*	2
	Pinus silvestris	2.5
	Populus alba	3
Forest-steppe and steppe	*Pinus silvestris*	3.5
	Betula verrucosa	4
Desert and semidesert	*Actinea acaulis*	0.3
	Artemisia spinescens	2
	Tamarix ramosissima	5
	Anabasis aphylla	5
	Quercus sp.	9
	Artemisia tridentata	10
	Haloxylon aphyllum	10
	Alhagi pseudalhagi	15
	Sarcobatus vermiculatus	18
	Atriplex canescens	20
	Pinus ponderosa	25
	Juniperus monosperma	60

Sources: Cannon [160] and Malyuga [563].

on the extent of the dispersion halo itself, which in turn depends to some degree on the fluctuation of the water table. The depth of the root system is largely a function of climate and, by inference, the type of soil involved.

Table 14.9 lists the depths of root systems of various plant species, and it will be seen that depths of up to 60 m have been recorded. The latter figure was for a live root found in a mine working on the Colorado Plateau [157]. The root depth of a plant is a function not only of climate but also of whether the species is a **phreatophyte** or a **xerophyte.** Phreatophytes are plants that depend on the zone of saturation below the water table for their moisture; xerophytes depend on surface water from rainfall and therefore have shallow root systems. Clearly, phreatophytes in general will be more useful than xerophytes in prospecting for deep deposits. In some cases plants with extremely shallow root systems can still indicate mineralization at depth. Cannon [157] has reported species of *Astragalus* that have indicated uranium deposits at depths of 25 m. To some extent the suitability of phreatophytes also depends on the type of root. For example, a species with a long tap root may well be able to give evidence for mineralization at depth but fail to pick it up at the surface.

Translocation of metals from the roots to the aerial parts of the plants has been mentioned as a factor affecting the usefulness of a species for

TABLE 14.10. Uranium and Vanadium Contents of Roots Compared with Aerial Parts of Plants

Species	Type of Root	Uranium in Ash (μg/g)			Vanadium in Ash (μg/g)		
		Roots	Tops	Ratio	Roots	Tops	Ratio
Juniperus monosperma	Deep	1600	7.8	200	3000	20	150
Juniperus monosperma	Deep	140	2.0	70	4000	50	80
Quercus gambeli	Deep	190	10.0	19	1700	90	19
Juniperus monosperma	Shallow	7	1.2	5.6	154	54	2
Atriplex confertifolia	Shallow	5	3.0	1.6	90	10	9
Astragalus preussi	Shallow	70	70	1.0	2600	3000	0.8

Source: Cannon [157].

biogeochemical prospecting. In general, the longer the root system, the less the enrichment of the trace elements in the upper parts of the plants. This is illustrated in Table 14.10, which gives the uranium and vanadium contents of roots compared with aerial parts of various species. The difference in values between the shallow and deep-rooted species is probably only a function of the greater difficulty of transportation of ions through a long root system.

In sampling vegetation, care should be taken to select leaves or twigs at several points around the circumference of the specimen studied. This is because root systems tend to translocate ions to aerial parts situated on the same side of the plant as themselves. It could well be possible that roots on one side of the plant could be in contact with an orebody whereas roots on the other side were in barren ground. Selection of vegetation from the wrong side of the species might result in the anomaly being missed entirely.

14.7.9. Drainage

Drainage is another factor that can affect prospecting with plants. Poorly drained soils, when compared with those that are well drained, can give misleading anomalies during an exploration project. This is because many elements are more mobile under waterlogged conditions and this will be reflected by greater uptake by plants. Mitchell [595] has studied this factor and carried out extensive trials on waterlogged soils. He found a greatly increased mobilization of cobalt, copper, iron, manganese, nickel, and vanadium as measured by extraction of the elements into acetic acid and ammonium acetate. Although increased elemental uptake due to drainage problems may not be as great as that caused by a genuine anomaly in the soil, any apparent anomalies associated with a change in the water content of the substrate should be examined with care and be reassessed critically.

14.7.10. Aspect

The question of aspect is of paramount importance in plant ecology. It also has significance in biogeochemical prospecting because accummulation of elements from the soil is controlled by photosynthetic processes. Uptake, therefore, will be influenced to some extent by the intensity and duration of the light received by the plant. Pot trials such as those of Ferres [292] have confirmed the influence of the photosynthetic process on the mineral uptake.

The daily variation in rainfall is also a factor that can influence mineral uptake by plants. Warren et al. [906] have shown that dry soils can increase the iron content of plants and decrease their manganese content. The combined effects of soil temperature and moisture content are illustrated by their findings and show that, in many cases, the elemental content of the same plant species differs markedly between sites of a northern and southern aspect.

14.7.11. Antagonistic Effects of Other Elements

The question of the availability of elements in soils has already been discussed at some length and will not be mentioned further beyond reaffirming its importance as a factor influencing the success of biogeochemical methods of prospecting.

The question of mutual antagonism of pairs of elements is also a factor that should be considered. In work in New Zealand, I have frequently noticed that if the concentrations of a pair of elements are plotted against each other, a linear function of negative slope frequently results. Examples of this are inverse relationships for calcium-nickel, nickel-magnesium, cobalt-calcium, and calcium-magnesium accumulated in serpentine plants [533]. The reason for this behavior is not clear; it may be linked to a breakdown of the plant's selection mechanism. The capacity of a plant to absorb mineral salts obviously has finite limits. If one constituent of the soil is accumulated to an excessive degree, this will have to be compensated by reduced uptake of another.

14.8. CONCLUSIONS

One of the most surprising features of biogeochemical prospecting is that apparently it can be used successfully in spite of all the numerous factors that can affect the reliability of the method. The inference is clear. Many of the variables must have a minor effect, though in some areas they can be more important than elsewhere. Other, more important variables may be reduced or even be virtually eliminated by use of certain preventative procedures.

TABLE 14.11.   Factors Affecting Elemental Uptake by Plants, and Methods of Reducing Their Effect

Factor	Importance	Method of Reducing Effect
Type of plant	Very great	Selection of suitable plants by testing
Organ sampled	Great	Selection of suitable organ by testing
Age of organ	Significant	Selection of suitable age by testing
Root depth	Significant	Use of ratios of two elemental concentrations in same sample
pH	Fairly significant	Use of ratios of two similarly affected elemental concentrations in same sample
Health of plant	Fairly significant	Select healthy specimens only
Drainage	Fairly significant	Avoid poorly drained areas
Availability of element	Fairly significant	Use concentration ratios of pair of elements with same availability to plants
Antagonism of other elements	Minor	None
Rainfall	Minor	Carry out work over short period of time
Variable shading	Minor	Avoid shady sites or use elemental ratios
Soil temperature	Minor	Carry out work over short period of time

Source: Brooks [97].

An attempt to measure the relative significance of some of the above variables was made by Timperley et al. [808] using a multiple regression technique (see Chapter 19) on biogeochemical data from a survey of copper and nickel in leaves of *Nothofagus fusca*. The regression equation had the form

$$y = k_1 a + k_2 b + k_3 c + \ldots + k_{15} o + c$$

where y was the dependent variable (BAC for the above species), k_1–k_{15} were regression coefficients, a–o were fifteen independent variables, and c was the regression constant. Ten of the independent variables were the concentrations of ten elements (Ni, Co, Cu, Zn, Cr, Fe, Ca, Mg, Mn, and K) in the plant material and the other five were slope of terrain, altitude of sample site, tree height, tree trunk diameter, and sampling height as a fraction of the total height of the tree. A computer program then determined whether the effect of each independent variable on the dependent variable

(y) was more significant than was a predetermined probability level. If the effect was more significant, the independent variable was retained in the regression equation and the other variables already selected were rechecked to see if they were still significant at the prescribed level of probability. Those no longer significant were removed. This procedure was repeated for all the remaining untested independent variables.

The above study showed that two variables (the nickel content of the plant and the altitude of the sampling site) were significant for influencing the BAC for nickel at the 0.1 level of probability. In the case of the BAC for copper, seven variables (the calcium, chromium, copper, magnesium, and nickel contents of the plant; the slope of the terrain; and the logarithm of the tree diameter) were significant at the same level of probability. Nevertheless, only 35% of the variance of the BAC for copper and 26.5% for that of nickel were explained by the regression equation. This seems to indicate that the influence of any one variable on the BAC for a given element is relatively minor.

Table 14.11 lists some of the factors influencing the accumulation of elements by plants in order of decreasing general importance. Suggested procedures for reducing the effects of these variables are also included. It is not claimed that the measures recommended will completely eliminate the adverse effects of all the variables concerned, but judicious use of them should help to ensure that any variations that do occur are less than the variations caused by ore anomalies in the exploration area.

Some of the problems can be overcome by using the ratio of a pair of elements instead of a single element only. The reference element is not an ore element and should vary in sympathy with the latter whenever fluctuations are due to variables other than the presence of ore elements in the substrate.

15

BIOGEOCHEMICAL PARAMETERS AND THEIR SIGNIFICANCE FOR MINERAL PROSPECTING

15.1. INTRODUCTION

A characteristic feature of the Russian language and culture is the preponderance of acronyms. It is not surprising, therefore, that many of these spill over into the sciences, and particularly into the biological sciences. A number of such parameters have been described for the field of biogeochemical prospecting during the past twenty-five years in an attempt to quantify a science that is somewhat inexact because of its reliance on naturally variable biological sample types. Sabinin [715] was one of the first to describe biogeochemical parameters and his work has been modified subsequently by various workers [99,458,647]. Five of the biogeochemical parameters are shown in Table 15.1. The table gives suggestions for a nonliteral translation of the parameters together with suggested abbreviations for them.

Each of the parameters will be discussed, and its significance for biogeochemical prospecting will be evaluated as far as possible in the light of my own personal experiences in prospecting in several countries. It should be emphasized that while the *magnitude* of each parameter has little signifi-

TABLE 15.1.   The Biogeochemical Parameters of the Russian Literature

Russian		English	
Name	Abbreviation	Suggested Name	Abbreviation
Coefficient of biological absorption	KBP	Biological absorption coefficient	BAC
Relative concentration of chemical elements in different species of plants	OSVR	Relative absorption coefficient	RAC
Relative concentration of elements in plant organs	OSDR	Acropetal coefficient	AC
Relative change of concentration of chemical elements in plants	OIS	Temporal absorption coefficient	TAC
Concentration of mobile form of chemical elements in plants	SPFR	Mobile element absorption coefficient	MAC

Source: Brooks [99].

cance for the field of mineral prospecting, it is the *variance* of it that can have an adverse effect on the biogeochemical method.

15.2. THE BIOLOGICAL ABSORPTION COEFFICIENT (BAC)

Kovalevsky [458] has defined the **Biological Absorption Coefficient** (BAC) as follows:

$$BAC = C_p/C_s$$

where C_p is the concentration of an element in plant ash and C_s is the concentration of the same element in the substrate. The original acronym used by this author was KPB (**koefizient biologicheskoi pogruzhyonnost** = coefficient of biological absorption); he later changed this to RPK (**rastitel'no pochvenni koefizient** = plant-soil coefficient). As I have transcribed both of these as BAC, this is not important in reading the English literature, but may cause confusion if the original Russian texts are consulted. There is a second point of possible confusion. Biogeochemical data can be expressed on either an ash weight or dry weight basis. The former standard gives concentration values that are about 20 times greater than when a dry weight basis is used. To avoid confusion, wherever possible I have used the dry weight standard throughout this book, using a conversion factor of 0.05 if the ash percentage was not quoted in the original reference.

BAC values can vary by several orders of magnitude (see Chapter 14), usually in the range 0.0001–10. However, there is evidence that even this upper limit can be exceeded. For example, Severne and Brooks [740] reported BAC values of over 25 for the hyperaccumulator of nickel, *Hybanthus floribundus,* and Cannon [160] has reported a BAC of about 130 for selenium in *Astragalus pattersonii.*

There is some relationship between high BAC values and whether or not the element concerned is essential for plant nutrition. Table 15.2 lists BAC values for a number of elements in vegetation and it will be readily seen that the essential elements tend to have the highest BAC values. The

TABLE 15.2. Biological Absorption Coefficients for Essential and Nonessential Elements

Perel'man Classification [647]	Element	Physiological Role	BAC $\times$ 10^4
Strong absorption	Boron	Essential	850
Intermediate absorption	Sulphur	Essential	480
	Zinc	Essential	450
	Phosphorus	Essential	440
	Manganese	Essential	200
	Silver	Nonessential	125
	Calcium	Essential	70
	Strontium	Nonessential	65
	Copper	Essential	65
	Potassium	Essential	60
	Barium	Nonessential	60
	Selenium	Essential	50
Weak absorption	Molybdenum	Essential	20
	Magnesium	Essential	17
	Nickel	Nonessential	15
	Cobalt	Nonessential	10
	Uranium	Nonessential	10
	Iron	Essential	6
Very weak absorption	Sodium	Essentials	3.5
	Rubidium	Nonessential	3.5
	Rare earths	Nonessential	1.5
	Chromium	Nonessential	1.5
	Lithium	Nonessential	0.75
	Silicon	Nonessential	0.30
	Vanadium	Nonessential	0.30
	Titanium	Nonessential	0.15
	Aluminum	Nonessential	0.15

Source: Brooks [99].

TABLE 15.3. Biological Absorption Coefficients for Copper and Nickel in New Zealand Trees

Statistical Data	*N. fusca*	*Q. acutifolia*	*W. racemosa*
	Copper		
Arithmetic mean (A.M.)	2.32	0.87	0.87
Standard deviation of A.M.	2.97	1.39	1.34
Relative standard deviation (%)	128	160	154
Correlation coefficient (r)	0.62	0.33	0.20
Geometric mean (G.M.)	1.37	0.86	0.66
Standard deviation of G.M.	3.07	2.69	1.77
Range	0–4.2	0–2.3	0–1.2
	Nickel		
Arithmetic mean (A.M.)	0.77	0.34	0.52
Standard deviation of A.M.	0.55	0.31	0.57
Relative standard deviation (%)	71	91	109
Correlation coefficient (r)	0.82	0.57	0.62
Geometric mean (G.M.)	0.62	0.48	0.81
Standard deviation of G.M.	1.91	1.48	2.82
Range	0–1.2	0–0.7	0–2.3

KEY: N—*Nothofagus*
Q—*Quintinia*
W—*Weinmannia*

Source: Brooks [99].

Note: Correlation coefficient is for the plant vs. soil data.
BAC values are for an ash weight basis and should be multiplied by 0.05 to convert to a dry weight basis.

criterion of essentiality, however, is a general one, and in cases such as selenium there will be many species of plants for which this element is not essential.

Timperley et al. [806] have studied the relationship between BAC values and the nature of the element absorbed. They concluded (see Chapter 14) that, in general, plants do not have a constant BAC for essential elements such as copper and zinc because values of the coefficient decrease logarithmically for increasing amounts of the element in the soil (see Figure 14.1). By contrast, nonessential elements tend to produce relatively constant BAC values in given species (see also Figure 14.4). The implications of this discovery are important for biogeochemical prospecting since it is obvious that good results cannot be expected when dealing with elements essential for plant nutrition. In practice, however, satisfactory results can be achieved provided that elemental concentrations in the substrate are sufficiently high to bring BAC values into the portion of the asymptotic curve parallel to the x axis.

Brooks [97] and Malyuga [563] have recommended that in biogeochemical surveys, two or three different species be used so that at least one of them is present at each point on a preselected sampling grid. To standardize the data, the concentration of each element in each species is divided by the appropriate BAC value. This technique of normalization is usually successful, provided that the BAC is accurately known and is not greatly affected by the variables listed in the preceding chapter.

Since the effectiveness of a biogeochemical survey will ultimately depend on the constancy of the BAC, some simple statistical test is needed to evaluate this. Table 15.3 shows means, standard deviations, and relative standard deviations for copper and nickel BAC values for three species of New Zealand plants sampled during the course of work carried out in Nelson Province. This table also shows that relative standard deviations are lower for nickel than for copper (i.e., BAC values for copper are more variable) and that this explains the fact that biogeochemical prospecting for nickel is usually more successful than it is for copper. To some extent, correlation coefficients (see Chapter 19) are inversely related to the relative standard deviations of the BAC values.

15.3. THE RELATIVE ABSORPTION COEFFICIENT (RAC)

Unless a given plant is well distributed in a test area, it is usually necessary to use more than one species during biogeochemical prospecting in order to have adequate coverage of the area. An approach to this problem is to use what I have termed the **Relative Absorption Coefficient** (RAC). Using Kovalevsky's terminology [458] for his own Russian acronym,

$$RAC = C_p/C_r$$

where C_p is the concentration of an element in one species and C_r is the concentration of the same element in another species used as a reference standard. The reference species is taken from a position as close as possible to that of the plant that is to be normalized. According to Kovalevsky [458] this position should be within 3–5 m. Since a ratio is used here, it does not matter greatly whether the original data were expressed on an ash weight or a dry weight basis.

When using RAC values, the same plant organ must be used throughout, or errors will arise due to the different element contents of different organs (see Chapter 14).

When the RAC has been computed for a number of species, the concentration of the element in one of the plants missing from a particular site can be calculated from a knowledge of the concentration present in the reference plant and by multiplying this value by the RAC for the missing species.

15.4. THE ACROPETAL COEFFICIENT (AC)

I have used the acronym AC to describe Sabinin's [715] **Acropetal Coefficient** (OSOR). This coefficient is defined as

$$AC = C_o/C_x$$

where C_0 is the concentration of an element in a particular plant organ and C_x is the concentration of the same element in a reference or standard organ of the same species.

In some cases, the Acropetal Coefficient will be constant and in others it will vary. In such a case there is said to be an **acropetal gradient**. This is illustrated in Figure 15.1 for three species of plant from Siberia [455]. In this example, the Acropetal Coefficients (leaves/twigs) decrease as the tungsten content of the twigs increases, except in the case of the willow. The plot is that of negative acropetal gradients and has been reported in the Russian literature [455,458–459].

Although I have investigated acropetal gradients in my own field studies, the AC values have tended to remain relatively constant. This may be because nearly all the species studied were evergreens, so that perhaps acropetal gradients are confined to deciduous plants.

Many workers have recommended that two- or three-year-old twigs should be used in biogeochemical prospecting [458,902]. This choice is well justified on theoretical grounds, as trace element levels in deciduous leaves usually have considerable seasonal variation [335] (see also Chapter 14).

It is clear that absorption of certain trace elements by plant organs depends on the structure and physiological function of their cells. The best plant-soil correlations appear to involve dormant cells (branches, twigs, and trunks). This principle has been carried further by the work of Cannon and Starrett [168] who recommended that dead material of trees and shrubs should be used in biogeochemical prospecting for uranium in North America.

Kovalevsky [458] in discussing Acropetal Coefficients has emphasized that leaves should never be used in biogeochemical prospecting because a limit of absorption is readily achieved for the element concerned. Even the use of twigs or branches is qualified by the recommendation that these be at least two or three years old.

It is worth mentioning that there is a further advantage in using twigs instead of leaves. The ash content of twigs is much lower (about 1%) than that of leaves (5%). If the dry weight concentrations of an element are about the same in both organs, the ashed twigs will give a level five times higher than that of the leaves. This may well be important for low concentrations near the limit of the analytical method used.

The Acropetal Coefficient serves as a useful warning to the field worker

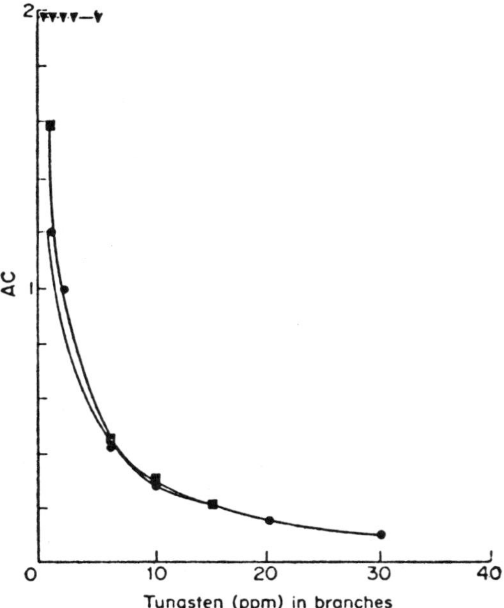

FIGURE 15.1. Variation of Acropetal Coefficients (AC) for tungsten in three species of plants from Siberia. Solid triangles = willow, solid squares = *Lonicera caerulea,* solid circles = *Ledum palustre*. The AC values (leaves/twigs) are plotted as a function of the tungsten content of the twigs. *After:* Kovalevsky [455].

that one of the sample types of potential use in the survey may have questionable value and should not be used.

15.5. THE TEMPORAL ABSORPTION COEFFICIENT

I have used the acronym TAC to describe the **Temporal Absorption Coefficient** initially referred to by the Russian acronym OIS [458]. This parameter governs the effect of time as a variable. It may be defined as

$$TAC = C_t/C_w$$

where C_t is the concentration of an element in a certain plant organ at a particular point in time, and C_w is the concentration of the same element in the same organ at a time corresponding to the winter rest period or at maturity. Variations in the TAC can be short-term or long-term. In the short term, we may consider the gradual accumulation or decrease of elemental concentrations in leaves of deciduous trees or shrubs. The TAC values increase or decrease to a limiting value of 1.0 over a 6-month period. For example, Guha [335] found that TAC values increased from 0.2 to 1.0 for

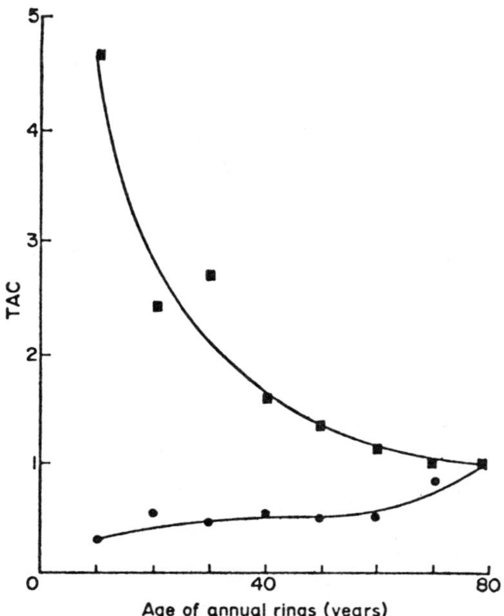

FIGURE 15.2. Variation in Temporal Absorption Coefficients (TAC) for copper (solid squares) and lead (solid circles) in annual rings of *Beilschmiedia tawa*. *After:* Nicolas and Brooks [618]. Copyright 1969, Australasian Institute of Mining and Metallurgy.

manganese in sycamore leaves and decreased from 12 to 1.0 for copper during the same period (May–October). It is clear therefore that sycamore leaves would be an unwise choice for biogeochemical prospecting for these two elements unless all the work was carried out over a short period of time.

Long-term variation in TAC values are found in slowly metabolizing tissues such as twigs or tree trunks. This is demonstrated in Figure 15.2, which is a plot of copper and lead in the ash of annual rings of a trunk specimen of *Beilschmiedia tawa* from New Zealand [618]. The implications of this plot are that tree trunks would be a most unsatisfactory biological sample type for prospecting unless care is taken to select only parts of similar age, such as the outside of the trunk section, as has been done successfully by Quin et al. [671] for tungsten exploration. The above discussion shows that the TAC serves as a useful guide as to which organs should be selected in biogeochemical exploration surveys.

15.6. THE MOBILE ELEMENT ABSORPTION COEFFICIENT (MAC)

The **Mobile Element Absorption Coefficient** (MAC) is the last of the five biogeochemical parameters to be considered. It is defined as follows:

$$MAC = C_m/C_p$$

where C_m is the concentration of the mobile (plant-available) form of an element in the soil solutions and C_p is the concentration of the same element in the plant. The value of this parameter for biogeochemical prospecting is much less obvious than the others. It is, in effect, the inverse of a BAC involving mobile forms of the element only. Insofar as it measures the latter quantity, it would seem to be more useful in agriculture than in mineral exploration.

15.7. THE OVERALL SIGNIFICANCE OF BIOGEOCHEMICAL PARAMETERS

Having discussed each biogeochemical parameter in turn, it is now possible to make a few general observations about their significance in mineral prospecting. Clearly it is not the absolute value of a parameter that is necessarily important, but rather its variance and rate of change as well as the factors that produce this change.

Variations in the BAC are perhaps the most important factors affecting the success of the biogeochemical exploration survey, even though many of these factors may be minor and easily corrected. Even when multiple regression of biogeochemical data is used to make corrections for BAC values (see Chapters 14 and 19), it is most unlikely that it will be possible to reduce the variance by more than about 50%.

The RAC is most useful for making allowances for missing species at predetermined sampling points.

The AC and TAC have great significance for biogeochemical prospecting by indicating the correct choice of biological sample types.

In contrast to the others, the MAC is not very important. This parameter usually remains relatively constant except where leaves are involved and where the climate is sufficiently wet to allow for leaching of elements from the leaves.

It is hoped that this review of the biogeochemical parameters of the Russian literature will have succeeded in highlighting this little-known aspect of the field of biogeochemical prospecting, and that it will result in a deeper understanding of the subject, thus resulting in a more quantitative approach.

16

A FIELD GUIDE
TO BIOGEOCHEMICAL
PROSPECTING

16.1. INTRODUCTION

The theoretical basis of biogeochemical prospecting having been established, consideration should now be given to practical applications of the technique. It is seldom possible to do a "blind" survey in a new area. That is to say, to venture boldly into exploration without some sort of preliminary orientation investigation. This would seem at first sight to be a disadvantage of the biogeochemical method, but on reflection it should be remembered that a pedological survey also requires preliminary orientation studies in order to establish which soil horizon or size fraction within that horizon should be sampled. The following discussion gives practical advice to the exploration geochemist on recommended procedures for biogeochemical prospecting.

16.2. SITE SELECTION

In selecting a site for an orientation survey, the field worker should be guided by a number of important considerations. The first of these is that the area should contain an anomaly or anomalies. It is not always necessary that the anomalies should be economic ore bodies, but they should contain concentrations of the element or elements sought, at levels at least two or three times background. Concentrations of the elements in the anomalies should approximate in magnitude to those to be expected in the "blind" area. Since by definition the latter will not usually be known in advance, it will only be apparent later whether or not the test anomalies were well chosen. If the blind area has elemental concentrations greater than the test area, the assessment of the potential of the biogeochemical method will have been conservative. If, however, the reverse is the case, the assessment will have been over-optimistic.

A second consideration for the selection of a test site is that it should be climatically, ecologically, geologically, and geomorphologically representative of the area that is to be surveyed ultimately.

The area of the test site should be such that it would allow for about 100 sampling points at spacing that would be employed for the main survey itself. Assuming a grid pattern of sites 30 m apart, this would allow for an area just under 1 sq. km.

Absence of known anomalies in a test area presents another problem. To some extent this can be overcome be selecting as a test site a contact between two geological formations with greatly different background concentrations of the element that is sought. There is, however, in this case the danger that different edaphic factors in the soils derived from these two formations may affect the BAC (Biological Absorption Coefficient; See Chapters 14 and 15) values of the species studied. If no contacts exist, it is still possible to use the vast amount of data now available for background levels of many elements in common plant species [194,883]. Indeed, in New Zealand, biogeochemical prospecting for copper and nickel has been carried out with species whose background levels had been established on similar geological formations 500 km apart.

A grid pattern of sampling may not always be appropriate, and in such cases the field worker may use a line transect or belt transect (see Chapter 3) which cuts across mineralization. One or several transects may be made. This type of approach is much favored by Russian workers and is certainly useful where only a small number of sample points is envisaged.

16.3. SELECTION OF TEST PLANTS AND SAMPLING PATTERNS

The first criterion for the selection of plants in a biogeochemical survey is that the species selected should be sufficiently widespread to enable good coverage to be made of the test area. Herbs and small shrubs should be avoided, if possible, because their root systems are likely to be so shallow that they probably will not penetrate to the dispersion halo of the anomaly. An assessment of the abundance of different species can be made by use of infrared photography (see Chapter 6), though an on-foot survey usually can give a good idea of relative abundances.

If possible, plants of similar size should be sampled. Very large trees should be avoided, if only because twigs or leaves are likely to be out of reach.

Although it has been suggested that a square grid sampling pattern is desirable, another possibility is to sample a given species at roughly equal distances and then mark the precise localities on the map. In this way, a single species can be used throughout a biogeochemical survey, though the sampling pattern may not correspond exactly with a square grid. It might be envisaged that the exploration biogeochemist would ultimately select two or three species for further testing as a result of application of the above criteria.

At this stage the field worker will have the option of carrying out a full-scale orientation survey with all preselected species or carrying out a mini-survey at, say, ten sites with varying elemental concentrations in the soil to determine the apparent constancy of the BAC values for each species. This will permit the less favorable species to be discarded.

If line transects are used for the sampling pattern, they should be oriented at right angles to the line of the mineralization and should be so arranged that at least two points fall within the anomaly. If a grid pattern is employed, a good distribution of sample points would be one third on background, one third on moderately mineralized ground, and one third on mineralization. If the terrain is extremely steep and dissected, the field worker may have to use a sampling pattern following spurs and ridges, so that all samples will have similar aspect and drainage. The criteria for determining the pattern of sample sites for vegetation are fairly similar to those used for soils. The reader is referred to Hawkes and Webb [347] for a discussion of this latter topic.

16.4. SAMPLING AND STORAGE OF VEGETATION

During the orientation survey the field worker might collect leaves, needles, twigs, or other biological sample types (see Chapter 17). Twigs should be at least one year old so that the material is young enough to have gone through the main processes of mineral accumulation and yet is not so old that senescence with attendant loss of minerals has started to occur to any extent. Other plant organs such as roots, flowers, fruits, bark, and trunk material may be collected also. The question of sampling in biogeochemistry has been discussed extensively by various workers [97,477,884,902,905].

The amount of vegetation collected should be sufficient to provide about 30–50 g of dry material. The samples should be removed from the trees or other plants with pruning shears or hedge trimmers, placed in prenumbered plastic bags, and sealed with rubber bands. The samples should be collected at various points around the circumference of trees or shrubs and preferably at a point as high as can be reached comfortably from the ground (i.e., about 2 m). Soil samples should be taken from various points around the base of each plant sampled and should be taken from the same horizon in each case. A total of about 200 g of soil should be more than adequate. The soil samples may be placed directly into kraft bags, finely-woven cotton bags, or polythene containers. As far as possible, collecting of plants and soils should be carried out over a limited time period.

At each sample site, observations on the topography and geology of the site, type of vegetation, existence of outcrops, and so on, should be entered in the field book. It is usually better to take prenumbered sample bags at random from a container and enter this number in the field book than to attempt to number bags in the field. Under adverse weather conditions, freshly marked numbers can soon become illegible. This also applies to the field book itself, and in wet areas it is advisable to construct a book from the tracing plastic used by cartographers.

The nature of the subsequent treatment of the samples will depend on whether they are to be analyzed in the laboratory or in the field. Generally, any field operation is much more difficult and less reliable than one carried out in the laboratory. Unless weight is a problem, it is better to take the samples back to the laboratory for further processing. Vegetation samples are initially collected in sealed plastic bags to keep them fresh before washing, which is the next stage of the process. Leaves stored in plastic bags will keep for about a week in a fairly fresh condition provided that temperatures do not rise much above 15°C. Twigs, of course, will keep almost indefinitely. Unless refrigeration facilities are available at the base camp, leaf samples will have to be processed further in the field if their onward despatch is to be delayed by more than a week.

On return to the laboratory, soil samples may be laid out on numbered pieces of paper and air dried for about two days. Alternatively the paper or cotton bags of samples may be dried in a drying oven for a few hours at 110°C. After drying, the soils are lightly disintegrated in a large mortar and pestle, and are then sieved through a 15-mesh nylon sieve before storage for further treatment. In arid areas, soils may be sieved immediately *in situ*.

When vegetation is removed from the containers, it should be washed vigorously under running water with a final rinse in distilled water. The samples should then be dried at 110°C in a drying oven. If the vegetation consists of leaves, crush the material in the hands after drying. This will remove the leaves from the supporting stems and small twigs. The material may then be placed in bottles or plastic bags until required again. Twigs should be cut into small sections with pruning shears before the drying process. Once the material is dry, it is much harder to cut.

The foregoing assumes that materials are to be taken to the laboratory for analysis. If field analyses are to be carried out, or if samples are to be reduced in weight for onward transmission, a different procedure should be followed. The soil bags should be suspended in a warm dry place at the base camp until dry. The samples are then gently disintegrated by hand or in a mortar and pestle and are sieved as before. The bulk of the samples is reduced by coning and quartering; a small representative sample is then available for onward transmission. In the field, vegetation samples should be washed in a stream or water tank and then transferred to kraft bags for drying in the sun. This procedure is not always possible since, in many areas, sun or water, or both, are lacking. If water is not available, the washing process has to be eliminated. This may cause contamination problems which can be overcome in the case of twigs or roots by peeling the specimens, a step that is laborious in the extreme. As a further step the samples may be calcined in the field by heating in aluminum pots held over a primus stove. There will be no need to ash the samples completely because this process can be completed in the laboratory. The ashed or semi-ashed material is then pulverized with a glass rod and stored in small plastic vials until required.

16.5. PREPARATION OF SAMPLES FOR CHEMICAL ANALYSIS

16.5.1. Vegetation

Many methods of chemical analysis (see Chapter 18) require that the sample be in solution. In the case of vegetation samples, this can involve either wet ashing or dry ashing. Wet ashing involves decomposition of the sample with a mixture of nitric and perchloric acids, whereas dry ashing involves heating the material in a muffle furnace for several hours at 500°C.

Due to the limitations imposed by furnace space, the main problem in calcining plant material is to strike the correct balance between the amount of material needed to give a representative sample and the amount that can be economically handled to give a fast output from the service laboratory. One way to improve the economics of the system is to pulverize the dried vegetation in a hammer mill, thus producing a fine powder from which a much smaller representative sample can be obtained. Even a small automatic coffee grinder can be quite effective for this purpose. Assuming that a hammer mill is used, about 10 g of dried material will be needed. This is placed in a 25 ml squat borosilicate beaker and ashed in a muffle furnace at 500°C for several hours. This will produce about 0.2 g of ash from woody material or about 0.5 g from leaves. A useful alternative procedure is to pre-ash leaves or twigs in 150-ml beakers over a hotplate. Charring will ensue and, if the material is kept long enough over the heat source, virtually complete calcination will result, so that all that is needed is to transfer the semi-ashed material into a much smaller container (10 ml) and complete the process in a muffle furnace. One problem that I have encountered is that if the heating by the hotplate is not uniform, some of the beakers will crack on cooling. As an alternative to this, use can be made of stainless-steel beakers or aluminum beakers for pre-ashing before determination of elements (such as uranium) that are not found in either metal container. The average muffle furnace will accommodate about 50 beakers of 25-ml size and about 100 of 10-ml size. Assuming that the muffle is left running continuously, about 200 samples can be processed every 24 hours. Ashing should be carried out in the presence of a small amount of air. If too much air is admitted, the material will catch fire; if too little is permitted access, there is a danger of volatilization of some constituents. Suddenly admitting air after a period of closure of the oven door can result in an explosion if organic matter is still present.

There has always been much controversy concerning the loss of elements during the ashing procedure. One of the best references on this subject is a book by Gorsuch [324]. Most workers [324,596] agree, however, that little loss of most elements will occur at 500°C provided there is adequate access of air. It is my own experience that slight loss of lead and significant loss of cadmium will occur. Surprisingly enough, arsenic can be heated to 800°C without significant loss, providing that excess chloride ion is not present [120]. Small amounts of chemical oxidants such as nitric acid or magnesium nitrate are sometimes favored as aids to trouble-free ashing. Very volatile elements such as mercury obviously cannot be determined in ignited samples.

Wet ashing is discussed in Chapter 18 and will not be mentioned further in this Chapter because the process is chemical rather than physical.

Further gentle grinding of the 15-mesh soil samples to pass a 60-mesh sieve is usually all that is required before chemical analysis of the material. If organic matter is to be removed, as in "dry" methods of chemical analysis (such as emission spectrometry; see Chapter 18), the soil may be ashed in small borosilicate beakers in exactly the same manner as vegetation.

16.6. THE WEIGHT BASIS OF EXPRESSING ANALYTICAL DATA

Perhaps one of the most irritating things to confront the student of biogeochemical literature is the extreme carelessness with which analytical data are sometimes expressed on both an ash weight or dry weight basis in the same publication, or what is even more annoying, are expressed without any indication of their weight basis. The ash content of twigs is about 1–2 percent for most species and is about 5–10% for leaves. A factor of anywhere from 10 to 100 represents the differences between reporting data on an ash weight and dry weight basis.

The above situation is much more serious for soils, because if the sample is very humic, reporting the concentration on an ash weight basis can give values a factor of ten or more higher than if the dry weight had been used. This is illustrated in Figure 16.1 where a false nickel anomaly appears in soils at station 4 of a traverse of a basic complex in New Zealand.

The Russian literature appears to favor expressing data on an ash weight basis for vegetation samples. I am of the opinion, however, that to be

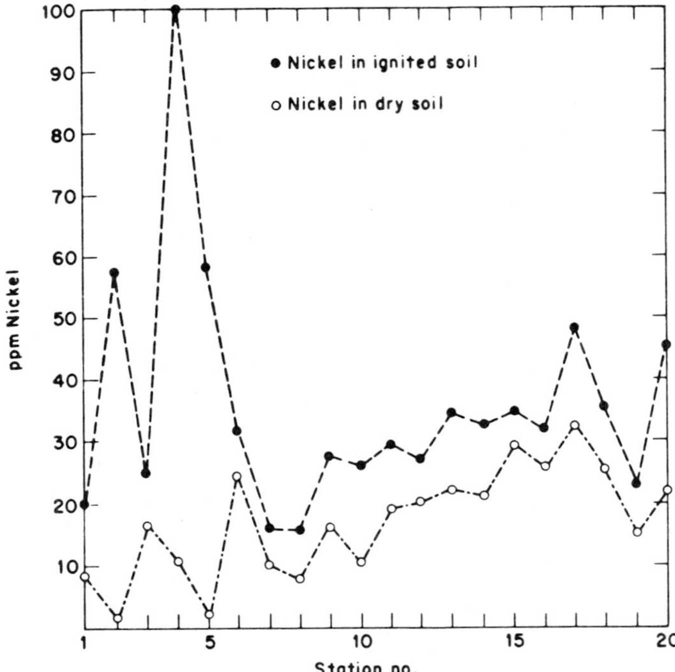

FIGURE 16.1. Nickel values in humus-rich soils sampled during a traverse of the Longwoods Range, New Zealand. The figure illustrates the importance of expressing soil analyses on a dry weight rather than an ash weight basis. Due to the high organic content at station 4, a false anomaly appears in ignited soils. *Source:* Brooks [97].

consistent with the usual method of reporting data for soils, a dry weight basis should be used for vegetation, not only for consistency but because this reflects the true-life situation. Therefore, throughout this work I have used this common standard and, wherever possible, have converted existing data previously reported for plant ash.

16.7. SAMPLE CONTAMINATION

The risk of contamination of vegetation from natural sources during a biogeochemical survey is a universal problem. A common source of such contamination is windblown dust from mine workings or exposed soils, and fumes from smelters. In New Zealand, with its dense vegetation cover, relative lack of heavy industry, and the presence of a heavy rainfall, contamination effects of this nature are rare and have never caused me many problems in field work. In arid regions, such as Australia and the western United States, this is an ever-present problem.

Cannon [151,157] has shown some of the contamination problems that can arise in vegetation and has stated that ore trucks can add 1–2 μg/g

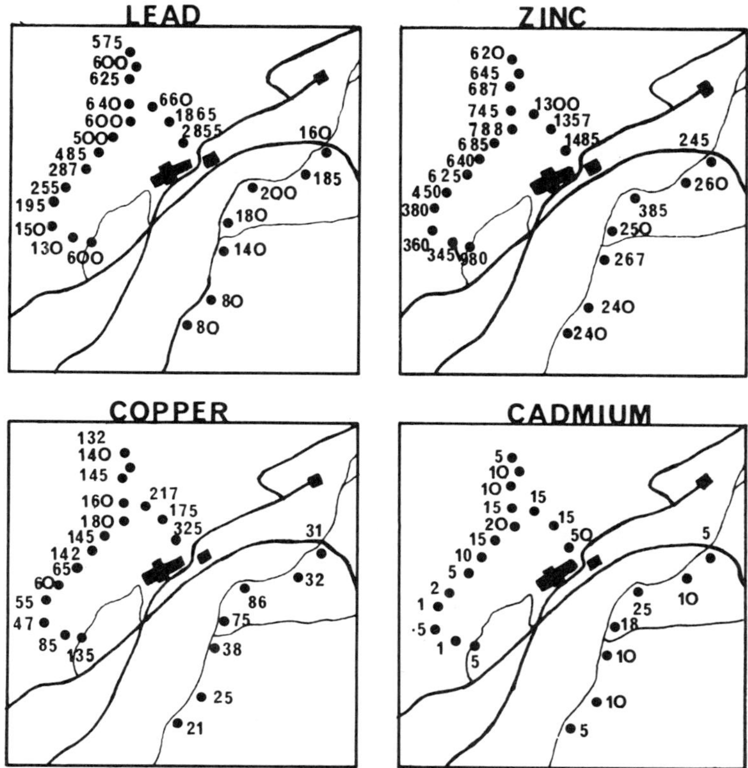

FIGURE 16.2.   Lead, zinc, copper and cadmium levels in the ash of leaves of *Beilschmiedia tawa* near an ore treatment plant at Te Aroha, New Zealand. *Source:* Ward et al. [880]. Reproduced by permission of the New Zealand Department of Scientific and Industrial Research.

uranium to vegetation and that a mill can increase the content of this element in trees by a factor of 1000. Similar findings were made by Ward et al. [880] who studied the effect of windblown ore upon leaves of *Beilschmiedia tawa* in the vicinity of a treatment plant at Te Aroha, New Zealand. The data are shown in Figure 16.2.

One of the most serious contamination effects is that which originates from the lead tetraethyl (TEL) of automotive exhaust fumes. Unlike other forms of contamination, the lead is absorbed so strongly by the environment that 80% of it is fixed virtually permanently in the upper soil horizons. Pollution of the biosphere by TEL has become a serious worldwide problem, particularly in the United States and other Western countries.

Warren and Delavault [895] were the first to report abnormal levels of lead in vegetation growing near major highways. Since then, a very considerable literature on the subject has appeared. Some of this work has been summarized by Brooks [100]. It would be tempting to assume that motor vehicle emissions are likely only to affect biogeochemical surveys for lead. In fact, as has been shown by Ward et al. [882], significant amounts of

TABLE 16.1.   Mean Concentrations of Heavy Metals in Vegetation Adjacent to the Auckland Motorway, New Zealand, at Sites Grouped According to Traffic Density

Mean Traffic Density (vehicles/24 hr)	Concentrations of Heavy Metals (μg/g dry weight)			
	Cr	Cu	Ni	Pb
>50,000	4.0	30	3.7	350
40,000–50,000	3.1	23	4.3	320
20,000–39,999	2.6	17	3.0	270
10,000–19,999	2.4	13	2.6	140
Background	0.6	10	0.9	5

Source: Ward et al. [882].

chromium, copper, cadmium, nickel, and zinc, as well as lead, are to be found in vegetation adjoining busy highways. These findings are summarized in Table 16.1.

The above discussion has clear lessons for biogeochemical prospecting. Surveys involving lead should not be carried out within 200 m of a major highway. Any lead anomalies found near highways should be treated with great circumspection. Since lead anomalies usually are associated with zinc, the zinc content of the vegetation should also be determined to confirm the reliability of the lead data.

There are other forms of contamination of vegetation apart from motor vehicle emissions and mining operations. Top-dressing with fertilizers or fungicides containing heavy metals can also be a source of error and should always be taken into consideration.

16.8.   PATHFINDER ELEMENTS

The concept of **pathfinders** is well established in mineral exploration, particularly in geochemical prospecting. Pathfinders may be defined as elements that, because of some property or properties, provide anomalies or halos more readily usable than the element that is sought and that are geochemically associated with that element.

Warren et al. [897] have shown that pathfinders may also be used in biogeochemical prospecting. The most useful species tested was the Douglas fir (*Pseudotsuga menziesii*). The normal arsenic content of growing tips of this species is <1 ug/g. In areas of sulfide mineralization where arsenopyrite is an important constituent, amounts up to 500 μg/g (dry weight) have been recorded in the twigs. Arsenic in soils and rocks already has a reputation of being a good pathfinder for gold and sulphide minerals, so that the possibility of using it for this purpose in biogeochemical prospecting is of great interest.

Warren et al. [906] have investigated the use of iron and manganese in vegetation as indicators for other elements. Tkalich [814] has used the iron content of plants for the same purpose. Talipov et al. [789] have explored the possibility of using arsenic in vegetation as a pathfinder for gold, and have obtained positive results.

The use of pathfinders in vegetation is a possibility to be explored during an orientation survey. Such data usually are obtained from the statistical analysis of interelemental relationships in vegetation alone, a procedure that should always be carried out at the end of the survey.

16.9. EVALUATION OF THE ORIENTATION SURVEY

After chemical analysis of the samples collected during the orientation survey, it is wise for the biogeochemist to carry out some sort of statistical procedure to establish the degree of confidence with which elemental concentrations in the substrate can be predicting from those in the vegetation. Such procedures are fully discussed in Chapter 19 and will be mentioned very briefly here. A useful technique is **correlation analysis,** used to deduce the degree of correlation between concentrations of specified elements in plants and soils. To evaluate this relationship, the correlation coefficient (r) is obtained. This function varies in magnitude between $+1$ and -1. A value of $r = 1.0$ for two sets of data implies a perfect direct relationship, whereas a value of -1.0 implies a perfect inverse relationship. As r approaches zero from either side, the trend toward nonsignificance increases. The value of r by itself is meaningless; it must be related to the number of sample pairs in order to establish the probability (P) that a relationship does or does not exist. This probability is itself based on a null hypothesis; that is, it measures the probability that a relationship does not exist. We therefore have the paradoxical situation that a low value of P implies a high degree of correlation between the two variables.

A second important statistical procedure that must be carried out is to evaluate the concentration level in plants, which represents the threshold between background and the presence of mineralization in the substrate. Various statistical procedures to achieve this are summarized in Chapter 19. If vegetation only provides a weak response to the presence of mineralization, a more subtle form of statistical analysis should be employed, such as discriminant analysis (see Chapter 19).

The statistical treatment should provide information not only on the best species to select for the biogeochemical survey, but also on which organ to select and whether the data should be expressed on a dry weight or ash weight basis.

Emphasis in this chapter has been on plant-soil relationships. A more meaningful evaluation of the potential of vegetation for prospecting would be achieved by comparison of plant data with bedrock data. This procedure is seldom carried out because of the difficulty of obtaining a sufficient number of bedrock samples to make a comparison statistically significant. If sufficient samples can be obtained, however, a useful exercise is to compare plant-soil data with plant-bedrock data.

Another approach to the orientation survey is a comparison of plant values for a particular element with data obtained by geophysical techniques such as induced polarization or geomagnetic measurements. Very little work of this nature has been carried out, although Marmo [573] compared hydrogeochemical and geomagnetic data with plant values for copper and zinc in Finland.

Although it is possible (and probably safer) to evaluate the significance of biogeochemical data completely by statistical methods, visual presentation of the data is more satisfying aesthetically.

If a grid basis has been used for the survey, separate plots may be used for each species employed, or alternatively, the data can be standardized if more than one species is used, by dividing the values for each species by the appropriate BAC (see Chapter 15). An illustration of this procedure is given in Figure 16.3.

There are many pitfalls in plotting contour maps of elemental isoconcentrations in plants and soils. Not the least of these is the fact that anomalies of different shape can be produced merely by selecting a particular threshold level for elemental concentrations. To avoid this difficulty, it is essential to use as a guide the threshold value indicated by an objective statistical procedure.

Contour maps can also be plotted by using elemental ratios in plants and soils such as the following: copper-zinc, nickel-cobalt, thorium-uranium,

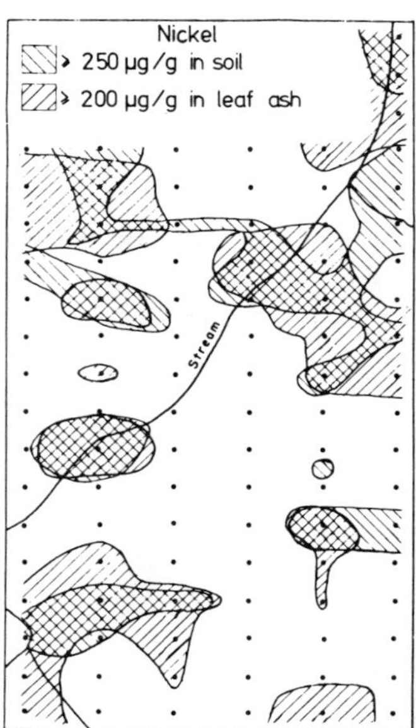

FIGURE 16.3. Degree of coincidence of nickel anomalies in soils and leaves of *Nothofagus fusca* sampled at the Riwaka Basic Complex, New Zealand. *Source:* Timperley et al. [805].

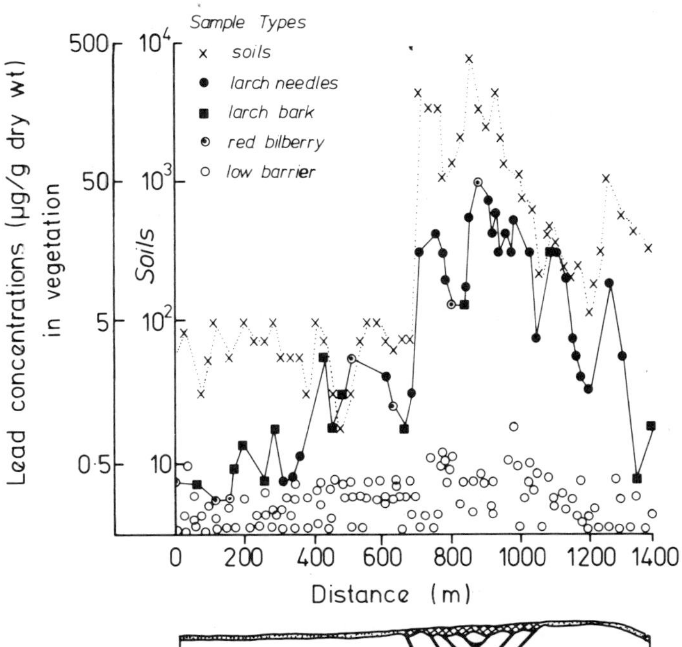

FIGURE 16.4. Lead in soils and various biological sample types at the Ozernoye base metal deposit, Siberia. The lead ores are shown as black markings beneath the overburden (cross-hatched) of the topographical sketch. *Source:* Kovalevsky [467].

uranium-radium. Use of such ratios carries the advantage that the use of BAC values is avoided and so are many of the variables that could affect the constancy of the BAC. Elemental ratios found in ore bodies usually are preserved in either plants or soils considered separately.

If a known anomaly in bedrock is not confirmed by both or either of the plant and soil data, a reason for this should always be sought. Anomalies common to plants only, particularly in steep terrain, may indicate absorption of soluble salts washed down from a true anomaly situated upslope, or may indicate bedrock mineralization covered with a transported nonmineralized soil. Exclusive anomalies in plants can sometimes indicate the presence of waterlogged soil (see Chapter 14). Anomalies common to soil only may indicate either a true anomaly in bedrock or that the soils themselves are part of a transported anomaly.

If line transects had been used for the biogeochemical survey, a somewhat different form of representation may be required [467]. Figure 16.4 shows data for lead in soils and various biological sample types over a base metal deposit in Siberia. This form of representation is much favored by Russian workers. It contains the maximum amount of information, though at the same time suffers some loss of clarity.

16.10.   BIOGEOCHEMICAL SURVEYS IN VIRGIN TERRITORY

With the completion of the orientation survey, the field worker will now wish to apply the method to virgin territory. In some areas only soil surveys would be feasible (i.e., in deserts) but in others the most effective procedure might be biogeochemical work. In the majority of cases, however, a joint plant and soil project might be advisable.

To decide the relative merits of soil or plant surveys, it is advisable to expose the soil profile by digging a pit, and to sample various horizons for analysis of their constituents. If the element that is sought is concentrated in one of the upper horizons, it is possible that there might be no advantage in using plants as a guide to mineralization, since soil sampling would be very easy. If the element is enriched at depth, however, and particularly if there is an impenetrable geochemical barrier such as siliceous hardpan or a carbonate layer between the surface and the zone of enrichment, plants probably will be more useful for prospecting. Most trees and shrubs should be capable of penetrating to the desired level in the soil profile.

Other situations in which the biogeochemical method could be used along with soil surveys (or even in place of them) are where the terrain is steep and eroded so that much of the soil is transported from elsewhere. Newly colonizing plants should presumably be able to sample bedrock and give a more accurate reflection of mineralization than the soil.

It is not fortuitous that much of the early biogeochemical work originated in Fennoscandia and North America. In areas such as these where there has been extensive glacial activity, the original soil cover has long since disappeared and has been replaced with the ubiquitous glacial till which is usually unrepresentative of bedrock. For this reason, soil geochemistry has been particularly unsuccessful in this part of the world. As an illustration of this, Figure 16.5 shows a profile of scintillometric readings, and of uranium concentration in soils and ashed twigs of *Picea rubens* (red spruce) across an area of uranium mineralization in Nova Scotia [107]. The plant data and scintillometric readings indicate quite closely the zone of mineralization but the soil is quite ineffective for this purpose.

In terrain with stable soils and gentle relief, and where elements are enriched in the upper, more accessible parts of the soil horizon, there may be little or no advantage in the use of the biogeochemical method. Nevertheless, for the sake of completeness and because no one method of exploration by itself may give the full picture, it is always a good idea to include both plant and soil surveys whenever economic conditions render this possible.

The working procedure in a biogeochemical exploration survey of virgin territory is little different from that of the orientation study, except that fewer plant species will be sampled.

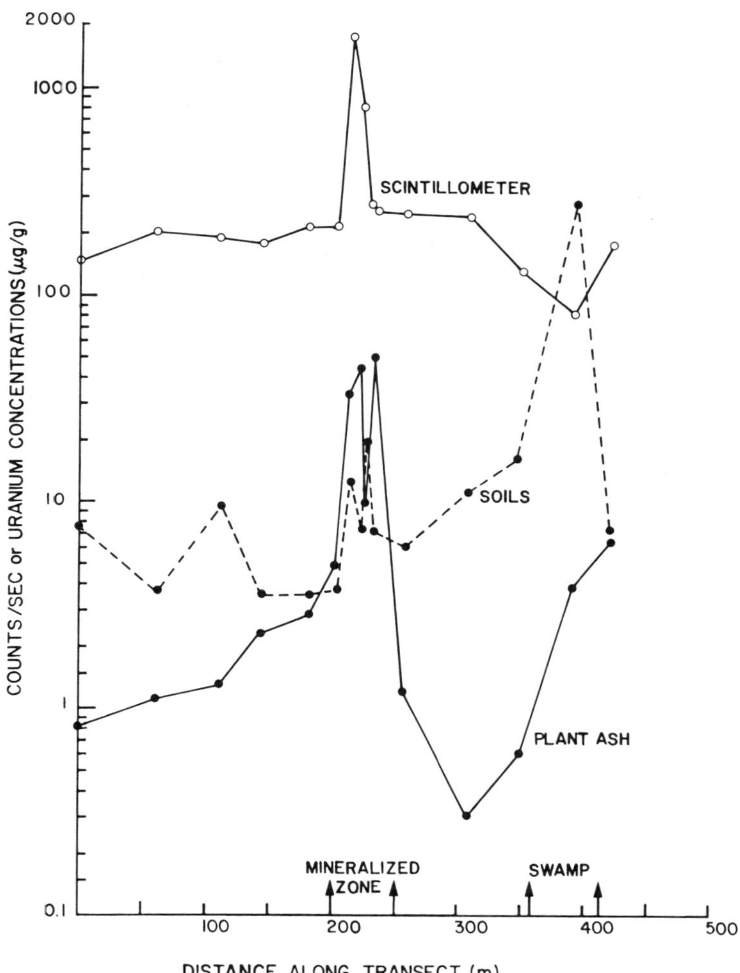

FIGURE 16.5. Scintillometric data (counts/sec) and uranium concentrations in plant ash and soils (μg/g) for a profile across mineralization in Nova Scotia. *Source:* Brooks et al. [107].

16.11. BACKGROUND DATA FOR BIOGEOCHEMICAL PROSPECTING

As a usable pool of knowledge of normal background levels for elements on certain plant species is accumulated, the need for extensive orientation surveys becomes progressively reduced. When the pool of knowledge includes species of continental distribution, it can be of great use for biogeochemical prospecting. A very large proportion of the existing knowledge of background data is derived from North America because of the work of Warren [883]. Available data from the Soviet Union are contained in works by Chukhrov et al. [194], Kovalevsky [477], and Malyuga [563].

17

ALTERNATIVE SAMPLE TYPES IN BIOGEOCHEMICAL PROSPECTING

17.1. INTRODUCTION

Although biogeochemists traditionally have concentrated their work on twigs and leaves of trees and shrubs, there are a number of alternative biological sample types that have been used in the past and which deserve more attention than is currently being paid to them. They include such diverse subjects as mosses, lichens, tree sap, roots, herbarium material, and many others. Some of these will now be considered in this chapter.

17.2. LICHENS

Uptake of trace elements by lichens has been described by Richardson and Nieboer [690]. Lichens are formed by an association of a fungus and an alga. The fungus with few exceptions belongs to the cup-fungi group (Ascomycotina) while the algae may be blue-green or green. The algae in most lichens form a layer enclosed on both sides by fungal tissue which has no protective cuticle so that the water content of the whole lichen thallus rapidly equilibrates with the surrounding environment. Such plants are termed **poikilohydric.** They dry quickly under bright sunny conditions and assume a state of inactivity, but revive quickly when remoistened. Water uptake occurs very rapidly and lichens may hold several hundred percent of their dry weight of water. As a consequence under damp conditions, toxic substances dissolved in rain water or flood waters have ready access to cells within the thallus. The presence of pollutants may be reflected in a reduced capacity of lichens for photosynthesis, or for nitrogen fixation in the case of those containing blue-green algae.

A second characteristic of lichens is their capacity to accumulate substances rapidly from their environment by processes which include the active uptake of anions and the passive absorption of cations by an ion-exchange process. Lichens have a considerable capacity for metal ions due to the presence of metal-binding sites on both the algal and fungal partners. The evidence is that the sites are largely extracellular and function in a comparable way to those of synthetic cation exchange resins. Laboratory studies have shown that lichens have molar exchange capacities close to those of synthetic ion-exchange resins.

Because of their ready uptake of trace elements from waters (and from aerial fallout), lichens have long been used as monitors of atmospheric pollution [690] from as many as 30 elements. For example, lichens within 10 km of an Ontario nickel smelter contained over 300 μg/g nickel (dry weight)[817]. The use of lichens for mineral prospecting is less well established. The first suggestion that they might be used for this purpose was from Rosenqvist [706], who was disappointed in the results. However, a fundamental paper by Leroy and Koksoy [508] suggested that lichens would be particularly suitable for biogeochemical prospecting because of their very great age (*ca.* 2000 years in some cases) which would facilitate absorption of many elements. They showed that the elemental content of lichens accurately reflected the composition of the sandstones upon which they were growing.

The use of lichens for biogeochemical prospecting has been reviewed by Richardson et al. [689]. Lichens growing on serpentine rocks in Finland contained up to 0.3% nickel in their ash [524]. The same author [526] also reported very efficient mechanisms for the selective absorption of zinc by *Umbilicaria pustulata.* Yliruokanen [961] and Erämetsä and Yliruokanen [276] have analyzed various trace elements in Finnish lichens. They found that elemental levels closely followed concentrations in the substrate, particularly for elements such as molybdenum and uranium. Up to 8 μg/g (500 μg/g in ash) uranium was found in *Cladonia alpestris* [276]. The same authors [275] have also reported values for the rare earths in six species of lichen. Brown [131] has studied lichens growing over old lead mines and has reported 0.46% of this element in the ash of *Stereocaulis pileatum.* An interesting example of unconscious biogeochemical prospecting by use of lichens is given by Tomassini et al. [816], who discovered high copper levels around Contwoyto Lake in Northwest Territories, Canada. At a later date, a mining company independently discovered a copper ore deposit in the area.

One of the most thorough studies of the potential of lichens for geobotanical and biogeochemical prospecting is by Easton [264], and has already been referred to in Chapter 4. He concluded that lichens were of much more use for geobotanical prospecting than for biogeochemical work. The reason for this is not hard to assess. Unlike bryophytes (mosses), lichens are not aquatic and, since their root systems are used only for anchoring and not for obtaining nutrients, they are not able to sample beneath the surface layer of the substrate. A final and very important factor is the extreme difficulty of identification. Although any reasonably intelligent person without botanical knowledge should have no difficulty in recognizing most higher plants, this is not the case for lowly plants such as lichens, and expert assistance therefore will be needed for studies involving them.

17.3. MOSSES

17.3.1. Introduction

Bryophytes are similar to lichens insofar that they can accumulate metals by both ion exchange and particle trapping [609]. The water relations of these plants correlate with their morphology and anatomy. Some understanding of the water relations is necessary when considering metal uptake by bryophytes.

Endohydric mosses with clearly differentiated water-conducting systems possess a cuticlelike surface on the leaves which impedes water and cation uptake from the leaf surface. *Polytrichum* and *Atrichum* are examples of this type of moss. **Ectohydric** mosses have no cuticlelike layer or internal conducting system, so that water and mineral absorption occurs over the entire surface, as in lichens. Examples of ectohydric mosses include **Pleurocarpous** mosses, some **Acrocarpous** genera (e.g., *Rhacomitrium, Grimmia,* and *Tortula*) and many liverworts. **Mixohydric** mosses combine some elements of each of the previous groups with weakly developed conducting tissues [781]. Thus, as might be expected, the ectohydric moss *Dicranella varia* shows a higher mineral content (20–200%) than the mixohydric moss *Philonotis fontana* which has a similar growth form [756]. Sphagnum, which is adapted for living in mires, has most of the characteristics of the ectohydric group.

Clymo [203] has carried out detailed studies on cation exchange in the genus *Sphagnum*. He developed a model that describes how H^+ ions on carboxylic acid groups in newly formed cell walls at the plant apex are exchanged for dissolved cations in the natural habitat in which the moss grows. The continuous source of H^+ ions, which is released by this process, maintains the pH as low as 3.0 in the mire environment in the case of

Sphagnum acutifolium. Hylocomium splendens, a species that grows on the ground rather than in mires, also takes up cations such as nickel via ion exchange [709]. The cation uptake of this species was in the order: Cu,Pb >Ni >Co >Zn,Mn, whether the cations were applied singly or as mixtures [709]. High cation exchange capacities have been observed in mosses. Thus a Ni^{2+} uptake of 80 μmol/g has been reported in *Hylocomium splendens* [709]. It has been estimated that the capacity for nickel uptake by *Sphagnum flexuosum* is approximately 1500 μmol/g. The ion-exchange capacity for singly charged ions in *Sphagnum* species is usually 900 to 1500 μmol/g and 600 to 1000 μmol/g in other bryophytes [203,668]. This is about one tenth to one third of the capacity of synthetic carboxylic-group cation exchangers. The high exchange capacity of mosses has been related to the carboxylic acid content of the cell wall [428]. Up to 20–30% of the dry weight of *Sphagnum* spp. may be uronic acid residues [226]. The cation uptake sequence mentioned above for *H. splendens* is consistent with carboxylic acid binding sites.

There has already been some discussion of the use of bryophytes for geobotanical prospecting (Chapter 4), and it is in this field that mosses probably are most effective. Some of the limitations that apply to the use of lichens in biogeochemical prospecting also apply to mosses; that is, their roots do not penetrate the substrate and there are problems with identification. However, unlike lichens, many bryophytes are aquatic and are surrounded by potentially mineralized water, so that they can be used in this way for prospecting.

17.3.2. Terrestrial Bryophytes

Although the best potential for biogeochemical prospecting lies with aquatic bryophytes, numerous investigations have been carried out on the relationship between elemental levels in the moss and in its substrate. For example, Lounamaa [524] used the nickel content of nine species of bryophyte to differentiate three rock types in Finland. Erämetsä and Yliruokanen [276] determined six elements in 22 Finnish mosses and found up to 180 μg/g uranium (dry weight) in *Rhacomitrium lanuginosum*. The same species was observed by Shacklette [745] to be an indicator of serpentine in Alaska. Unusual accumulation of chromium (an element seldom accumulated to any extent by vegetation) was observed by Lee et al. [506] in New Caledonian specimens of *Aerobryopsis longissima*. They found up to 300 μg/g (dry weight) in this species growing on *Homalium guillainii,* a hyperaccumulator of nickel. Ward et al. [881] observed up to 5 μg/g cadmium, 34 μg/g copper, 202 μg/g lead, and 156 μg/g zinc in specimens of *Hypnum cupressiforme* from a mining area in New Zealand.

One of the best reviews of trace elements in bryophytes is by Shacklette [746]. The same author [747] has published an excellent review of the role of copper mosses in mineral exploration.

17.3.3. Aquatic Bryophytes

Sampling and Analysis.   Aquatic bryophytes are often found in swiftly running streams draining deeply forested areas, and are usually attached to rocks and stones in the stream beds. The rhizoids have little penetrating power and serve as anchors rather than as means of nourishment. Clumps of bryophytes, however, do serve as natural traps for finely divided sediment in the stream waters, and a major problem arises in separating the plant material from this sediment.

When samples are collected from streams, they should be washed as carefully as possible *in situ* before the material is placed in plastic bags for onward transmission. The material should be washed further in the laboratory and should be hand picked where necessary to remove portions containing the greatest amount of entrapped sediment. The cleaned material is then ashed in borosilicate beakers at 500° C for several hours, and the residue is pulverized gently with a glass rod. Grading the material through a 20-mesh nylon sieve removes a part of the residual inorganic contamination. The plant ash (20-mesh size) is then weighed.

The main difficulty in the analysis of bryophytes is to decide to what extent contamination problems are present. One way to circumvent this difficulty is to analyze the substrate or contaminants entrapped in the clumps and to make an allowance for this in the final calculations. It may be assumed that the ash of clean bryophytes should be about 10% of the dry weight and that proportions greater than this represent contamination. During the analysis of 44 bryophytes, Whitehead and Brooks [925] found that the mean ash weight was 16%, with 37 specimens giving values in the range of 7–27%. The mean contamination was probably about 6% and the range was 0–17%. The same authors noted that concentrations of uranium and several other elements in the entrapped sediment were significantly lower than in the bryophyte material itself. The effect of contamination, therefore, was to lower the apparent elemental content of the bryophytes rather than to increase it.

Prospecting with Aquatic Bryophytes.   As most bryophytes are fully or partially aquatic, their great ion-exchange capacity can be useful in providing a prospecting method that is an alternative to the notoriously unreliable hydrogeochemical method. Recent work by Wenrich-Verbeek [920] showed that mosses growing in water containing 0.005 μg/ml uranium contained up to 1500 μg/g of this element (an enrichment factor of 300,000) and there appeared to be a good correlation between the two variables. A positive correlation also existed between the concentration of uranium in the moss and the level of this element in the stream sediments.

One of the highest accumulations of any element by mosses has been found in *Pohlia nutans*. Boyle [84] has reported up to 2.4% copper (dry weight) in this species growing near a spring in a cupriferous bog at Sackville, New Brunswick, Canada. This probably represents the highest copper level ever recorded for any plant and reflects the extreme tolerance of *P. nutans* to an element that is normally very toxic to plant life.

Shacklette and Erdman [749] have studied uptake of several elements by aquatic mosses in spring waters of central Idaho. They reported anomalous levels of uranium (up to 1000 µg/g) in the ash of *Brachythecium rivulare*. A case history from New Zealand will now be described.

Mosses were collected from streams draining an area of uranium mineralization in the Lower Buller Gorge region of South Island, New Zealand. Collection of the bryophytes was somewhat of a problem because they grew in closely intertwined clumps containing several species together. At first an attempt was made to collect individual species, but the separation and identification of these species was so difficult and laborious that such a procedure was impractical. Ultimately the species composition of the bryophyte clumps was ignored completely and the samples were treated as composites of various species including: *Bryum blandum, Dicranella vaginata, Distichophyllum pulchellum, Fissidens rigidulus, Lophocolea planiuscula, Plagiochila deltoides, Pterogophyllum dentatum, Riccardia* sp., and *Thamnium pandum*. The bryophytes were analyzed for beryllium, copper, lead, and uranium. The first three elements were determined because previous work had shown that they were good pathfinders for uranium in that area. The aim of the work was to compare the uranium content of bryophytes with the uranium content of the stream waters and, hence, indirectly with the presence or absence of mineralization in the catchment area. The latter comparison was rendered possible by the fact that the area had already been investigated thoroughly by geochemistry and geophysics.

Data for the uranium content of the streams were available from previous work by Wodzicki [946], but an attempt was made to obtain further data by determining the uranium of peat absorbers treated in the manner suggested by Horvath [368]. Samples of peat were placed in cotton bags and anchored in each stream for a period of one week. At the end of this period, the bags were removed, dried, and the contents ashed in a muffle furnace. The samples were then analyzed and the data are shown in Table 17.1 together with data for water, which had been previously been recorded by Wodzicki [946]. Although the data for uranium did not indicate absolute amounts of this element in the waters, they were at least comparative for the streams concerned. The table also contains details of flow rates of each stream.

Statistical interpretation of the data was rendered difficult by the small number of sites (14) involved. Of the 14 streams sampled, 8 were known to have uranium anomalies in their catchment areas. The statistical procedure adopted was the well-known Games Theory [616] and involved assigning a score of $+2$ to those streams where a known anomaly was confirmed by higher than average uranium levels in the sample type. Similarly, a score of $+2$ was assigned where absence of mineralization was confirmed by low uranium levels in the samples. Values of -2 were assigned where the

TABLE 17.1.   Elemental Concentrations in Stream Waters (ng/ml) and Ashed Bryophytes (μg/g) from the Lower Buller Gorge Region of New Zealand

Stream Number	U in Peat	U in Water	Elemental Concentrations in the Bryophytes				Nature of Known Mineralization in the Region
			Be	Cu	Pb	U	
1	10.5		82	37	910	14.3	None
2	18.6	3.25	69	45	296	19.5	None
3	18.6	2.30	41	14	120	16.8	Weak
4	9.3		33	20	230	8.8	None
5	9.4		36	17	345	12.1	None
6	9.0	1.40	30	6	140	4.7	None
7	57.0		109	78	510	11.2	Weak
8	324.0		33	143	260	52.0	Strong
9	15.6		92	42	410	12.8	Strong
10	18.3		93	30	293	18.7	Strong
11	33.0		53	14	690	68.0	Strong
12	36.0		65	60	650	86.0	Strong
13	75.0	0.80	43	140	180	6.0	Weak
14	4.0	0.65	13	61	150	0.7	None
Reliability Index (max. = +25)	+17	−2	+8	+3	+5	+13	

Note: The very high value for stream #8 was due to disturbance caused by recent mining activities.
Source: Whitehead and Brooks [925].

analyses did not confirm presence or absence of mineralization. The above values were assigned for cases where the known mineralization was strong. In cases where the mineralization was weak, values of +1 or −1 were assigned. The above procedure gave a total possible maximum score of +25 and a minimum of −21. From Table 17.1 it can be seen that a score of +13 was obtained for the uranium content of bryophytes as indicators of mineralization. The peat absorbers had an even higher reliability index (+17). The pathfinders were far less effective and gave indices ranging downwards from +8 for beryllium. Although the uranium content of the five streams studied by Wodzicki [946] was apparently no guide whatsoever to uranium mineralization, the number of samples was too small to have much statistical significance.

The flow rate of the water is a factor that will control its uranium content. Thus if all uranium concentrations in bryophytes are normalized to a constant standard by dividing each by the flow rate of the stream, a better basis of comparison can be achieved. Once the existence of mineralization in the catchment area has been suspected, the tributaries should each be tested by the bryophyte method until the source has been discovered.

Although the above case history has concerned uranium, there is no reason why the same principle should not be applied to other elements. The reliability of the procedures will probably never be as good as that for stream sediments except perhaps in the case of uranium and other elements which are very mobile and are not retained for long periods in these sediments.

17.4. HUMUS

Humus has long been considered as an acceptable biological sample type in biogeochemical prospecting [563]. Although it might be considered to be an integral part of the soil profile (A_0 horizon), it is of such recent origin that it is better considered as a biological sample type. In many parts of the world that have suffered extensive glaciation, soil geochemistry is often ineffective in the search for ores. This problem is particularly serious in Fennoscandia where humus sampling has been pioneered as a prospecting tool. An excellent account of case histories of such work in Fennoscandia has been given by Nuutilainen and Peuraniemi [628]. They reported positive response of humus to mineralization at depth in three areas of Finland where glacial till gave little or no response.

Toverud [819] has demonstrated the effectiveness of humus as a sampling medium for tungsten. Two areas previously surveyed by sampling the heavy mineral concentrates of glacial till and trenching and diamond drilling were chosen for the investigation. One was in a mountainous region of northern Sweden and the other was a forested area of central Sweden where tungsten mining is now taking place. In a third area, humus samples were collected from the neighborhood of a tungsten mine in central Sweden, which had been previously sampled for humus in 1950. The results showed that humus material can be a useful complement to, or replacement for, heavy metal mineral sampling on a local scale in tungsten prospecting. The cost of the humus sampling program was three to four times lower than that of determining heavy metals in glacial till. The disadvantage of humus sampling was that no mineralogical studies could be carried out.

In prospecting for molybdenum, Isohanni et al. [381] found humus to be a good alternative to glacial till in the Aittojärvi area. One of the most successful case histories for the use of humus lies in the search for gold in Colorado [230]. Gold anomalies in soil, float pebbles, and cobbles poorly reflect gold deposits in bedrock beneath an extensive layer of alluvium and glacial drift in the Empire District. The correlation between the gold content of the mull (forest humus) and the existence of known mineralization at depth is shown in Figure 17.1, and is really quite remarkable. Gold has a ready tendency to form organic complexes and is one of the elements that usually provides a barrier-free system (see Chapter 14) for many biological sample types in biogeochemical prospecting.

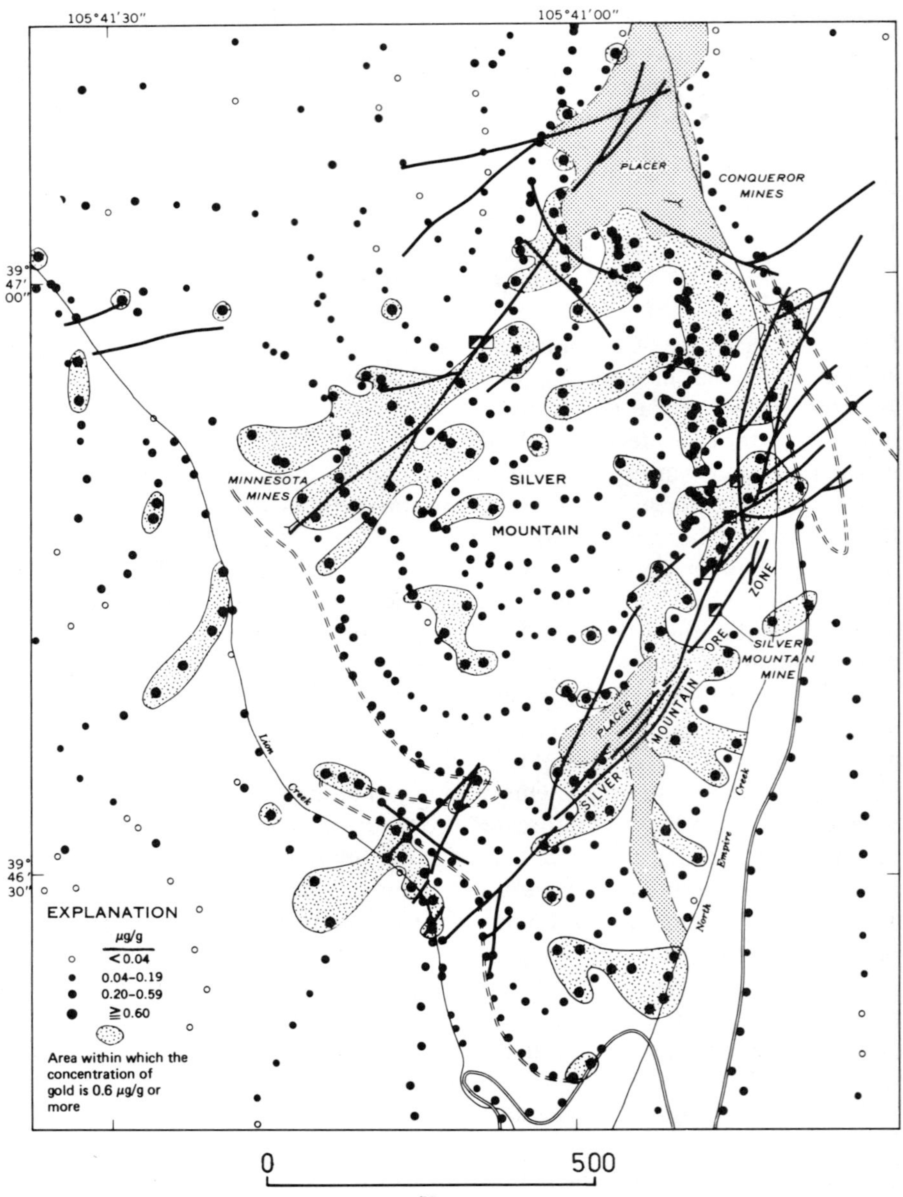

FIGURE 17.1. Distribution of gold in the ash of mull (humus) in the Empire district of Colorado. *Source:* Curtin et al. [230].

17.5. PEAT

17.5.1. Introduction

A large area of the land in northern latitudes is covered with water-saturated organic terrain known variously as bog, muskeg, peatland, swamp, and so on. In Canada alone, about 15% of the total area of the country is covered with this type of formation. The peatlands form a dense covering above the bedrock and are a formidable barrier to prospecting for minerals in the substrata beneath the organic layer. The magnitude of the problem in Canada and the Soviet Union has been discussed by Hawkes and Salmon [346] and Shvartsev [759], respectively. An extensive review by Usik [832] is devoted to the use of geobotanical and biogeochemical methods of prospecting in peatlands, and is the source of much of the basic geobotanical information contained in this section.

Peatlands can be classified into two main groups depending on their hydrological condition. The first of these groups is said to be **minerotrophic** in origin and derives its water from drainage or percolation through rocks, till, soil, and so on. **Ombrotrophic** peatlands, however, receive water solely from local precipitation and are deficient in mineral-influenced drainage waters. Biogeochemical or geochemical prospecting methods will tend to be more successful in minerotrophic peatlands, as movement of water will increase the chance of detecting mineralization. Even in minerotrophic formations, however, there are still many problems that render prospecting difficult. The first of these is the ready tendency of peat to complex with a number of elements, particularly copper, iron, manganese, uranium, and zinc. The result of this chelatory effect is to increase background levels for a number of elements so that anomalies are harder to detect. Movement of water in the bog can also have the effect of producing a displaced dispersion halo away from the original source of mineralization.

In the broadest definition of the term, biogeochemical prospecting includes the analysis of all organic material, living or dead, so that analysis of peat can justifiably be discussed, albeit briefly, in this section. Extensive pioneering work on the relationship between peat or peat vegetation and the environment has been carried out by Salmi [717–720].

17.5.2. Factors Affecting Adsorption of Elements by Peat

Like most humic material, peat has an extraordinary capacity to adsorb cations. Szalay [782] has undertaken extensive investigations on this property, principally with regard to uranium. Adsorption of this and other elements is highly pH-dependent and is at a maximum in the pH range of 4–10. Below a pH of 4, adsorption decreased rapidly because of the competition from free H^+ ions. In the optimum pH range, the **distribution coefficient** (concentration in the peat divided by the concentration in the water) for the UO^{2+} ion is about 10,000 (fully saturated peat can contain up to 10% uranium on a dry weight basis) and for Th^{4+} is about 20,000 [782]. In general, distribution coefficients increase strongly with the valence and atomic mass of the cation. The various cations compete with each other, and those of higher valence and atomic mass displace those with lower values of these two variables.

The principle agents for fixation of elements by peat are the humic acids. Szalay [782] has shown that the adsorption of ions decreases to a negligible amount if the humic acids are removed or neutralized. Pratt et al. [664] reported on the chelation of copper and nickel by the same material. The association of molybdenum and uranium with humic acids has been reported by Szalay and Szilagyi [784], whereas for uranium alone, this association has been reported by Kranz [482], Manskaya et al. [572], and Titaeva [810].

There is a close association between the adsorption of ions by peat and by bryophytes, since one of the major sources of peat is sphagnum moss, whose cation-exchange capacity has been studied extensively by Puustjärvi [668]. Since adsorption of cations is highly pH dependent, pH measurements should always be made whenever peat samples are taken. In the critical range of pH 2–4, even small changes in the value can greatly affect the adsorptive properties of peat.

Elemental distributions in peat profiles usually are not uniform. Work by Hvatum [376] and Mitchell [594] has shown that lead, molybdenum, and zinc tend to concentrate in the upper layers of peat profiles, whereas copper, cobalt, and manganese are sometimes concentrated at the surface and bottom of the peat bogs. It is not known to what extent surface enrichment is due to biogenic migration (biogeochemical cycle; see Chapter 14), but the general distribution of elements within the past profile is probably a function of a large number of factors including Eh and pH. Erämetsä et al. [274] found that uranium tended to concentrate about three quarters of the way down the bog profile. A partial solution to controlling some of the factors influencing elemental distributions in peat profiles has been proposed by Shvartsev [759], who suggested that concentration ratios of pairs of elements should be taken as an indication of the presence or absence of mineralization within the bog. This is analagous to the classical work of Warren and his co-workers in the use of element ratios in plant material.

17.5.3.   Mineral Bogs

The concentrations of elements in peat bogs sometimes become so high that the peat itself becomes an economic source of minerals. Records of mining these bogs extend well back into the early nineteenth century, where references can be found to "copper bogs" in Ireland [820] and Wales [348]. Henwood [348] reported that the peat fields of Merioneth at one time gave an annual yield of copper worth $100,000. Unfortunately, extraction of

TABLE 17.2. Occurrences of Mineral Bogs in Various Parts of the World

Element	Locality	References
Copper	Canada	84, 300, 301, 769
	Finland	718, 719
	Ireland	820
	Soviet Union	570
	United States	266, 296, 528
	Wales	348
Iron	Canada	346
	Finland	718
Manganese	Canada	129, 829
Uranium	Australia	239
	Soviet Union	431, 516, 599
	Sweden	30
	United States	82
Vanadium	Finland	718
Zinc	Canada	87, 281
	Finland	719
	United States	153

Source: Brooks [97].

elements from many bogs presents as yet unresolved economic and technological problems [82].

Elements found in mineral bogs are few in number but are all of economic importance. Table 17.2 lists some of the better-known mineral bogs in various parts of the world and gives the appropriate references. One of the best known of these is at Otanmäki in Finland [718]. The ore is mined from dry areas of the region but extends beneath several large bogs. Investigations by Salmi showed anomalous amounts of iron, titanium, and vanadium in the peat material at points corresponding to mineralized outcrops in the substrate. This is illustrated in Figure 17.2. The association of relatively high pH values (6–7) with mineralization at Otanmäki, is probably due to the presence of lenses of limestone, which are commonly associated with zinc in Finland.

Mineralization can be detected in peat bogs even when the orebody is covered with an overburden of sand or till up to 15 m in depth [832]. Procedures in prospecting with peat involve sampling with a bog drill and immediate measurement of pH and Eh. The material is then dried, ashed at 450° C and analyzed for its elemental content by some suitable method such as atomic absorption or emission spectrometry (see Chapter 18). Usik [832] has recommended that detailed local investigation on peat bogs should involve the following five basic steps.

1. Studies of physical conditions such as groundwater flow, diffusion of elements, and so on

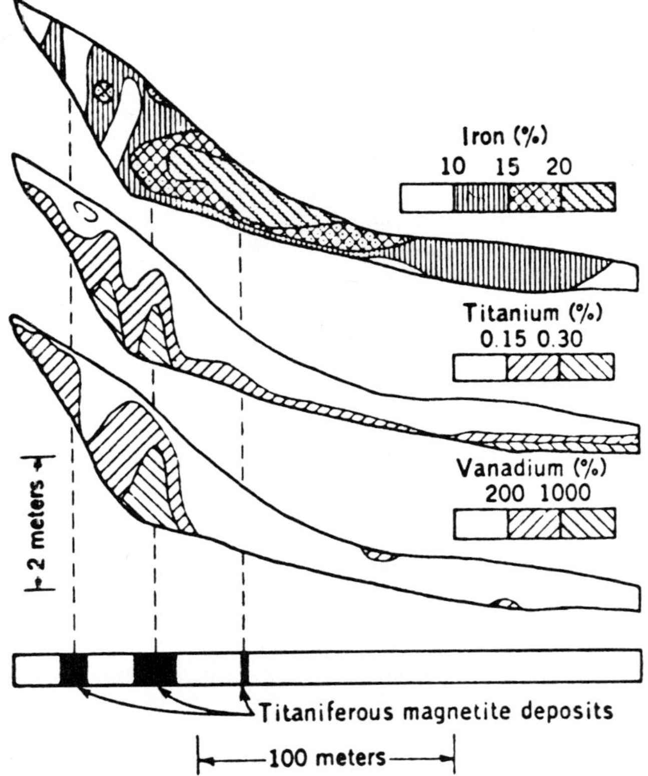

FIGURE 17.2.   Section through the Malmisuo Bog, Finland, showing the relationship between iron, titanium, and vanadium in peat and in the underlying ore. *Source:* Hawkes and Webb [347], after Salmi [718].

2. Studies of elemental distributions in vertical profiles including the underlying clays and tails.

3. Studies of the areal and vertical distribution of Eh, pH, and elemental concentrations.

4. Studies of the nature of chelation of elements to organic material

5. Botanical and ecological studies.

17.5.4.   Prospecting in Peatlands

Apart from the above example from the Otanmäki orefield, there are relatively few case histories concerning the use of peat for prospecting. This is, of course, apart from the use of live plants growing in peat bogs, for which there is a fairly extensive literature. However, Yliruokanen [961] studied 130 sphagnum peat bogs in Finland and found anomalous metal concentrations in 15 of these bogs. Some of the anomalies indicated nearby mineralization. It is likely that much of the future work in this field will continue to be carried out in Fennoscandia, where some 25% of the surface is covered with sphagnum bogs.

17.6. TREE SAP AND PLANT JUICE

In an extensive review of the geochemistry of gold, Boyle [85] has reported that the gold content of plant juice has been recommended as a method of prospecting for gold in Soviet Armenia [495]. It was observed that the gold content of the juice was significantly higher than that of either aqueous extracts of the plant material or of the soil itself. A related method (also for gold) was reported by Krendelev and Pogrevniak [484]. They utilized birch sap as a biological sample type. One of the advantages claimed over the conventional biogeochemical method was that the data required no correction for the time of sampling (see Chapters 14 and 15) as the spring begins and terminates at about the same time over large regions of similar latitude. Another advantage that was cited was that birch forests that occur so widely in the Soviet Union also occur in other boreal countries, such as those in Fennoscandia and in North America. Further cited advantages were the simplicity of obtaining and processing samples and the fact that sap is a ready-made, natural metal-bearing solution that can be subjected to immediate neutron activation analysis (see Chapter 18) without any preliminary preparation whatsoever. Samples of sap obtained by tapping the birch *Betula platyphylla* were taken during the spring sap run along profiles across a pyritic gold deposit in western Transbaikal. The gold content of the sap varied from 0.0015 to 0.33 ng/ml (ppb). The higher values clearly coincided with anomalous gold contents present in the secondary halos in soils over the deposits. The same technique was used for base metal deposits in the Soviet Union [485].

The utilization of sap as a sampling medium should be applicable to regions that have a widespread distribution of certain species of tree. The maple and the various conifers in Canada seem obvious targets in Europe and North America.

17.7. ORGANIC SAMPLES FROM STREAMS

N. H. Brundin, one of the originators of the biogeochemical method in the late 1930s, has suggested that the dead organic matter in streams could be used for prospecting purposes [133]. Favorable initial results by use of this sample type led to full-scale testing in northern Sweden. The organic material was sampled from the banks of streams below the usual water level and consisted of dead organic matter in different stages of humification mixed with varying amounts of inorganic material, and often penetrated by living roots. Statistical procedures were used to compensate for the presence of interfering minerals such as limonite in the organic samples. It was concluded that organic samples were more suitable than stream sediments in prospecting for uranium, copper, lead, and molybdenum. A further advantage of this sampling medium was that it could be collected everywhere along streams in the region, whereas stream sediments were much more difficult to find. Another advantage of the procedure was the higher elemental content of the ashed material compared with that of stream sediments so that analytical problems were reduced. A disadvantage of the sample type was that a sophisticated statistical procedure was necessary to correct for the presence of limonite and other interfering constituents in the samples.

17.8.   HERBARIUM MATERIAL

Herbarium material represents a vast reservoir of already collected plant material. In some cases this material may be used for prospecting in countries that the prospector has never even visited. This field is very new and will not be discussed further, since Chapter 21 is devoted completely to it.

18

CHEMICAL ANALYSIS OF PLANT MATERIAL

18.1. INTRODUCTION

During the past few decades, progress in instrumental methods of analysis has greatly outstripped that of previous years. This has had far-reaching effects on the various types of geochemical prospecting that rely so heavily on chemical analysis of a wide range of sample types. The development of these new instrumental techniques makes it possible to carry out analyses cheaply, accurately, and at great speed. This has had a marked stimulatory effect on geochemical exploration techniques, and there is no doubt that the current level of activity in these fields could not be maintained without these analytical procedures.

Although it is perhaps not true to say that the days of "wet chemistry" are numbered, it certainly is true that the working time of the analyst is becoming increasingly occupied with mere preparation of samples for subsequent instrumental analysis. This factor has also had a cost benefit in exploration work, because it is now possible to use unskilled personnel to prepare the solutions. These lower-paid workers can now process a much greater volume of work than did the relatively highly paid "wet chemists" of 30 years ago. So cheap has analysis become indeed, that it is now cheaper in both absolute and relative terms than it was in, say, 1939 to analyze a single element such as copper in a sample of rock or soil.

Because of the paramount importance of instrumental methods of analysis in mineral exploration programs, the main emphasis in this chapter will be on four of these: neutron activation analysis (NAA), emission spectro-

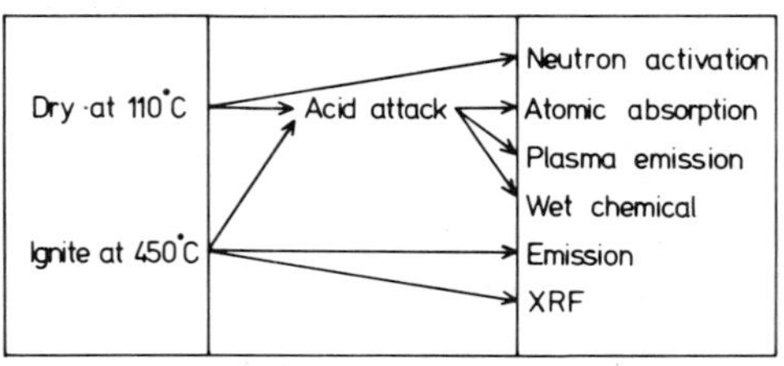

FIGURE 18.1. Schematic representation of preparation, dissolution, and analysis of vegetation.

metry (ES), inductively coupled plasma emission spectrometry (ICP), and atomic absorption spectrometry (AAS). Other techniques such as X-ray fluorescence spectrometry (XRF) and colorimetry will also be mentioned, although they are not as important as the above methods.

The analysis of material for biogeochemical prospecting involves a combination of all or some of four basic steps: preparation and dissolution of the sample, followed by separation and instrumental determination of the element or elements. Whereas all methods of analysis involve the first and last of these steps, most modern methods of analysis do not require a preliminary separation of the analyte, and many do not even require a dissolution stage. Figure 18.1 gives a schematic representation of the various processing stages required for various instrumental techniques as well as for classical "wet chemical" methods. The term "wet chemical" includes such procedures as colorimetry, fluorimetry, gravimetry, and titrimetry. Figure 18.2 shows the usual ranges of concentration levels determined by various methods of analysis. Since biogeochemical methods of prospecting usually involve trace elements, it is only the more sensitive of the techniques shown in Figure 18.2 that are generally applicable to this method of prospecting.

This chapter does not include information on the analysis of rocks and soils. The reader is referred to Reeves and Brooks [681] for a full treatment of this subject.

18.2.   SAMPLE PREPARATION

Plant material for chemical analysis usually is required in one of three forms: pulverized dry material (for NAA) ashed material (ES and XRF), or in solution (AAS and ICP).

Samples for neutron activation analysis merely have to be dried at 110°C for a few hours and pulverized in a hammer mill or other suitable disintegrator such as a rotary blender. Even this latter stage is not always essential, since leaves, twigs, or needles only need to be chopped up into suitable lengths and placed in a vial for irradiation.

The problem of ashing plant material already has been discussed in Chapter 16, and little remains to be said except that it is usually preferable to wet ashing with explosive and highly corrosive perchloric acid. Some

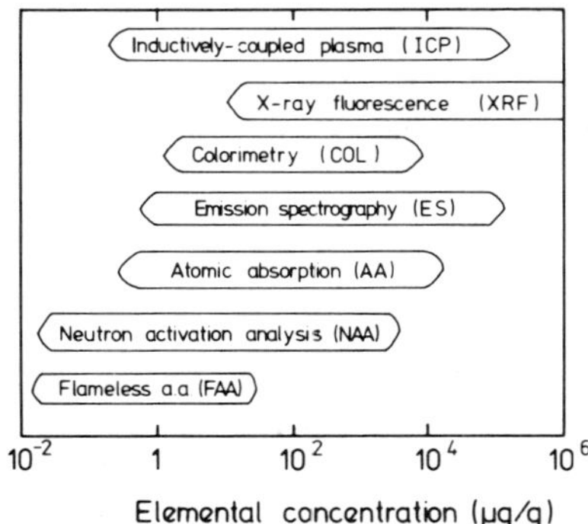

FIGURE 18.2. Working ranges of elemental concentrations in vegetation analyzed by various methods.

elements such as cadmium, however, are too volatile for dry ashing so that wet ashing has to be used. Dry ashing at about 500°C with adequate access of air is a very satisfactory procedure. The work of Gorsuch [324] is perhaps one of the best monographs on this important subject.

It has frequently been asserted [169] that a disadvantage of biogeochemical prospecting compared with soil sampling procedures is the necessity for the extra step of ashing of the plant material. This argument is not always relevant, however, since soil samples also need to be ashed for methods such as XRF or ES. A further point to remember is that it is much easier to dissolve plant ash than soils or rocks (gentle heating with $2M$ hydrochloric acid is usually sufficient) so that the combined ignition-dissolution for plant material may require less operator time than does the single dissolution step for rocks or soils using hydrofluoric acid and oxidizing acids. Many service laboratories are not used to analyzing plant material and the prospect of doing so often causes them anxiety, particularly with regard to the economics of the operation. Samples of vegetation placed in 100-ml squat borosilicate beakers will ash to completion in a few hours over a conventional laboratory hotplate. In my own laboratory we routinely pre-ash vegetation on hotplates and transfer the small volume of material to 25-ml beakers for final ashing in a muffle furnace at 500°C.

If the plant ash is now to be dissolved, a recommended procedure is as follows: weigh about 0.1 g of plant ash into a 12 × 150 mm borosilicate test tube and add 10 ml of 2M hydrochloric acid. Place the tube in a water bath of boiling water for about 15 minutes and decant or filter off any insoluble residue, unless it settles readily and can usually be ignored during the instrumental stage. If high concentrations of lead or silver (insoluble chlorides) are expected, $2\,M$ nitric acid should be used instead. Occasionally an appreciable amount of siliceous residue remains in the plant ash (e.g., *Equisetum* species). In such cases, attack with hydrofluoric acid may become necessary.

Wet oxidation of organic matter carries with it the problem not only of use of dangerous oxidants but also the possibility of trace element contamination by the large volume of reagent relative to the sample. Nevertheless, for some volatile elements such as selenium and mercury, there is no alternative. Fuming nitric acid may also be used, but the most usual attack is with a 4:1 mixture of nitric and perchloric acids with a volume (ml) least 20 times that of the weight of sample (g). The samples are heated until copious white fumes of perchloric acid are evolved. It is often necessary to add an antifoaming agent to the reaction mixture. If the analyte is too volatile even for wet oxidation with an open system, recourse will have to be made to a robust Teflon bomb containing the sample and oxidants, and which is heated for an hour or so at 110°C in a drying oven. The pressure is so great in this operation that the lid of the bomb is sometimes removed with explosive force. From my own experience, it is wise to further reinforce the bomb by encasing it in a stainless steel cage to prevent loss of the lid.

A number of common instrumental techniques will now be described and their potential for chemical analysis of plant material will be assessed.

18.3. ATOMIC ABSORPTION SPECTROMETRY

18.3.1. Flame Methods

The development of instrumental methods of atomic absorption spectrophotometry by Walsh [875] has been without doubt one of the most important advances in analytical chemistry of all time. The technique is of great simplicity and is relatively free of serious interference problems, so that analyses for many elements can be carried out in the same solution without the need for tedious separation procedures.

The basic theory of atomic absorption was known as long ago as the middle of the nineteenth century. In essence it concerns the observation that certain wavelengths of radiation emitted by excited atoms (emission spectrum) will be strongly absorbed by unexcited atoms of the same element. Hence radiation emitted by these excited atoms will be reduced in intensity if allowed to pass through a region in space containing a population of unexcited atoms of the same element. A spectral line absorbed in this way is said to be a **resonance line.** The extent of the attenuation of the emission spectrum is determined by a detector, and is a direct measurement of the concentration of the analyte. The reader is again referred to Reeves and Brooks [681] for further details of the theory and practice of this technique.

In summary, atomic absorption spectrometers consist of four main parts: the source, the absorption cell, the monochromator, and the detection system.

Sources comprise hollow cathode lamps which are usually designed for a single element, although multielement sources are obtainable. The emission spectrum is produced by passing a d.c. discharge which ionizes the low-pressure filler gas (argon or neon). The gas ions strike a hollow cathode made of the analyte element and release a cloud of excited atoms which produces the emission spectrum.

The absorption cell comprises a nebulization system which sprays the analyte solution into a long narrow flame fed by a gas mixture such as air/acetylene. The flame decomposes the chemical compounds into atoms and provides a long absorption path since it is fed by a stream of gas passing through a long narrow slit in the same plane as the energy emitted from the source. The temperature is sufficiently high to decompose the compounds into atoms but not so high that a significant proportion of the ground-state atoms become excited.

The monochromator, usually of the grating type, serves to isolate the particular spectral line that is required.

The detector system amplifies the signal from a photomultiplier which receives the energy from the monochromator. Because of the inherent stability of a.c. amplification systems, they are preferable and are a feature of most commercial instruments. The result is obtained as a digital readout, meter reading, or chart recording.

Table 18.1 lists detection limits for various methods of analysis. The values are only approximate, and in all cases refer to the original solid sample, not to the solution prepared for analysis. It is extremely difficult to assess true detection limits because they are influenced by so many factors other than the concentration of the analyte in the sample. For example, gold in arsenic-free plant material might be determined down to 0.01 μg/g (10 ppb) without difficulty by neutron activation analysis, whereas the same procedure might only provide a detection limit of 1 μg/g if much arsenic were present in the sample.

The question of sensitivity is of great importance for biogeochemical prospecting since care must be taken that the correct method of analysis is used at all times. For example, a field worker used to having his samples analyzed by AAS might request that molybdenum also be determined by this method. If, as is likely to be the case, the results were to lie near the limit of detection of the method, the accuracy of the data would be in question. The field worker would have been better advised to have the analyst use a method such as ES.

The sensitivity of AAS techniques can be improved by greater amplification of the signal. However, this also increases the background noise and is self-defeating unless a very stable signal is obtained from the source. Double-beam commercial instruments are available to increase the stability of the signal and hence to allow for greater amplification.

One of the early problems with AAS was the omnipresent background which gave a signal even when the analyte was entirely absent in the sample. This was easily the greatest cause of erroneous data in the early days. The background is due to solid particles in the flame which physically impede the source signal and appear to indicate atomic absorption. Most modern instruments have a continuous background corrector in which an auxiliary (deuterium) lamp is used to subtract nonatomic absorption from the total absorption signal.

TABLE 18.1. Detection Limits of the Elements for Various Instrumental Methods of Analysis (values calculated on original solid sample)

Element	AAS	FAAS[a]	ICP	ES	XRF	NA[ab]
Ag	0.5	0.002	0.15	0.05	8	0.001
Al	6	0.03	1	2	40	0.001
As	30	1	2	100	3	0.0006
Au	0.5	0.1	0.5	10	16	0.0001
B	300	—	0.5	10	—	—
Ba	5	—	0.02	5	13	0.0002
Be	0.1	0.01	0.01	10	—	—
Bi	1	0.07	2	20	19	—
Ca	0.1	0.003	0.01	2	4	0.2
Cd	0.1	0.001	0.4	10	9	0.03
Co	0.3	0.06	1.5	10	2	0.0002
Cr	0.3	0.05	0.2	1	2	0.1
Cu	0.3	0.07	0.6	0.5	5	0.001
Fe	0.5	0.03	0.6	5	2	20
Ga	3	0.02	1.5	3	3	0.001
Ge	50	—	2	5	3	0.002
Hg	10	0.001	1	100	17	0.002
In	3	—	2	1	10	0.00001
Ir	200	—	0.8	50	15	0.0004
K	0.3	0.01	3	2	20	0.04
La	400	—	0.25	10	12	0.0008
Li	0.3	0.05	0.20	0.5	—	—
Mg	0.02	0.001	0.004	2	300	0.03
Mn	0.3	0.005	0.05	10	2	0.00001
Mo	1	0.4	0.2	5	5	0.03
Na	0.5	0.001	1	0.5	600	0.001
Nb	1000	—	1	30	4	0.2
Ni	0.5	0.1	0.4	5	2	0.04
Pb	0.5	0.05	1.5	5	18	0.04
Pd	25	2	1.5	10	7	0.001
Pt	25	2	2	50	15	0.01
Rh	2	—	1.5	10	7	0.0001
Ru	15	—	1	10	6	0.002
Sb	10	0.3	1.5	20	11	0.001
Sc	10	—	0.04	2	4	0.0003
Se	25	1	3	—	3	0.01
Si	5	—	0.5	20	35	30
Sn	5	0.6	1.3	10	11	0.02
Sr	0.5	0.005	0.01	5	3	0.001
Ta	300	—	1.5	100	11	0.03
Te	15	—	2	200	11	0.004
Ti	5	0.03	0.1	10	2	0.01
Tl	10	—	2	1	18	—
U	600	—	5	100	30	0.0004
V	1	1	0.2	5	2	0.0001

TABLE 18.1. *(Cont.)*

Element	AAS	FAAS[a]	ICP	ES	XRF	NA[ab]
W	150	—	1	20	12	0.0004
Y	15	—	0.1	10	12	—
Zn	0.1	0.001	0.1	100	3	0.01
Zr	250	—	0.1	10	4	0.3

Key: AAS—flame atomic absorption spectrometry
FAAS—flameless atomic absorption spectrometry
ICP—inductively coupled plasma emission spect
ES—emission spectrometry
NAA—neutron activation analysis
Note: For solution methods of analysis such as AAS, FAAS, and ICP, a $50\times$ dilution of the sample has been assumed.
[a]Assuming a 5 μL sample.
[b]Assuming one-hour activation with a flux of 10^{13} thermal neutrons/cm²/s and measurement by gamma spectrometry.

Atomic absorption with flames certainly provides the advantages of specificity, sensitivity, relative freedom from interferences, multielement determinations on the same sample, freedom from elapsed time requirements (e.g., cooling times required for some NAA procedures), and the provision of data in direct readout form. Nevertheless, there are some disadvantages. Ward et al. [876] have shown that quite severe matrix effects can be produced in solution of high ionic strength. To some extent this problem can be overcome by dilution of the sample, but this in turn reduces sensitivity. Even in dilute solutions, other interference effects are present, and the above authors have suggested that the ultimate solution to the problem is to have standards with the same matrix as the analyte solution. Despite the above cautionary note, there is little doubt that AAS is continuing to prove to be of immense use for biogeochemical exploration work and is still perhaps the most favored method of analysis in use at present for this purpose.

18.3.2. Flameless Methods

Flameless atomic absorption spectrometry (FAAS) replaces the flame cell with a system in which ground state atoms are either generated outside of the cell and are fed through it with a carrier gas, or are generated in a small enclosed space by means of an electric current. An example of the former is the determination of mercury, in which the sample is dissolved in acid and the mercury is reduced to the metallic form by tin(II) chloride. The mercury is carried through a glass cell with quartz end windows. The ground state atoms have a long residence time in the cell and their absorption of emission from the mercury source is therefore very much greater than with conventional flame methods. Mercury, which has poor sensitivity by flame methods, can be detected quite easily down to 0.001 μg/g (1 ppb) by this procedure. A good introduction to the subject is a review of more than 400 papers on the determination of mercury by flameless methods [831].

Atomic absorption techniques using electrothermal methods have already been mentioned very briefly. The basic principle (with many variants) involves introducing a small volume of solution (ca. 2–20 μl) into a graphite furnace [496] or on to a graphite rod [923]. An electric current is passed through the furnace or rod in three stages: low current to dry the sample at about 120°C, an intermediate current to char the sample at about 800°C to remove organic matter, and a high current to atomize the analyte at a temperature approaching 3000°C. The beam of the hollow cathode source passes through the cell and the ground state atoms of the analyte. These forms a cloud around the point of application of the sample, readily absorb the incident radiation, and, because of their long residence time, produce an absorption very greatly in excess of that produced by the flame method.

It will be noted from Table 18.1 that flameless methods are of the order of 100–1000 times more sensitive than are conventional flame procedures. In spite of this, they have not been used widely in biogeochemical work except in the case of mercury, which really cannot be determined at these levels by any other method. There are several reasons for this. In the first place, AAS methods are usually sufficiently sensitive for most elements. Second, the FAAS techniques are much slower than conventional AAS procedures. Once samples are in solution, they can be put through an AAS instrument at the rate of one every two or three seconds, compared with one every three minutes for FAAS. In spite of this, and apart from the obvious case of mercury, there are one or two elements (such as gold) whose concentrations [878] are so low in vegetation that the only alternative to FAAS would be the expensive and more time-consuming method of neutron activation analysis.

18.4. EMISSION SPECTROMETRY

Emission spectrometry is one of the oldest of the instrumental methods of analysis, and dates back to 1861 when Bunsen discovered caesium by the spectroscope. The technique remained qualitative, or at best only semi-quantitative, until the pioneering work of Goldschmidt and his co-workers at Göttingen in the early 1930s. After this period there was a rapid increase in the use of the technique, and it became very widely used for the routine analysis of a number of elements in plants, rocks, and soils. Until the advent of AAS in the early 1960s, ES was the most favored method of instrumental analysis used for prospecting purposes and, with colorimetry, accounted for nearly all analyses made.

The theory and practice of emission spectrography has been described by Ahrens and Taylor [10]. The sample is excited in a graphite or carbon electrode by means of a d.c. arc or an a.c. arc or spark. The sample produces excited atoms whose emission spectra are refracted or diffracted in a prism-type or grating-type monochromator. The separated spectral lines are recorded either on a photographic plate or by some form of electronic system. The intensities of the spectral lines are proportional to the original concentration of the analyte. Limits of detection for various elements are shown in Table 18.1.

Early instruments using photographic plates involved a great deal of tedious work in converting the optical density of the photographic images to concentration units for the analytes in the original samples. However, direct-reading instruments are much superior in regard to speed of analysis, but are much more expensive. The source and optical arrangement are the same as for the photographic instruments, but the detection system incorporates a number of photomultipliers positioned at various exit slits and corresponding to specific analysis lines. Electrical impulses from the photomultipliers are stored in condensers during excitation and are released sequentially at the end of the exposure. The result is displayed on a scale, printed on a typewriter, or punched on data cards for computer processing. Spectrographs are now available with up to 22 channels so this number of elements can be determined simultaneously. ES cannot compete with AAS either in speed, precision, or costs when a limited number of elements is to be determined. The technique is very useful, however, for elements such as boron, beryllium, and molybdenum which are determined more easily by ES than by other procedures.

One great disadvantage of emission analysis is that the line intensities of most elements vary with the bulk composition of the sample. This **matrix effect** can be reduced by having standards of the same bulk composition as the samples. A further disadvantage of ES, particularly if photographic instruments are used, is that they demand a higher degree of operator skill than do direct readers or atomic absorption spectrometers.

The reproducibility of direct-reading ES instruments is comparable with that of AAS (ca. 2%), but for photographic instruments a 5–10 percent precision might be expected for most biogeochemical work.

18.5. X-RAY FLUORESCENCE SPECTROMETRY

X-ray fluorescence spectrometry is a nondestructive method of analysis, but has not been used very widely for the analysis of plant material because of certain basic factors. There are the high capital cost of equipment, the relative slowness and insensitivity of the method, and the fact that relatively large samples are required for analysis. This latter factor is particularly disadvantageous for the analysis of plant ash, where samples tend to be small. X-ray fluorescence does have the advantage that the simultaneous determination of many elements can be carried out. Relatively precise values can be obtained for the concentration of the major constituents of rocks, soils, or plant ash. Table 18.1 lists the limits of detection of the elements by this method and demonstrates its poorer sensitivity than that of other methods.

The basis of XRF is the bombardment of the sample with **primary X-rays** produced from a tungsten or molybdenum X-ray tube. These primary X-rays expel electrons from the atoms in the sample, leaving a "hole." This vacancy is filled by electrons from higher energy levels and the net result is a characteristic X-ray emission spectrum. The **secondary x-rays** thus produced pass through a rotating crystal where, by Bragg's Law, only certain wavelengths are diffracted for a specific angle between the X-ray beam and the axis of the crystal. The diffracted X-rays are recorded by a detector (Geiger or scintillometer) rotated at twice the speed of the crystal. The output of the detector is plotted automatically as a function of the angle of rotation. From the magnitude of this angle, the identity of the element can be deduced and its concentration calculated from the intensity of the X-rays received by the detector.

As mentioned previously, XRF is hardly applicable for speedy analysis of a large number of samples. It can be useful, however, in providing a total analysis of major constituents of the material in a fraction of the time that would be required for "wet chemical" methods.

18.6. COLORIMETRY

With the advent of AAS and direct-reading emission spectrometers, there has been a rapid decline in colorimetric methods of analysis as applied to mineral exploration. Nevertheless, the pioneering work of Sandell and his co-workers [721] is still useful for the colorimetric determination of elements such as tin, tungsten, and so on, which are difficult to determine by many instrumental methods when their concentrations are low. There will be a continuing need for colorimetric methods in field analyses, although even here the demand will diminish because of the trend toward speedier analysis in service laboratories. The advent of the helicopter has meant that sample data can be obtained almost as quickly from the service laboratory as from *in situ* measurements.

18.7. RADIOMETRIC METHODS OF ANALYSIS

18.7.1. Neutron Activation Analysis

Neutron activation analysis is one of the better-known radiometric methods of analysis. The technique depends on irradiating a sample with a flux of particles such as neutrons, protons, or deuterons. The absorbed particle in most cases turns the original stable isotope into a radionuclide as in the following example:

$$^{75}_{33}\text{As} + {}^{1}_{0}n = {}^{76}_{33}\text{As} + \gamma$$

The arsenic isotope of mass 76 produced above has a half-life of 26.4 h and decays to give a stable selenium nuclide of the same mass number. A process such as this with absorption of a neutron and emission of gamma radiation, is known as an n, γ reaction. Limits of detection (assuming a 1 g sample) are shown in Table 18.1. As mentioned previously, detection limits are very difficult to assess because they depend on many factors other than the concentration of the analyte in the sample. These factors include irradiation times, cooling times, and the concentration of any interfering element or elements.

The concentration of the analyte is determined by comparing its radioactivity (usually gamma) with that of standards treated in exactly the same way as the analyte sample. Neutron activation analysis is perhaps overall one of the most sensitive of all analytical techniques. It does have a few disadvantages, however. The first of these is its relative slowness. Assuming a one-hour irradiation period and a counting time of 15 min, only about 20–30 samples a day can be analyzed. There is the additional problem of the capital cost of the equipment, which in 1982 was approaching $1,000,000. Against these disadvantages must be set the undeniable advantages, not only of sensitivity, but also of reliability and precision. For some elements such as uranium, there are virtually no other suitable methods for determining very low concentrations of this element with both precision and accuracy.

Perhaps one of the best examples of the use of NAA for biogeochemical prospecting is in the determination of gold at very low concentrations [315,680]. Other good examples are the determination of uranium in plant material [107,252].

18.7.2. Direct Measurement of Natural Radioactivity

The measurement of alpha activity in plants has been used in exploration work [21,321,926,927]. Whitehead and Brooks [926] compared the method with other techniques for detecting uranium and concluded that the alpha activity of plant ash is a reliable indicator of mineralization.

The beta activity of plants is a far less reliable indicator due to the presence of large concentrations of potassium-40 in plant ash. Figure 18.3 shows the alpha activity of plant ash and soils expressed as a function of the beta activity, which decreases to a limiting value of about 300–400 counts/100 min/25 mg as alpha activity decreases to zero [926]. Grodzinsky and Golubkova [333], in making the same observations, concluded that 45–71% of the beta activity of plant ash is due to radioactive potassium.

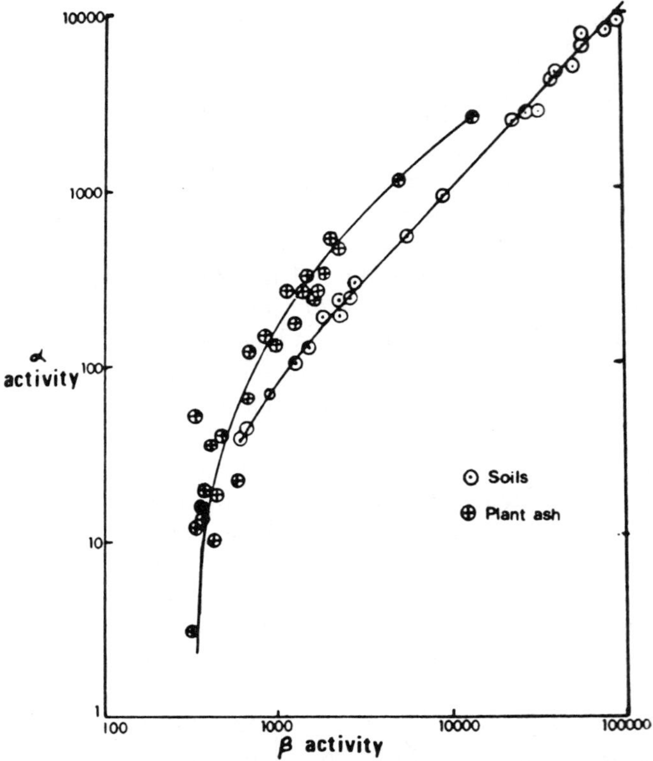

FIGURE 18.3. Alpha activity of plants and soils as a function of their beta activity. *Source:* Whitehead and Brooks [926].

18.8. INDUCTIVELY COUPLED PLASMA EMISSION SPECTROMETRY

Inductively coupled plasma emission spectrometry (ICP) is a relatively new technique and is finding increasing favor for chemical analysis of many trace elements in a wide variety of materials including plants. It incorporates a special source that depends on the interaction between the magnetic field of an oscillating radiofrequency current and the charged species present in the plasma. A stream of argon gas passed through the radiofrequency field produces a toroidal plasma in which temperatures of over 5000°C can be produced. The sample solution is introduced as an aerosol into the inert gas stream and is efficiently vaporized and atomized in the plasma. The high stability of the plasma permits precisions approaching 1%. The plasma source is combined with a monochromator and readout system similar to that of a direct-reading spectrograph and provides facilities for about 20 different channels.

Advantages of ICP are low limits of detection (Table 18.1) for most elements, coupled with an extremely wide range of usable concentrations. The procedure is very speedy and can process one sample every few minutes. A big disadvantage is the high capital cost of the equipment not only for the basic instrumentation, but also because of the necessity for on-board computer facilities to make corrections for extensive interferences which are a feature of the technique.

18.9. CHEMICAL FIELD TESTS

Field methods used for the analysis of plants, rocks, and soils are nearly all colorimetric and are essentially only semiquantitative. Many of the procedures were developed by the U.S. Geological Survey and at the Applied Geochemistry Department at Imperial College, London. Most techniques designed originally for soils can be used for plant ash with minor modifications. One method for doing this is to digest plant material with a 1:10 mixture of perchloric and nitric acids heated over an alcohol burner or gasoline stove. As an alternative, the samples may be ashed in the field (see Chapter 16) and the ash dissolved in dilute (2 M) hydrochloric acid. Field methods of analysis are justified primarily on the grounds of *in situ* determinations, but if the contrast between background and anomalous concentrations of a particularly analyte is small, field tests may not be sufficiently reliable to make the distinction.

As mentioned previously, field tests are becoming less popular because of increased costs and because of quicker turn-around of analytical data by modern service laboratories, coupled with speedier transport between the field and the laboratory. There is still a place, however, for field tests under some circumstances. My own research group uses test papers specific for nickel and cobalt, in which the vegetation sample is pressed between moistened sheets of this paper. In the case of nickel (using the reagent dimethylglyoxime) a red coloration indicates a nickel content in excess of 500 $\mu g/g$. There is no reason why other test papers specific for other elements should not be developed. Biological samples are particularly well suited for this type of test because the analyte element is usually in a soluble form. This is not so in soils and rocks, where some sort of preliminary acid digestion usually is required before a test can be made.

18.10. MOBILE LABORATORIES

It is obvious that there is a great difference between the quality of data obtained from field tests and those obtained from a properly equipped laboratory. The advantage of greater accuracy and precision in the first case is counterbalanced by the greater elapse of time needed for the data to be in the hands of the field worker. The solution to this problem is obviously to take the laboratory into the field.

Pioneering work in the development of mobile laboratories has been carried out in the Soviet Union [563], in the United States at the U.S. Geological Survey, and in Canada [298]. Debnam [239] has reviewed the development of mobile laboratories and suggested that miniaturized direct-reading spectrographs could be installed in them. This suggestion was made as far back as 1969, and these instruments are now a standard feature of such laboratories.

A mobile laboratory specifically designed for biogeochemical work in Canada has been described by Fortescue and Hornbrook [298]. The unit consisted of two 10-m house trailers, one for sample processing and the other for chemical analysis. The units were transported near the operating site on flat-top rail cars and then towed the rest of the way by truck or tractor. The trailers were stored in heated garages during the winter. In the biogeochemical laboratory, a crew of three was responsible for the collection, subsampling, and preparation of the samples; a second crew of three, was responsible for the analyses. Daily output of the laboratory was about 13 elemental concentrations in each of 30 samples using an optical spectrograph.

The U.S. Geological Survey (F. N. Ward, personal communication) uses truck-mounted spectrographic units housed in large van-type trucks and in smaller step-van vehicles. Electric power can be obtained from gasoline-powered generators or from a temporary installation by the local power company. Operations such as mineral separations, colorimetric tests, and atomic absorption analysis are also carried out in camper laboratories which are hauled to temporary operation sites in the field. In addition to housing a spectrographic laboratory, the step-van vehicle is used for sample preparation. At present these mobile laboratories are used largely to furnish analytical data to field projects connected with geochemical assessment programs.

Other advances in mobile laboratories have occurred in the area of data handling. At selected field locations, a portable terminal equipped with telephonic access to a centralized time-shared computer facility has been set up. Data can be fed into the computer by telephone and statistical parameters and graphical analysis can be transmitted back to workers in the field.

Mobile laboratories have been of value recently in the exploration of several areas with difficult access. For example, units equipped with AAS instruments have been shipped to sites in the Solomon Islands and have been flown to within a few hundred kilometers of the North Pole. The use of mobile laboratories will continue to increase because of the advent of rapid data acquisition during exploration programs and because of their convenience in the exploration of remote areas. The advent of ICP should make a big contribution to greatly improved mobile laboratory facilities. These instruments are already in use in helicopters (see Chapter 20), and their use in mobile laboratories seems a logical extension of this and has probably already been carried out.

18.11. A FINAL ASSESSMENT OF METHODS OF ANALYSIS OF VEGETATION

The success of a biogeochemical program, as well as any other project in mineral exploration, depends to a great degree on the reliability of the data and the speed at which they can be obtained. Economic factors are also a consideration in choosing a suitable analytical method. The problem facing the field worker or analyst is to consider these three variables and to select a technique that is the best compromise between speed, reliability, and cost.

There is little doubt that for very large numbers of samples, AAS, ICP, and direct-reading emission spectrometers will have a wide appeal. The capital costs of the last two instruments (particularly ICP) are very high, but so is their output. The cost of an optical spectrograph is much less than that of a direct-reading instrument, but its output is less, as well. AAS equipment requires a relatively modest capital outlay and is probably the most cost-effective procedure where only a limited number of elements are to be determined. The use of colorimetric procedures is probably still justified where instrumental facilities are lacking or where the number of samples is so small that a heavy capital outlay is not justified. At the time of writing (1982) the heaviest capital cost is for ICP instruments, but relative costs may fall as the instruments are developed further.

Apart from the criteria listed above, another important consideration in the selection of a suitable analytical procedure is its sensitivity and the range of concentration values to be expected. Ideally, the normal working range of an analytical method should be matched with the concentration range of the element in the sample. A lack of communication between the field worker and the analyst can result in the selection of an unsuitable procedure. For example, the cryptic message "do Mo by aa" given by a field worker to the service laboratory for samples containing less than 25 μg/g of this element would be a most unwise instruction, since all the data would be at or near the limit of detection of the method and at best would be unreliable. The correct choice in this case would probably have been ES or ICP.

It is hoped that this review of chemical methods of analysis applied to biogeochemistry will have been instrumental to some degree in bridging the gap between the field worker and the analyst. A meaningful dialogue between the two can only be of benefit for mineral exploration.

19

STATISTICAL INTERPRETATION OF DATA FROM BIOLOGICAL PROSPECTING METHODS

19.1. INTRODUCTION

A sound statistical interpretation of data is a *sine quae non* for most successful biogeochemical investigations. The advent of high-speed computer facilities has allowed increasingly sophisticated statistical procedures to be carried out. Such facilities have also been an important factor in statistical manipulation of the very large amounts of data that can now be provided rapidly by the newer methods of chemical analysis (see Chapter 18).

The biogeochemist of today is faced with the need to detect orebodies that are becoming increasingly difficult to find, and with the need to identify anomalies in vegetation that are less and less differentiated from the background. The following summary of statistical procedures is formulated largely with the needs of the exploration biogeochemist in mind, but is equally applicable to other geochemical methods of exploration. Definitions and symbols for the statistical terms used in the discussion are set out in Table 19.1.

19.2. POPULATIONS AND DISTRIBUTIONS

An existing distribution of data is known as a **population.** It is assumed that all components of this population are linked by some generic factor. For

Table 19.1. Statistical Notation

Symbol	Interpretation
n	Number of observations
μ	Population mean $= \dfrac{1}{N}\displaystyle\sum_{i=1}^{N} x_i$
$\bar{x}$	Sample mean $= \dfrac{1}{n}\displaystyle\sum_{i=1}^{n} x_i$
σ^2	Population variance $= \dfrac{1}{N}\displaystyle\sum_{i=1}^{N} (x_i - \mu)^2$
$\hat{\sigma}^2$	Best estimate of population variance calculated from sample $= \dfrac{1}{n-1}\displaystyle\sum_{i=1}^{n} (x_i - \bar{x})^2$
σ or $\hat{\sigma}$	Standard deviation $= (\sigma^2)^{1/2}$ or $(\hat{\sigma}^2)^{1/2}$
ϕ	Number of degrees of freedom on which statistic is based
H_O	Null hypothesis
P	Probability

example, they may be biogeochemical abundance data for vegetation samples drawn from a particular geological substrate. Such samples may represent one population only, or they may represent several populations within a particular biogeochemical province. For example, if copper is measured in vegetation in the vicinity of an orebody, some values may belong to the background population and others to a population growing over the dispersion halo of the orebody. There may well be a third population representing plants growing on the orebody itself. A knowledge of the **distribution** of the population is of importance in interpreting biogeochemical data, as the reliability of many statistical procedures is dependent on the nature of these distributions.

19.2.1. Normal and Lognormal Distributions

Nearly 30 years ago Ahrens [9] suggested that the concentrations of most trace elements in felsic and mafic rocks are distributed lognormally. His original paper provoked a lively discussion [33,183,590,752,854] which still persists today. Assumptions about the nature of the distribution are important in expressing the "average" composition of rocks, soils, and biological materials. If all the data for a particular element are plotted as a histogram with linear coordinates, a symmetrical or a skewed distribution arises if only one population is present. If the histogram is symmetrical, the distribution is said to be normal and the median, arithmetic mean, and mode of the data are coincident. If the histogram is asymmetric with a pronounced skew, it is possible that a normal distribution will be obtained if the logarithms of the abundance data are used. The original distribution is then said to be lognormal. In such a case the median of the data corresponds to the geometric mean.

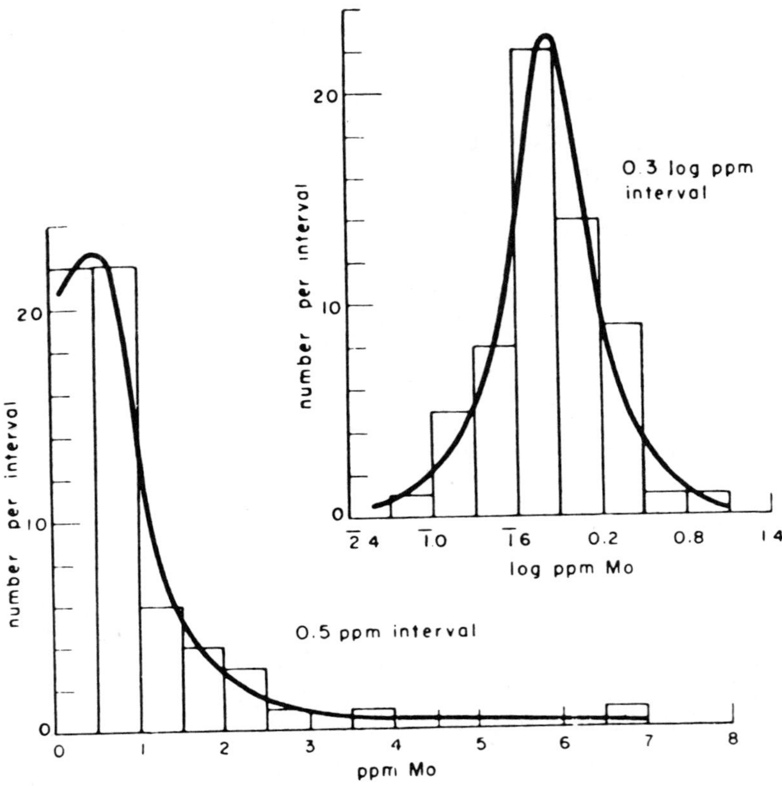

FIGURE 19.1.   Molybdenum concentrations in granites illustrating a lognormal distribution. *Source:* Ahrens [9].

Figure 19.1 shows two histograms from a single lognormal distribution plotted on linear and logarithmic scales. Alternatively, a simple calculation can replace this comparison of histograms. Agreement between the median and arithmetic mean indicates normality. For the data shown in Figure 19.1, the median is 0.70 μg/g and the arithmetic and geometric means are, respectively, 1.2 and 0.95 μg/g. It is clear that the data approximate more closely to a lognormal distribution.

The most important reason for deciding the nature of a distribution is that the most sensitive statistical procedures (such as those for deciding whether two sets of samples are taken from the same or different populations) assume that the data are distributed normally. If the data are lognormally distributed, the statistical procedure should be applied to the logarithms of the elemental concentrations.

19.2.2.   Mixed Populations

Homogeneous populations are somewhat rare in biogeochemical data collected in the vicinity of an anomaly. In a mineralized area, there may be

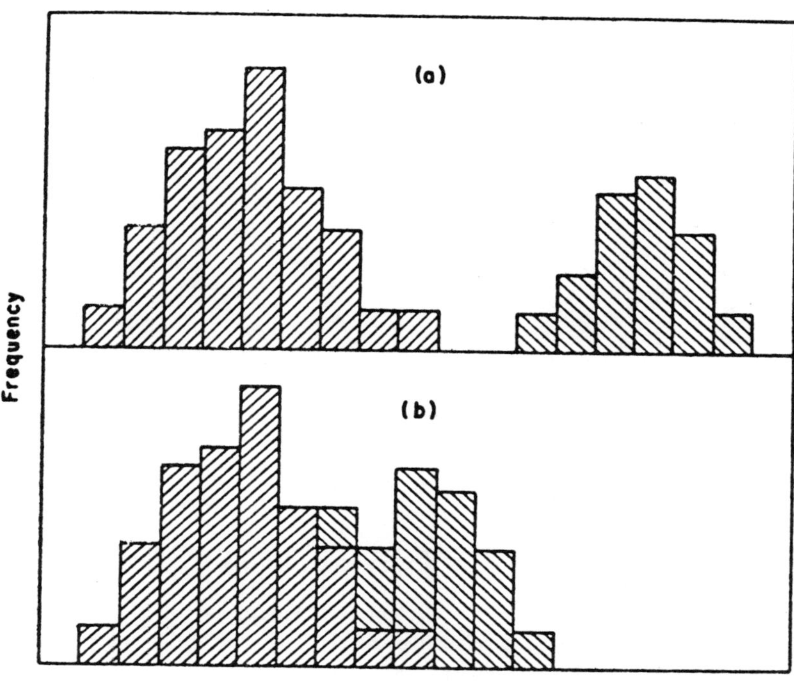

FIGURE 19.2. (*a*) Histogram for separate populations representing anomalous and background concentrations of an element in vegetation. (*b*) The same as (*a*) but with overlap of the two populations. *Source:* Brooks [97].

two or more different populations of an elemental concentration in plant or soil samples. There probably will be a population representing the normal background values, and another representing anomalous values from an orebody or its dispersion halo.

It is common practice to use histograms for biogeochemical data to indicate the concentration that should be taken as anomalous for the area. This procedure is quite straightforward when the two populations are quite separate, as shown in Figure 19.2*a*. In Figure 19.2*b*, however, the two populations are nearly coincident, and there is some difficulty in deciding which concentration levels should be taken as representing mineralization and which should indicate the threshold values between the background and anomalous values.

It has been shown [800] that if the cumulative frequency of the concentrations is plotted on probability paper, each population will be represented by a straight line or curve. The intersection of the lines or curves represents a level that might be taken somewhat arbitrarily as being the threshold level between background and anomaly. On this plot a single distribution appears as a straight line if its type (normal or lognormal) is matched with the type of scale (linear or logarithmic) used as the ordinate of the probability paper.

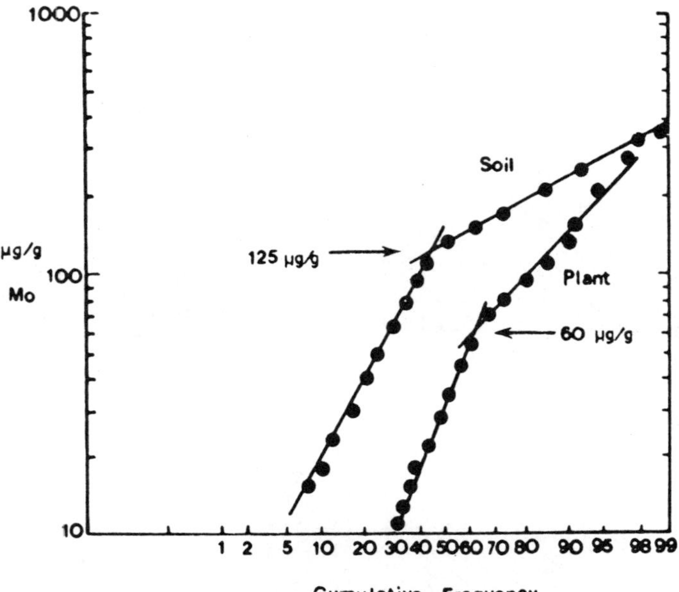

FIGURE 19.3.   Threshold values for plant and soil anomalies for molybdenum obtained from cumulative frequency plots on probability paper. *Source:* Brooks [97].

Figure 19.3 shows a cumulative frequency plot of molybdenum concentrations found in a study of soils and plants in a mineralized area. Two separate lognormal distribution are evident in both cases.

The intersection of two cumulative frequency lines or curves does not necessarily indicate a sharp discontinuity between two populations, and there can be considerable overlap on each side of the intersection. The two distributions may be separated [171] and the degree of overlap calculated. Williams [943] has given worked examples of this method of separation. A more detailed interpretation of cumulative frequency curves and their use in mineral exploration has been given by Sinclair [763].

In using cumulative frequency plots, the data are arranged in increasing (or decreasing) order and points are plotted with the cumulative frequency (expressed as a percentage of the total number of samples) along the x axis and the concentration values along the y axis. If there is a large number of data points, the data may be grouped and the highest number in each group cited on the plot.

19.3.   PARAMETRIC AND NONPARAMETRIC METHODS

Statistical methods may be classified into two groups, **parametric** and **nonparametric,** according to the nature of the data used. Parametric methods use raw data (e.g., elemental concentrations, masses, particle sizes, etc.), whereas nonparametric methods use the ranks or other classifications of the raw data. Univariate or multivariate tests, which are discussed in this chapter, may be carried out on either type of data.

Parametric methods of statistical analysis are often only valid if the data are distributed normally. It is therefore necessary to know the type of distribution beforehand and to transform to logarithms if necessary. Nonparametric methods are independent of the nature of the distribution but seldom have the same degree of predictive power at a given confidence limit. Therefore they require more samples to give the same degree of significance as a parametric procedure.

The various measurement scales that may be used for parametric and nonparametric data have been described by Marshall [579] and Siegel [761].

19.4. SIMPLE TESTS OF SIGNIFICANCE

19.4.1. The Null Hypothesis and the Concept of Probability

A frequent problem in biogeochemistry is the decision as to whether a given set of data originates from a common source or from several sources. This problem arises because it is hardly ever possible to measure an entire population. Instead, reliance has to be made on samples from this population. Elemental concentrations measured in these samples vary, partly because of the process of sampling from an inhomogeneous material and partly because of experimental error involved in the methods of measurement.

As an example of the above, we may consider the case where plants from two adjacent areas were analyzed for copper. The means of the two groups were 23.9 and 27.0 μg/g, respectively. One person might judge that the groups represent different populations because the means seem to be appreciably different, whereas another might have decided that they were essentially similar. By using sample tests of significance, judgments can be placed on a quantitative basis and their reliability stated in quantitative terms.

The first step in applying a test of significance is to adopt a **null hypothesis** (H_0). With the null hypothesis the assumption is made that a parameter being considered does not differ from some specified value. A determination is then made of the probability of obtaining the specified value by chance. Taking the above example of copper values in plant material, the null hypothesis would be that there is no significant difference between the means, and that both groups are derived from the same population. We can now calculate the probability (P) that the above statement is true; that is, that there is no significant difference and that H_0 is accepted. Suppose that the probability is very small (e.g., 0.001 = 0.1%). Then two conclusions may be drawn: either (a) the 1 in 1000 chance has occurred and H_0 is true, or (b) H_0 is false.

Before the data analysis is begun, the risk we are prepared to take of wrongly asserting a difference should be decided. The level of this risk will vary according to the nature of the experiment being carried out.

Table 19.2.   Levels of Significance

Probability (P)	Conclusion	Symbol
> 0.1	Not significant	NS
0.1–0.05	Possibly significant, some doubt cast on H_0	PS
0.05–0.01	Significant, H_0 is rejected, though with reservations	S
0.01–0.001	Highly significant, H_0 confidently rejected	S^*
< 0.001	Very highly significant, H_0 very confidently rejected	S^{**}

Source: Brookes et al. [96].

As a general rule in the geological and biological sciences, the null hypothesis is rejected only if there is a less than 10 percent probability ($P < 0.1$) of its true being. Brookes et al. [96] have listed the generally accepted levels and divisions for tests of significance (Table 19.2).

19.4.2.   The *F* Test

The F test is used to make a comparison of variability. For example a biogeochemist might want to assess which of two species growing in a test area had the more variable copper content or he might want to compare the variability of two laboratories analyzing the same sample supplied as replicates with different code numbers (an old trick to weed out dubious service laboratories). We may therefore imagine that a single sample of plant ash or soil is divided into 20 replicates and that one set is sent to each of two laboratories.

The F test is based on the ratio of the variances (see Table 19.1) of two sets of data. The variance estimate (σ^2) is given by

$$\hat{\sigma}^2 = \frac{1}{n-1} \sum_{i=1}^{n} (x_i - \bar{x})^2 \tag{19.1}$$

although the equivalent expression

$$\hat{\sigma}^2 = \frac{1}{n-1} \left[\sum_{i=1}^{n} x_i^2 - \frac{1}{n} \left(\sum_{i=1}^{n} x_i \right)^2 \right] \tag{19.2}$$

is often more convenient to calculate.

Suppose we have two variance estimate $\hat{\sigma}_1^2$ and $\hat{\sigma}_2^2$, based on ϕ_1 and ϕ_2 degrees of freedom, respectively. We adopt the null hypothesis H_0 that $\hat{\sigma}_1^2$ and $\hat{\sigma}_2^2$ estimate the same population variance, and then test the probability that $\hat{\sigma}_1^2 > \hat{\sigma}_2^2$, where $\hat{\sigma}_1^2$ is the larger estimate of variance. The greater the ratio of the two variance estimates, the less likely that H_0 is true.

Suppose the results of the above testing of laboratories give the following

data for copper levels in replicates of a single plant ash sample: Lab. A—
20.8, 19.5, 22.1, 21.0, 21.8, 17.4, 21.5, 18.2, 19.0, 18.0 μg/g. Lab. B—17.5,
18.8, 19.3, 18.9, 19.2, 20.0, 18.8, 19.5, 19.2, 19.6 μg/g. The variances are $\hat{\sigma}_A^2$
= 2.971 and $\hat{\sigma}_B^2$ = 0.451. Hence

$$F = \frac{\hat{\sigma}_A^2}{\hat{\sigma}_B^2} (\phi_A, \phi_B) = \frac{2.971}{0.451} (9, 9) = 6.58(9, 9)$$

where ϕ_A refers to the set of data with the greater variance. Reference to
Appendix I shows that the experimental value for F (6.58) exceeds the value
(5.35) that corresponds to the 1% probability level (P = 0.01) and the null
hypothesis is rejected confidently. There is clearly a highly significant (S^*)
(see Section 19.4.1) difference in the results produced by the two laboratories; that is, there is less than a 1% chance that there is no difference in
variability.

The above use of the null hypothesis points to a somewhat illogicality in
this statistical concept. Most people would instinctively prefer to have asked
the question, "What is the probability that there *is* a relationship?" rather
than the converse negative proposal. We have the situation that a low value
of P implies a high probability that there is indeed a difference.

19.4.3. Student's *t* Test

If we return to the case of the two groups of plant samples analyzed for
copper above, we may now ask ourselves the question whether there is a
difference between the means and, if so, with what degree of confidence
can we assert that this difference exists. This is answered by use of the
well-known Student's *t* test. The following layout is based on an example
by Hinchen [353].

The concentrations of copper (μg/g dry weight) in plants from two adjacent areas were as follows: Population A—25.4, 26.1, 20.9, 18.4, 27.1, 30.9,
21.2, 23.4, 22.4, 23.7; Population B—27.1, 26.3, 28.9, 27.7, 29.3, 20.7, 21.6,
28.9, 29.8, 30.1.

1. Calculate the respective means ($\bar{x}_1$ and $\bar{x}_2$). These are 23.94 and 27.04
   respectively.
2. Calculate the standard deviations ($\hat{\sigma}_1$ and $\hat{\sigma}_2$) for each set from the
   formula

$$\hat{\sigma} = \left[\sum \frac{(x_i - \bar{x})^2}{n - 1} \right]^{1/2} \tag{19.3}$$

where n is the number of samples in a population and $x_i - \bar{x}$ is the
deviation of each sample value from the appropriate mean. In this
case, the standard deviations are 3.58 and 3.33, respectively.

3. Use the F test to verify that the two sets of data have the same variance. Only if the F test shows the variances to be insignificantly different, can the t test be used to examine the significance of any differences in the means. In the present example, the variances are not significantly different.

4. Obtain a better estimate of the variance, σ_p^2 by pooling the information from the two sets of data as follows:

$$\sigma_p^2 = \frac{(n_1 - 1)\hat{\sigma}_1^2 + (n_2 - 1)\hat{\sigma}_2^2}{n_1 + n_2 - 2} = 11.97 \qquad (19.4)$$

5. Calculate Student's t for the comparison of the two means using

$$t = \frac{\bar{x}_1 - \bar{x}_2}{\sigma_p(1/n_1 + 1/n_2)^{1/2}} = -1.99 \qquad (19.5)$$

This value of t is based on $(n_1 + n_2 - 2)$ degrees of freedom.

6. Refer to Appendix 2 and note that a value of $t = -1.99$ gives a probability between 10% and 5% for 18 degrees of freedom (ϕ).

We may now assert that H_0 is rejected at the 10% level and that there is a possibly significant (PS) difference between the means of the two sets of data. However there remains the possibility ($> 5\%$) that the observed difference in the means has occurred by chance from populations with the same mean.

The significance of a given value of t depends on the number of degrees of freedom involved. A value of $t = -1.99$ arising from a situation where $\phi = 120$ (i.e., two sets of 61 analyses), implies that the means are significantly different (S) and that the probability of H_0 being true is less than 0.05.

This test is only valid if the data are distributed normally. If the raw data have a lognormal distribution, they should be converted to logarithms before making the calculations. The t test is nevertheless comparatively "robust"; that is, small deviations from normality can be permitted without greatly affecting the validity of the conclusions.

19.4.4. The Chi² χ^2 Test

The χ^2 test compares the variation of a sample of n random values (normal distribution) with the variance of the distribution:

$$\chi^2 = \sum_{i=1}^{n} \frac{(x_i - \mu)^2}{\sigma^2} \qquad (19.6)$$

Where data have been classified into groups, we may compare frequencies observed in these groups (f_0) with the expected frequencies (f_e) based on the normal distribution. To do this an alternative version of the test is used:

$$\chi^2 = \sum \frac{(f_0 - f_e)^2}{f_e} \qquad (19.7)$$

where the summation is carried out over all groups.

The Chi^2 test is not used very often in the course of prospecting work. However, it can be used to test whether or not a given distribution is normal or lognormal. The data are divided into class (i.e., concentration) intervals. The observed number is then compared with the theoretical number for a normal distribution (obtained from standard tables) and is calculated as above. If a significant difference is shown to exist, the data may then be converted to logarithms and the calculation repeated to see if a better fit results. Tables for the Chi^2 test are given in Appendix 3. More comprehensive tables of the normal distribution, F, t, and χ^2 distributions can be found in standard statistical works of reference [66, 293, 645].

19.5.   BIVARIATE ANALYSIS

19.5.1.   Introduction

Relationships between two variables are studied by the techniques of **regression** and **correlation.** In regression we study the relationship by expressing one variable (y) in terms of a linear (or more complex) function of the other (x). This functional relationship finds particular use in (a) reporting hypotheses about the possible causation of changes in y by changes in x, (b) predicting values of y for given values of x, and (c) using changes in x to explain some of the observed variations in y. In correlation analysis we study principally the extent to which the two variables are associated, or vary together, without any suggestion that one of the variables is dependent on the other.

19.5.2.   Regression Analysis

If the assumption can be made that one variable (x) is independent and is measured without error, and the other variable (y) is dependent and subject to error, it is possible to seek a relationship between the two by calculating a **least squares line** through the data points. In the simplest case, that of a linear relationship, this is done by comparing the data points (x_i, y_i) with a line of the form

$$y = bx + a \qquad (19.8)$$

and minimizing the sum of the squared deviations of the dependent variable.

Table 19.3.   Calculation of the Regression Line for a Plot of Nickel Concentrations (μg/g) in Plant Ash as a Function of the Nickel Content (μg/g) in the Soil

x_i	y_i	x_i^2	x_iy_i
10	15	100	150
20	30	400	600
30	45	900	1350
40	30	1600	1200
45	45	2025	2025
50	55	2500	2750
55	65	3025	3575
65	45	4225	2925
70	70	4900	4900
70	75	4900	5250
$\Sigma x_i = 455$	$\Sigma y_i = 475$	$\Sigma x_i^2 = 24575$	$\Sigma x_iy_i = 24725$

Source: Brooks [97].

This quantity, $\Sigma (y_i - y)^2$, is minimized by partially differentiating it with respect to a and b and equating each derivative to zero.

The parameters a and b are found to be

$$b = \frac{n \sum x_iy_i - \sum x_i \sum y_i}{n \sum x_i^2 - (\sum x_i)^2} \qquad (19.9)$$

$$a = \frac{1}{n} (\sum y_i - b \sum x_i) = \bar{y} - b\bar{x} \qquad (19.10)$$

The slope of the line, b, is decreased as the **regression coefficient** for the regression of y on x and the function 19.8, with the least squares value of a and b inserted, is the regression equation.

As a worked example it may be seen how regression analysis can be used to calculate the regression line for a set of biogeochemical data representing the nickel content of plants, expressed as a function of the levels of this element in the supporting soils. The data are shown in Table 19.3.

If values from the table are substituted in Equations 19.9 and 19.10, the following results are obtained:

$$b = \frac{(10 \times 24{,}725) - (455 \times 475)}{(10 \times 24{,}575) - (455)^2} = 0.80$$

and

$$a = 47.5 - (0.80 \times 45.5) = 11.1$$

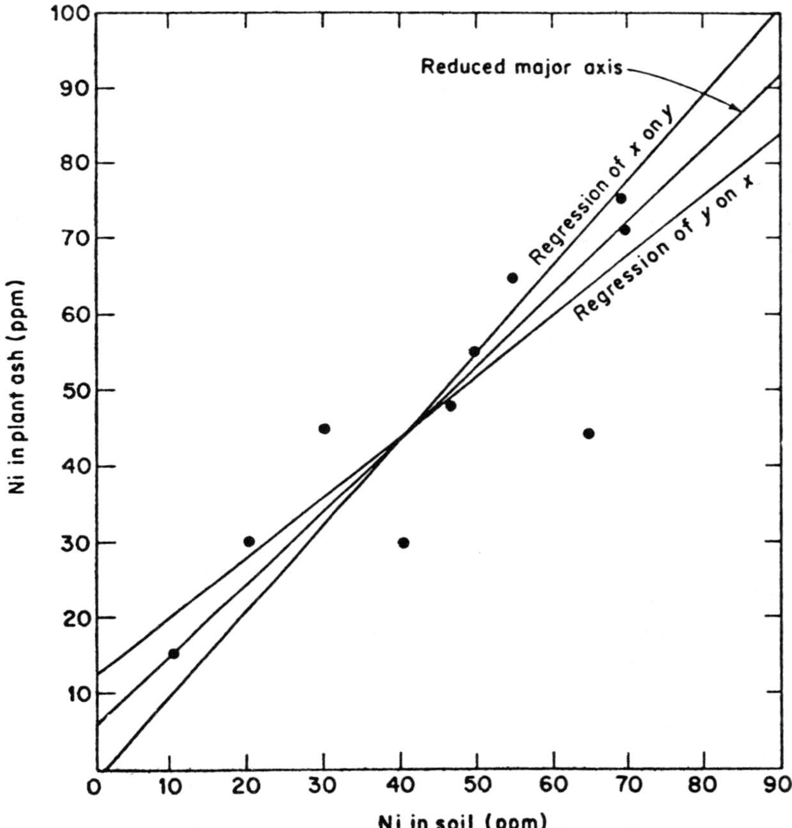

FIGURE 19.4. Regression lines and reduced major axis for plot of nickel concentration in plant ash (*y* axis) as a function of the nickel content of the soil (*x* axis). *Source:* Brooks [97].

hence

$$y = 0.80x + 11.1$$

This line (the regression of *y* on *x*) is shown in Figure 19.4, which is a plot of the data of Table 19.3.

To obtain the regression of *x* on *y* it is only necessary to interchange the values of *x* and *y* in Table 19.3 and recalculate as before. This gives the following equation:

$$x = 0.91y + 0.77$$

This line is also shown in Figure 19.4.

The two regression lines are seldom the lines that would have been drawn by visual assessment. There is, however, a fallacy in the use of regression lines for the interpretation of plant-soil data. The assumption is made that x is an independent variable free of error. If x represents elemental concentrations in soils, there is no reason to suppose that soil values are any more free of error than are those for vegetation. Soil data are subject to analytical as well as many other sorts of error.

The problem of this deficiency in regression analysis applied to data has been discussed by Imbrie [379] who has recommended that use should be made of the **reduced major axis** in placing a line through a set of points. This, according to Imbrie "(i) makes no assumption of independence; (ii) is invariant under change of scale; (iii) is simple to compute; and (iv) produces results intuitively more reasonable than corresponding results obtained using regression analysis." Figure 19.4 also shows the reduced major axis which lies between the two regression lines, and is probably closer to the line that would have been drawn by visual assessment only.

19.5.3. Correlation Analysis

In the previous subsection it was shown how a regression line can be fitted to a series of points on a graph. One vital question remains unanswered, however: How closely do the points cluster around the line? In other words, how good is the linear relationship at predicting y from x? This is a very important question in biogeochemical prospecting, where concentrations of elements in plant material often have to be compared with values in the substrate.

A good approach to the problem is use of the so-called Pearson product moment correlation coefficient (r). This may be expressed as follows:

$$r = \frac{\Sigma xy - (\Sigma x \Sigma y)/n}{\{[\Sigma x^2 - (\Sigma x)^2/n][\Sigma y^2 - (\Sigma y)^2/n]\}^{1/2}} \qquad (19.11)$$

The following is a worked example for a calculation of r from the data in Table 19.3. The number of samples is fewer than normally would be employed, but space limitations preclude use of more data. The method of calculation is based in Hinchen [353] as follows:

Calculate Σx ($=455$), Σy ($=475$), Σx^2 ($=24{,}575$), Σy^2 ($=25{,}875$), Σxy ($=24{,}725$), $(\Sigma x)^2/n$ ($=20{,}700$), $(\Sigma y)^2/n$ ($=22{,}560$), and $(\Sigma x \Sigma y)/n$ ($=21{,}610$). Substituting these values into Equation 19.11 we obtain the relationship

$$r = \frac{24{,}725 - 21{,}610}{[(24{,}575 - 20{,}700)(25{,}875 - 22{,}560)]^{1/2}} = 0.87$$

The correlation coefficient can vary between the limits of $+1$ and -1. Positive values of r indicate that the two variables are directly related; negative values imply an inverse relationship. Values approaching $+1$ or

Table 19.4. Correlation Coefficients for the Relationship Between Copper, Lead, and Zinc Concentrations (Logarithmic Coordinates) for Plants and Soils at Te Aroha, New Zealand

Species	Organ	State	No.	Copper		Lead		Zinc	
				r	Sig.	r	Sig.	r	Sig.
Schefflera digitata	Leaves	Dried	36	0.05	*NS*	0.27	*NS*	0.45	*S**
	Leaves	Ashed	36	−0.03	*NS*	0.28	*NS*	0.41	*S*
Beilschmiedia	Leaves	Dried	35	−0.05	*NS*	0.26	*NS*	−0.04	*NS*
tawa	Leaves	Ashed	35	−0.03	*NS*	0.31	*PS*	0.06	*NS*
	Wood	Dried	28	0.26	*NS*	0.45	*S**	0.22	*NS*
	Wood	Ashed	28	0.24	*NS*	0.48	*S**	0.29	*NS*

Source: Nicolas and Brooks [618].

−1 indicate increasing linearity between the two variables, whereas 0 implies no linear relationship whatsoever.

In the above calculation, a value of 0.87 was obtained for r and implies that a direct relationship exists. To establish the degree of confidence that may be placed on the assertion that a relationship does indeed exist, the probability value should be obtained from Appendix 4 for n-2 degrees of freedom, where n is the number of pairs of data. For 8 degrees of freedom, as in the above example, the value lies just above 0.001. This is the probability that a relationship does not exist and implies that there is almost a 99.9% chance that there is a linear correlation and that this is not due to random choice. Examination of Appendix 4 will show that the significance of r is a function of the number of degrees of freedom, and that in this respect it is similar to the t test.

Correlation coefficients and levels of significance are important statistical parameters for biogeochemical prospecting. This is illustrated in Table 19.4. The table gives a great deal of information. It appears that the best sample type for determining zinc in vegetation would be the leaves of *Schefflera digitata* expressed on a dry weight basis, whereas for lead, the best results would have been obtained with the wood ash of *Beilschmiedia tawa*. Neither species was suitable for copper. A common error in presenting data of this nature is to report correlation coefficients only. The numerical values of these coefficients are not even necessary. The values of the probability function P, and perhaps a symbol of significance such as S or S^* (see Table 19.2) are all that is required.

The correlation coefficient may only be calculated from normally distributed data. If either or both sets of variables have a lognormal distribution, they should be transformed logarithmically before making a calculation. The sort of erroneous result that can be obtained by not making this transformation is illustrated in Figure 19.5, which shows hypothetical values for the concentration of an element in plants and soils. There are 10 points

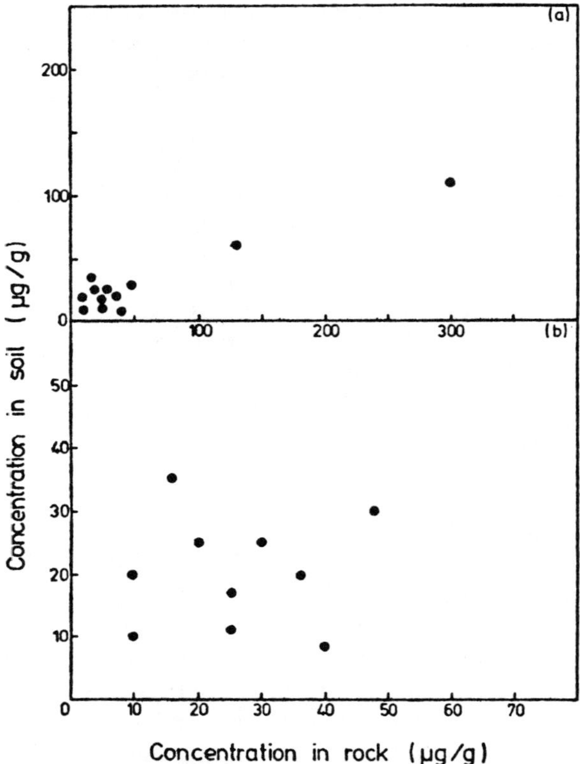

FIGURE 19.5. Hypothetical values for concentrations of an element in plant material as a function of the concentration of the same element in the soil. (*a*) Includes two abnormally high values and gives a highly significant value for the correlation coefficient between the two variables. (*b*) Ignores the two high values and gives a value for the correlation coefficient which indicates a nonsignificant relationship. *Source:* Brooks [97].

clustered near the origin in plot *a* and two with much higher values of both variables. Statistically this distribution behaves as virtually three data points with the cluster acting as a single point. The value for *r* is very high (0.95). If the two high values are omitted as in plot *b*, the value of *r* is reduced to a nonsignificant 0.06. The effect of outliers of this kind is reduced by a logarithmic transformation. Even then, however, a few points might still appear as outliers and are better omitted. It is far better to err on the conservative side than to accept uncritically, data that only appear to be favorable.

If an approximately normal distribution of the data cannot be achieved even with logarithmic transformation, use may be made of the Spearman rank correlation coefficient (r_S). This is a nonparametric parameter and is not influenced by the type of data distribution. The mode of calculation is as follows:

1. The pairs of data are arranged in two separate columns, with the values in the first column ranked with the highest values at the top.
2. The values in the second column are then so arranged that they match horizontally with those of the first column. If there is a complete correlation, then the values in the second column will also be ranked perfectly, with the highest values first.
3. Calculate the value of r_S from the equation

$$r_S = 1 - \frac{6\sum d^2}{n^3 - n} \tag{19.12}$$

where n is the number of pairs of values and d is the difference of rank of each horizontal pair in the two columns. When there are tied values in either column, a modified formula must be used:

$$r = \frac{A + B - \sum d^2}{2(AB)^{1/2}} \tag{19.13}$$

where

$$A = \frac{n^3 - n}{12} - \sum T_a \qquad B = \frac{n^3 - n}{12} - \sum T_b \tag{19.14}$$

The terms T_a and T_b are corrections for the presence of ties in columns a and b, respectively. If t is the number of values tied for any given ranking.

$$T = \frac{t^3 - t}{12} \tag{19.15}$$

The summations are carried out for all ties in each column.

The following is a worked example using the data from Table 19.3. The value of r_S hence obtained will be directly comparable with that obtained for r above.

The data in the first column of Table 19.3 are ranked and the results are 1, 2, 3, 4, 5, 6, 7, 8, 9.5, 9.5. The last two values represent a tie, and in such a case it is customary to assign to each value the means of the ranks that would have been assigned if no tie had existed. The ranking of the second column as a consequence of rearrangement of the first then gives values of 1, 2.5, 5, 2.5, 5, 7, 8, 5, 9, 10. Note that in this column there is one tie of two values and one of three. Hence,

$$A = \frac{10^3 - 10}{12} - \frac{8 - 2}{12} = 82$$

and

$$B = \frac{10^3 - 10}{12} - \frac{8 - 2}{12} - \frac{27 - 3}{12} = 80$$

The rank differences for each pair taken horizontally across the columns are:

$$0,\ -0.5,\ -2,\ 1.5,\ 0,\ -1,\ -1,\ 3,\ 0.5,\ -0.5$$

From this,

$$\sum d^2 = 18$$

Hence,

$$r_S = \frac{82 + 80 - 18}{2(82 \times 80)^{1/2}} = 0.89$$

The above value of 0.89 for r_S compares with 0.92 for the Pearson coefficient. The significance of the Spearman parameter is, however, only 91% that of r. If the data of Appendix 4 are used to obtain significance values, and if, for example, 100 degrees of freedom were involved, a value of 0.321 for r, should be equivalent to 0.337 for r_S (91 degrees of freedom). Appendix 4 can therefore be used for rank correlation coefficients merely by reducing the number of degrees of freedom to 91 percent. Although the Spearman coefficient does not depend opn the type of distribution of the data, this advantage is somewhat offset by the fact that more computer time is needed for the calculation because of the time required for ranking. The rank coefficient is particularly useful when only a small number of data pairs is used, so that the true type of distribution cannot be ascertained with much certainty.

19.6. MULTIVARIATE ANALYSIS

19.6.1. Multiple Regression and Discriminant Analysis

In many cases it is appropriate to attempt to express a dependent variable y in terms of more than one independent variable: $x_1, x_2, \ldots, x_m$. One general model frequently used is that in which data are fitted to an equation of the form

$$y = a_1x_1 + a_2x_2 + a_3x_3 + \cdots + a_mx_m + c \qquad (19.16)$$

An estimate of a coefficient a_i in the regression model can be found from the rate of change of the dependent variable with respect to variable x_i, with all other variables being held constant. These estimates, the sample partial regression coefficients, are generally different from the simple regression coefficients of y on x. The multiple regression can be expected to account for more of the variation in y than does any simple regression. It should be noted that, because the variance of only one variable (y) is subjected to analysis, multiple regression is multivariate only in the sense that several variables are measured on each sample.

As in the two-variable case, the method of least squares can be used to provide the best estimates of the parameters a_1, a_2, a_3, . . . , a_m, and c. By minimizing the sum of the squared deviations of the observed values of y from the line, a set of $m + 1$ simultaneous equations is obtained.

Computational procedures for dealing with situations where m is large are discussed in several works on statistics [234, 267]. One such procedure is described as stepwise regression [267, 587]. At each stage of the computation, this procedure selects the independent variable that increases by the most significant amount the variance of y that is explained by the regression equation. If, and only if, this amount is more significant than a predetermined probability level, this independent variable is inserted into the regression equation. Following the insertion of a variable into the regression equation, the other variables already in the equation are rechecked to find if they are still significant at the predetermined level of significance. If not, they are removed from the equation. On completion of the analysis, the percentage of the variance in the dependent variable that has been explained by the regression equation is computed.

An example of a stepwise regression procedure as applied to biogeochemical data has already been given in Chapter 14, and was used to assess factors affecting the BAC of ashed leaves of *Nothofagus fusca* during the search for nickel and copper in New Zealand [808]. Another example was referred to briefly in Chapter 3 and was used for geobotanical work in Australia [620]. It involved using a linear combination of several variables and selecting values for the constants a_1, a_2, and so on (see Equation 19.16), so that the differences between the various groups were maximized. The degree of discrimination (discriminant analysis) can be tested by computation of the Mahalanobis D^2 statistic [540]. This statistic is calculated from the means, variances, and covariances of the variables measured in replicate samples from the various groups. It is, in effect, a measure of the "distance" between the groups. The larger the value of D^2, the greater the differences between the groups. The same pocedure was used for a biogeochemical survey in the same area. Details are as follows:

Bark samples of *Eucalyptus lesouefii* were sampled in a belt transect across amphibolites and ultrabasic rocks at Spargoville, Western Australia.

**Table 19.5.   Values for the D^2 Statistic and the Associated Degree of
Discrimination for Elemental Concentrations in the Bark of *Eucalyptus Lesouefii*
Used to Predict the Nature of the Substrate**

Variables Used	D^2	Number of Correct Predictions	
		A (34 sites)	UB (29 sites)
Ca	0.122	17	17
Co	1.66	23	15
Cr	9.36	24	14
Cu	2.18	24	15
Mg	0.789	24	13
Mn	0.165	21	10
Ni	3.15	24	14
Pb	1.22	20	13
Zn	3.76	26	14
Ca, Co, Cr, Cu, Mg, Mn, Ni, Pb, Zn	16.76	29	17
Cr, Ni	10.14	27	15
Cr, Ni, Zn	10.62	25	14
Co, Cr, Ni	13.49	26	16
Cr, Cu, Ni	10.17	27	15
Cu, Ni	3.56	23	16
Cr, Mn, Ni	10.15	27	14
Cu, Cr, Ni	11.93	27	13
Cr, Mg, Ni	13.92	25	18
Cr, Ni, Pb	10.51	27	16
Co, Cr, Ni, Pb	13.49	26	16
Cr, Mg, Ni, Pb	14.17	25	19
Co, Cr, Mg, Ni, Pb	16.54	29	16
Cr, Cu, Ni, Zn	10.63	25	14
Cr, Cu, Mn, Ni	10.17	27	15
Ca, Cr, Cu, Ni	11.93	27	13
Cr, Cu, Mg, Ni	14.13	26	17
Co, Cr, Cu Ni	13.68	27	17
Cr, Mg, Mn, Ni	14.03	25	18
Cr, Cu, Mg, Mn, Ni	14.15	26	18
Ca, Cr, Cu, Mg, Mn, Ni	14.20	26	17
Cr, Cu, Mg, Mn, Ni, Zn	14.34	26	18

Source: Nielsen et al. [620].
KEY: A—amphibolites
        UB—ultrabasics

The samples were then analyzed for calcium, chromium, cobalt, copper, lead, magnesium, manganese, nickel, and zinc. The object of the experiment was to see whether or not the chemical composition of the bark indicated the geological nature of the substrate.

In the case where only two populations are considered, the Mahalanobis D^2 statistic is given by

$$D_p^2 = \sum_1^p \sum_1^p W^{ij}(\bar{x}_{i1} - \bar{x}_{i2})(\bar{x}_{j1} - x_{j2}) \qquad (19.17)$$

where $\bar{x}_{i1}$ and $\bar{x}_{i2}$ are the sample means for the ith character for the first and second samples, respectively; (W^{ij}) is the reciprocal of the covariance matrix (W_{ij}); and p is the number of variables used.

To test the hypothesis that there is no difference in mean values of the p characters for the two populations, the following statistic was used as a variance ratio with p and $(n_1 + n_2 - 1 - p)$ degrees of freedom:

$$F = \frac{n_1 + n_2 - p - 1}{(n_1 + n_2 - 2)} \frac{n_1 n_2}{n_1 + n_2} - D^2 \qquad (19.18)$$

where n_1 and n_2 are the number of samples in each population. When $p = 1$, the D^2 statistic reduces to a simple t test.

Table 19.5 summarizes the results using different combinations of the elemental concentration variables. The best discrimination of ultrabasic rocks was obtained from the determination of chromium, magnesium, nickel, and lead in the plant tissue. The degree of discrimination using only four elements was slightly better than when all the elemental concentrations were used. Taking both rock types together, the highest number of overall correct predictions (36 out of 63) was achieved using all of the variables. In this case, 29 out of 34 amphibolites and 17 out of the ultrabasic sites were assigned correctly. However, only a small degree of discrimination was lost when only cobalt, lead, magnesium, and nickel were used. This is understandable as three of these elements are enriched in ultrabasic rocks compared to amphibolites. It will be noted that in this area, geobotany (see Chapter 3) gave much better discrimination than did biogeochemistry.

19.6.2.   Trend Surface Analysis

Trend surface analysis is a mathematical method of analyzing map data using techniques similar to those of multiple regression. In geochemistry it is used to separate the values of a variable such as an elemental concentration in rocks, soils, or plants into various components to facilitate interpretation of the geological environment. Trend analysis leads to an equation from which an estimated value of the variable may be obtained for each point on the survey area.

Mapped biogeochemical data may be considered as having three main components: (a) regional trends due to large-scale geological processes such as changes or rock type; (b) local deviations from the trend arising from mineralization; (c) random variation over the sampling site, usually called noise, and caused by very local geological effects or even sampling or analytical error.

It is necessary to postulate a suitable function to describe the regional trend. For example, a linear trend surface is an equation of the form

$$y = a_1x_1 + a_2x_2 + c \tag{19.19}$$

where the variables x_1 and x_2 are geographical coordinates. In some cases it is necessary to try to fit the regional trend to a more complex function; for example, a second-degree trend surface:

$$y = a_1x_1 + a_2x_2 + a_3x_1^2 + a_4x_2^2 + a_5x_1x_2 + c \tag{19.20}$$

The trend surface is found by a least squares procedure minimizing the sum of the squared deviations of observed values of y from the surface. This best-fit surface is considered to be the regional component of the data, and the local component and noise are contained in the deviations. The deviations, or **residuals,** may be separated into local and noise components by observing the sign of adjacent residuals. Where a group of adjacent residuals all have the same sign, they are said to be **autocorrelated.** Positive autocorrelated residuals reflect regions of anomalous local influence. If the dependent variable is a metal concentration in vegetation, soils, or rocks, these anomalous features may be mineralized zones. Residuals whose signs vary in a random fashion contain the noise component of the data.

In practice, it is often difficult to decide at which stage refinement of the best-fit surface begins to fit small-scale geological effects as well as the regional trend. If the geology of the sampling area is known, refinement can be made until the surface of best fit agrees with the known geology. If, however, the geology of the region is unknown, one technique is to retain in the trend surface equation only those terms that result in a significant reduction in the residual sum of squares. The degree of significance is specified for a particular analysis. The method usually is designed to give a regular surface with few maxima. In general, the higher the level of significance specified, the fewer maxima the surface will contain. Because the prescribed significance level is controlled by the operator, the surface of best fit that is obtained will not necessarily correspond to the actual regional trend over the sampling area. Similarly, the trend residuals may not arise solely from small-scale geological effects, but may include some of the

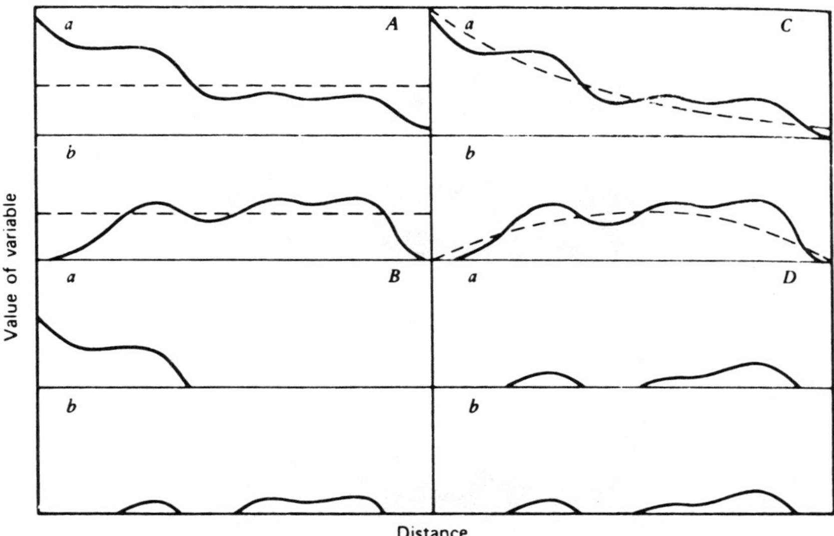

FIGURE 19.6.   Trend analysis of hypothetical data for biogeochemical data from a transect across mineralization. A—raw data, B—anomalies indicated by isoconcentration contours only, C—estimated trends (broken line), D—residuals. *Source:* Timperley et al. [807].

regional trend. It is the operator who must decide what the components of the trend analysis represent.

A simple example of the use of trend analysis is shown in Figure 19.6, where the curves represent biogeochemical data obtained in a transect across a suspected mineralized area [807]. Data sets *a* and *b* represent copper and molybdenum values, respectively, in vegetation.

It can be seen from plots A and B of Figure 19.6 that data sets *a* and *b* are each influenced by different regional processes and that apparently they do not indicate the same anomalous areas if only isoconcentration contours are used. However, if the trends due to the regional processes can be computed (plot C) and the residuals plotted, it is clear from plot D that the data sets do indeed indicate the same anomalies.

A good example of the use of trend surface analysis for interpretation of biogeochemical data is given by the work of Timperley et al. [807] in a survey of nickel in New Zealand. The data are shown in Figure 19.7.

Nickel was determined in leaf ash of *Nothofagus fusca* and in three size fractions of the supporting soils. The concentration data were distributed lognormally and therefore were transformed to logarithms before being plotted as contours representing anomalous regions (2.146–2.398 for the logarithmic values). Agreement between the soil and biogeochemical data was poor. Comparison between the trend of the plant data and the trend of the −120 mesh soil fraction indicated that as the nickel concentration in the soil increased, the concentration in the plant also increased, but to a

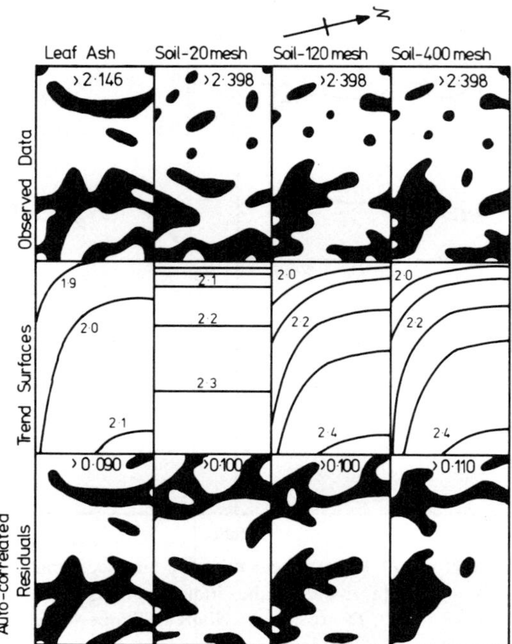

FIGURE 19.7. Contour maps for nickel showing observed data, computed trends, and the autocorrelated residuals for plant and soil data. Concentrations are logarithms of µg/g values. *Source:* Timperley et al. [807].

larger extent. This indicated that the accumulation of nickel by the plant, relative to the soil, decreased as the nickel concentration in the soil increased, an observation that had been made previously by a different statistical approach [808]. From these findings it seemed apparent that the concentration of nickel in the leaf tissue of the plants was determined mainly by the concentration of this element in the soil and partly by some other factor or factors.

The autocorrelated positive residuals for the nickel data are contoured in Figure 19.7. Two adjacent residuals of the same sign were considered to be autocorrelated. It can be seen that the residuals for each set show a greater similarity with each other than do the raw data contour plots. It had been shown previously [808] that the nickel content of plant leaf tissue was determined primarily by the nickel concentration in the soil, and therefore the regional process influencing the plant data is the same as that influencing the soil data. However, it is also noted that an additional process influences the plant data. This is interpreted as possibly arising from some mechanism such as the exclusion of nickel from the vascular system when large concentrations of exchangeable metal ions are present in the soil. The influence of these processes on the plant data is present in the raw data but is removed by trend analysis, resulting in better agreement between the residual values for the different data sets.

A recent example of the application of trend surface analysis to a more complex three-dimensional situation can be found in the work of Putnam [667]. For a more comprehensive discussion of the technique and for other applications, the reader is referred to standard works [8, 146, 218, 233, 234, 488, 617].

19.6.3. Factor Analysis

Factor analysis is a technique used for the interpretation of relationships between the variables in a multivariate collection of data. It allows the data to be expressed in terms of a number of factors that is fewer than the number of original variables. The factors created are new variables having the form of linear combinations of the original variables. Each factor extracted successively accounts for a decreasing proportion of the data variance.

The analysis may be carried out by examining the relationship between the samples (**Q mode**) or the relationship between the variables (**R mode**). R-mode factor analysis is particularly useful in clarifying the relationship between variables in a complex array of data. Where the variables are elemental concentrations in a set of geological or biogeochemical sample types, the analysis can lead to explanations of the data variability in terms of geochemical or biogeochemical interelemental associations. If factor analysis is appropriate for a given data (i.e., if the variances observed are the result of correlations between variables and underlying factors), only a few factors should be needed to account for a high proportion of the variance.

As an example of the use of factor analysis for the interpretation of biogeochemical data, details are given of work carried out on Coppermine Island, New Zealand [40]. Factor analysis was used in this work to study intercorrelations between the variables determined at each sampling site. It was hoped that the intercorrelated geochemical and biogeochemical data of each species and of soils at these sites coulds be related to the distribution of the vegetation communities which had previously been identified by infrared aerial photography and by ground work. R-mode factor analysis was used for the above study. Data were transformed logarithmically where necessary and the axes were rotated orthogonally, so that the variance of the loadings on each factor was minimized and the variance of the loadings between each factor was maximized. Table 19.6 shows the percentage of the variance accounted for by each variable for one of the plant species analyzed (*Melicytus ramiflorus*), and gives the factor score regression coefficients from which the factor scores at the sampling sites can be evaluated. For convenience these coefficients were expressed in terms of the normalized variables; that is, each variable had been transformed in such a way as to have zero mean and unit standard deviation.

In selecting the number of factors for each species, the principle used was either to accept only those with **eigenvalues** greater than unity or those

Table 19.6.   Factor Analysis of Biogeochemical and Soil Concentration Data from Coppermine Island, New Zealand Using *Melicytus ramiflorus*

					Factor Score Regression Coeff.		
Factor	Eigenvalue	% Variance Explained	Cumulative Percentage	Variables	Fac. 1	Fac. 2	Fac. 3
1	3.68	30.68	30.68	Cu in soil (tot.)	0.02	−0.25*	0.03
2	1.97	16.39	47.07	Cu in soil (sol.)	−0.08	−0.66*	0.03
3	1.77	14.75	61.84	Zn in soil (tot.)	−0.04	−0.03	0.02
4	1.10	9.20	71.04	Zn in soil (sol.)	−0.06	−0.06	−0.02
5	0.91	7.63	78.66	Cu in leaves	0.20*	−0.04	0.12*
6	0.67	5.62	84.27	Zn in leaves	−0.12*	0.03	−0.26*
7	0.60	5.04	89.32	Fe in leaves	0.32*	0.07	0.17*
8	0.44	3.67	92.99	Mn in leaves	0.15*	−0.06	−0.08
9	0.38	3.17	96.16	Cu in twigs	0.14*	−0.02	−0.01
10	0.20	1.70	98.87	Zn in twigs	−0.03	0.04	−0.58*
11	0.16	1.35	99.22	Fe in twigs	0.23*	0.02	0.00
12	0.09	0.77	99.99	Mn in twigs	0.18*	−0.09	−0.13*

Source: Yates *et al.* [955].
* major contributors

for which there was a natural break in the sequence of eigenvalues. Having selected the number of factors, an iterative procedure was used to determine the factor loadings (factor score regression coefficients). The selected factors in Table 19.6 were as follows:

Factor 1. A general plant factor with the main contribution from iron in the leaves and twigs.

Factor 2. A copper-in-soil factor having a large contribution from soluble copper in the soil and a smaller contribution from total copper in the soil.

Factor 3. A zinc-in-plant factor having a large contribution from zinc in leaves and twigs, with smaller contributions from copper and iron in leaves and from manganese in twigs.

The factor scores have been plotted against each other in Figure 19.8 and attempt to distinguish between the kanuka forest, the mixed forest, and the coastal belt communities.

In all three graphs, there is an observable variation between the factor scores of the kanuka forest and mixed forest. The best separation is due to Factor 3 (zinc in plant) which indicates a positive tendency for score values from sites within the kanuka forest. Factor 2 (soluble and total copper in soil) also shows separation of factor scores of sites from these two communities, though to a lesser extent. The coastal belt sites are again separated from all other communities and this is due to an iron-in-plant factor (Factor 1).

In assessing the use of factor analysis as a tool in mineral exploration, it

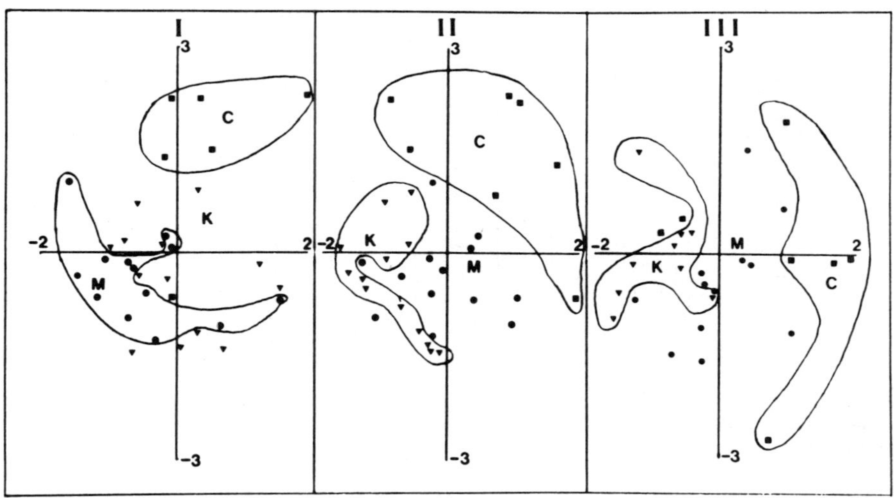

FIGURE 19.8. Factor analysis on sites including *Melicytus ramiflorus* expressed as combinations of plots for three factors. I—Factors 1 and 2, II—Factors 1 and 3, III—Factors 2 and 3. Plots show different groupings of the following communities: C—coastal, K—kanuka forest, M—mixed forest. *Source:* Yates et al. [955].

is clear that in the above test area, four plant communities could be distinguished on the basis of chemical data alone. It is obvious that in this case, factor analysis could hardly have replaced aerial photography as an exploration tool. However, if geochemical and biogeochemical data were available anyway as a result of an independent survey, it would be possible to use such data to obtain geobotanical information (i.e., distribution of plant communities) without the need for the expense of aerial photography or time-consuming plant mapping on the ground. Plant mapping on the ground in New Zealand with its very dense bush cover is seldom feasible economically so any technique that can replace or supplement it is of some value. The reader is referred to other references [147, 172, 235, 312, 341, 427, 499, 617, 714] for a further discussion of factor analysis.

19.6.4. Cluster Analysis

Cluster analysis [325, 489, 641] is a technique for classifying objects into related groups and subgroups (taxonomic classification). Many methods of computation and presentation are available, and the data can be correlated by comparing variables (R mode) or samples (Q mode). As with most other multivariate techniques, the method is most practical with a computer.

The simplest way of correlating multielement data is to generate a matrix that contains the correlations between all possible pairs of elements considered. However, such a matrix is difficult to interpret and impossible to plot meaningfully on a map. With use of cluster analysis, however, it is possible to group variables according to their mutual correlations; that is, to choose

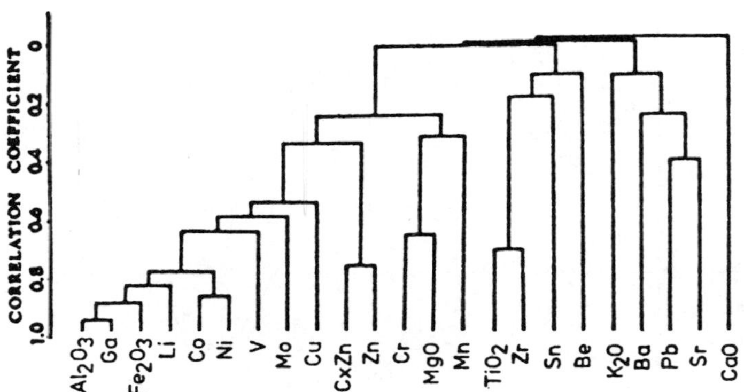

FIGURE 19.9.   Dendrogram of cluster analysis of the elemental content of stream sediments in the United Kingdom. Source: *Obial* [629]. Copyright 1970, Institution of Mining and Metallurgy, London.

mutually correlated variables that exhibit the greatest within-group correlation relative to the between-groups correlations, taking into account all possible combinations of the given elements. Obial [629] has used cluster analysis to interpret the major, minor, and trace element contents of 170 stream sediments collected in the United Kingdom. The data are shown as a dendrogram in Figure 19.9.

Clearly the best correlation ($r = 0.95$) is for Al_2O_3 and Ga. Both are correlated well with F_2O_3 ($r = 0.90$). In this particular case, six elemental groupings were recognized: (a) Al_2O_3, Ga, Fe_2O_3, Li, Co, Ni, V, Mo, Cu; (b) Zn and extractable Zn; (c) Cr, MgO, Mn; (d) TiO, Zr, Sn, Be; (e) K_2O, Ba, Pb, Sr; (f) CaO. Although the above example was based on geochemical data, there is no reason why biogeochemical data could not be used in the same way.

20

AERIAL BIOGEOCHEMICAL PROSPECTING

20.1. INTRODUCTION

Sampling of vegetation from the air presents an obvious way of overcoming access problems in many parts of the world (such as New Zealand and New Guinea) where a major cost of any ground operation comprises cutting of trails and access routes.

An Australian company working in New Guinea has attempted to sample stream sediments from the air by means of a tubular missile dropped from a helicopter into stream beds and retrieved with sediments trapped within the tube. Unfortunately, success has been meager due to the relatively small amount of suitable sediment in the streams concerned.

In New Zealand in the 1960s, an early attempt was made to devise a foliage reaper that would cut vegetation by means of a rotating cutter approaching from the side of the tree and suspended from a helicopter operating at tree height. Unfortunately, practical problems were encountered in making the operation sufficiently secure to comply with air safety regulations since there was a tendency for the tail rotor to become entangled in the vegetation. A solution to this problem might have been to operate at a higher altitude to increase the safety of the operation and then to cut the vegetation with apparatus that would convey the chopped material into the cockpit by means of a suction device. However, this presented further problems because the higher the aircraft above the sampling point, the greater the power needed to convey the chopped material to the cockpit, and the less the control of the guidance of the cutter.

An early method for aerial sampling of atmospheric particulates was devised by Weiss [918], who employed a filter system for sucking in air and an X-ray analytical technique (see Chapter 18) for analyzing the trapped particulates. The same worker [917] referred to the particulates as "mineralized aerosols" and stated that the method was not applicable to terrain that had heavy vegetation.

In recent years a new technique has been developed that involves collecting these atmospheric particulates and analyzing them *in situ* by a sophisticated analytical technique, which is described later in this chapter. First, however, the very important subject of the relationship between mineralization and atmospheric particulates will be discussed.

20.2. MINERALIZED ATMOSPHERIC PARTICULATES DERIVED FROM VEGETATION

Sources of atmospheric particulates are both anthropogenic and natural. The former sources are associated with industrial activity and other man-made sources, such as automotive emissions. These particulates, however, tend to be relatively large (> 10 μm) and have a low residence time in the atmosphere, so that they are readily precipitated near the source. Smaller particles (usually of natural origin) have a residence time of about 4 days and can be transported over much longer distances.

One of the earliest workers to suggest that plants exude mineralized particles was Nemeryuk [614] who condensed the volatile exudates from various plants and analyzed them for sodium, potassium, calcium, and magnesium. He concluded that water vapor evolved into the atmosphere during transpiration contains salt particles of 0.1 nm diameter.

The presence of soluble salt particles on the surface of vegetation was also confirmed by the well-known observation that inorganic ions are leached from vegetation by rain water [939]. Kotolov et al. [436] collected oak leaves along traverses of several base metal deposits in the eastern provinces of Siberia. The leaves were rinsed in deionized water which then contained concentrations of base metals (copper, lead, and zinc) which were a factor of 100 greater than before. These workers coined the term "atmospheric dispersion aureoles" to describe their observations.

Perhaps the most detailed of early investigations into the movement of elements into the atmosphere from plants was by Curtin et al. [228]. Exudates from conifers, presumably consisting largely of volatile materials, were sampled at 19 subalpine localities in Colorado and Idaho where anomalous amounts of several elements had been determined in vegetation and mull during previous geochemical testing. The trees sampled were lodgepole pine (*Pinus contorta*), Engelmann spruce (*Picea engelmannii*), and Douglas

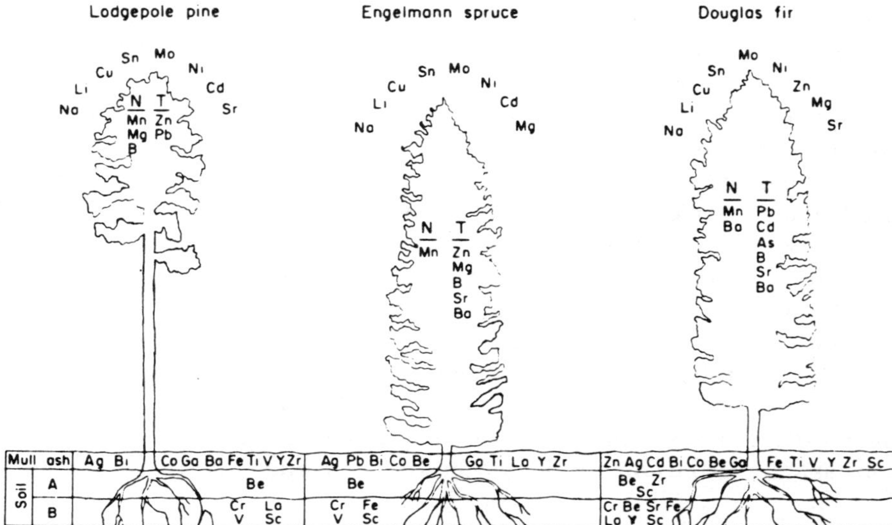

FIGURE 20.1.   Diagrammatic representation of the principal findings of a study of the highest elemental concentrations in soils, humus (mull), vegetation, and the ash of elemental halos above trees. N indicates ash of needles and T denotes ash of twigs. *Source:* Curtin et al. [228].

fir (*Pseudotsuga menziesii*). The condensed exudates were passed through paper filters and then successively through millipores of 0.05–5 μm size. The collected material was ashed and analyzed for Li, Be, B, Na, Mg, Ti, V, Cr, Mn, Fe, Co, Ni, Cu, Zn, Ga, As, Sr, Y, Zr, Mo, Ag, Pb, Bi, Cd, Sn, Sb, Ba, and La. The presence of these elements suggested that volatile exudates from vegetation are a medium for the transport of elements in the biogeochemical cycle in subalpine environments. These workers suggested that air sampling and analysis of these exudates could be a useful prospecting tool. The expeimental data are summarized in Figure 20.1. One or more species showed the highest concentrations of Na, Li, Mo, Ni, Sn, Cd, Cu, Zn, Bi, Cr, Ag, Mg, and Sr in the ash of the exudates. The elements of Pb, Sb, As, Mn, Ga, and Ba were most highly concentrated in ashed mull and vegetation, whereas in soils and ashed mull, the most highly concentrated elements were Fe, Ti, Co, Be, La, V, Y, Sr, and Zr.

The predominant organic constituents emanating into the atmosphere from vegetation are terpenes such as isoprene, pinene, and myrcene [677]. Curtin et al. [228] suggested that metal complexes with these terpenes [824] may be the agency by which the atmospheric halos are formed.

Later work by Beauford et al. [54, 55] indicated, however, that the initial release of metals takes place in particulate form. Field studies and laboratory investigations [55] using radioactive zinc and lead showed release of both elements to the atmosphere. The loss mechanism wa associated with growth conditions, the concentrations of the elements in the leaves, and with meteorological factors. Loss of zinc was about 20 pg/h/cm^2 of leaf surface and corresponded to about 9 kg of zinc per sq. km of vegetation cover per year. The release of lead was much less at about 5 g/sq. km.

A thorough survey of the composition of particulate matter from the surface of *Ulex europaeus* (gorse) was carried out by Horler et al. [358]. Samples were collected in dry weather from the foliage of the bushes of 20 background and mineralized sites in southwest England, together with soil and herbage samples. The foliar dusts were separated into four size fractions and analyzed for 16 elements by ICP (see Chapter 18). Regression analysis of the data (Chapter 19) showed that the analytical values obtained for foliar dust particles were correlated ($P > 0.05$) in at least one particulate fraction (> 45 μm) with soil concentrations of the following elements: Cu, Pb, Mn, Zn, Al, Ca, Cr, Fe, Be, Sr, Ba, P, and Cd. The multielement data supported by electron microprobe analysis and visual and microscopic examination of the particles showed that plant material, as well as soil dust, contributed substantially to the composition of the surface particles on the vegetation.

20.3. METEOROLOGICAL FACTORS AFFECTING ATMOSPHERIC SAMPLING

A review of meteorological factors affecting atmospheric sampling has been given by Barringer [42]. In such operations it is obviously essential that the particulates collected are of local origin and have not been transported long distances. This problem can be solved to some extent by sampling only the larger particles (> 10 μm) that have a short residence time in the atmosphere. A highly localized sample can be obtained, provided that there is strong vertical mixing of the material, as for example under sunny conditions where the temperature of the atmosphere decreases with increasing altitude. Inversion layers suppress mixing and produce erroneous results.

Under favorable conditions, and when the winds are light, convection plumes are formed, which are nonrotating columns of rising warm air that travel at a lower velocity than the mean wind speed. These convection plumes have a high loading capacity for particulate material that appears to move only a short distance from its origin. The elemental composition of particulates within the plumes is usually different from that of particulates in the cooler surrounding air. Barringer [42] has described a method of discriminating between particulates from plumes and from the surrounding air. This technique, involving selective sampling of air particulates, is very important because of the presence of three background populations of material in the air. There is a small global or continental background that occurs entirely in the < 10 μm fraction and that is derived from fallout of material from the stratosphere and upper troposphere. There is also a regional background from the middle troposphere with a size range exceeding 10 μm. Finally there is a local background generated in the zone of mixing and which covers the entire size range up to 100 μm. The ratio of

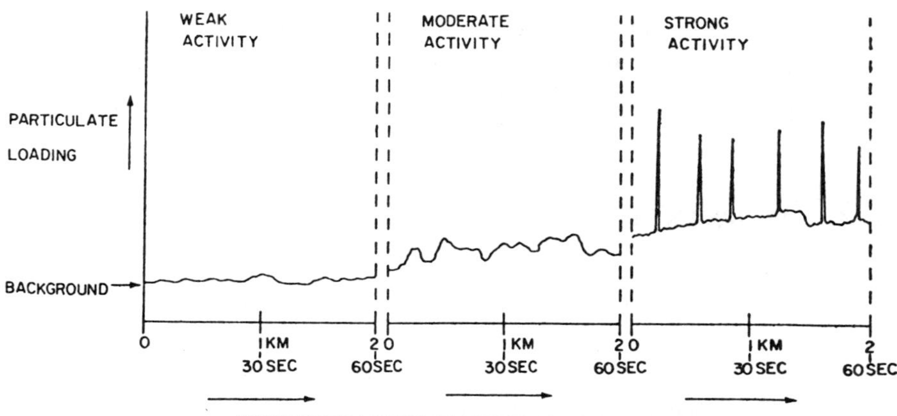

FIGURE 20.2. The effect of mixing activity of the atmosphere upon particulate loadings. *Source:* Barringer [42].

regional to local background is strongly controlled by meteorological conditions. In covering the climatic range from stagnant air through moderate mixing to strong mixing, the contribution of local background to the total burden increases dramatically (see Figure 20.2). At the same time, the increasing thermal plume activity creates columns of particulates rich in very local material which can be sampled from the air by equipment known as AIRTRACE [42].

In addition to day-to-day variations in mixing conditions, there is also a marked diurnal cycle. At night, when convection mixing ceases and still air conditions prevail, there is a steady settling of particulates within the lower troposphere. In the morning, if the sun is shining and adiabatic or super-adiabatic conditions exist, active mixing will commence, building up to high levels by noon. This will raise particles into the air in thermal plumes at velocities of up to several meters per second and there will be a continual increase in the particulate burden. The upward flux of particles during active mixing is much greater than the downward flux, since the latter is controlled substantially by sedimentation. By early afternoon even some of the giant particles may have moved considerable distances laterally, making it imperative to reject the background material if sampling resolutions of a kilometer or so are to be achieved.

In general, the great variability of the conditions that affect the contribution of regional material to the local particulate burden indicates the difficulties of gathering reliable data unless a thermal plume-sampling procedure is used.

There are also seasonal effects which place constraints on aerial sampling of particulates. Whereas strong mixing can occur at all times of the year in warm climates, there are areas where much of the winter is unsuitable. However, satisfactory results have been achieved in Canada at temperatures well below $-20°C$ and with total snow coverage on the ground. It appears that if sunshine is present, strong mixing can occur even under these winter conditions, and an adequate particulate burden for airborne biogeochemical surveys can be provided by evergreens and branches of deciduous species.

20.4. EQUIPMENT FOR SAMPLING AND ANALYSIS OF ATMOSPHERIC PARTICULATES

20.4.1. AIRTRACE

Equipment for atmospheric sampling of biogeochemical particulate matter was devised by Barringer [42] and his co-workers. The system is known as AIRTRACE and is usually mounted in a helicopter with surveys being carried out about 10 m above treetop height. Navigation is from a photomosaic. The samples are collected by means of light weight cyclones handling a hundred or so cubic meters of air per minute. Louvered collectors can handle up to four times this volume. Particles are concentrated into an air stream and are passed to a device that transfers them into an argon carrier gas. Spectroscopic analysis can then be carried out *in situ* or at a later date.

Fast-response thermal detectors at the air scoop inlets in conjunction with solenoid valve switching ensure that the air intake is switched off whenever the colder air surrounding the convective plume is encountered.

The collected particulates are impacted on to a special adhesive tape and then vaporized by means of a pulsed carbon dioxide laser. The vapor is then fed directly into an inductively coupled plasma (ICP) emission spectrometer (see Chapter 18) and analyzed immediately for 14 elements. The ICP system combines good sensitivity with multielement analytical capability. Since it is not feasible to weigh the sample material, anomalous values are detected by recording the analyte/titanium ratio. Titanium is a major constituent of most materials and has a fairly constant abundance in most geochemical and biogeochemical samples.

Another approach to the problem of analyzing samples that cannot be weighed, is to normalize the data by assuming that the total of the oxides of Si, Al, Ca, Mg, Fe, Mn, and Ti constitutes 90% of the dry mass of the material. The method of normalizing against a computed mass is quite satisfactory for analysis of soils, but less so for plant material, which is much more variable in its main constituents.

To use the ratio of a pair of elements (or to normalize against bulk composition of common oxides) is a somewhat simplistic approach. A more sophisticated procedure involves using linear regression (see Chapter 19) to plot the variation of a single element (such as copper) against six other preselected elements. The ICP system can provide determination of 25 elements in each of 3000 samples in the course of a single day of flying.

It should be emphasized that AIRTRACE analyzes all particulate matter, whether of biological origin or not. However, various tests such as density measurements have shown that many anomalies are mainly of biological origin.

20.4.2.   SURTRACE

A modification of AIRTRACE has been described by Barringer et al. [43]. This technique is known as SURTRACE and depends on analysis of the surface layer of vegetation rather than the particulates in the ambient air. SURTRACE equipment utilizes a helicopter with specially designed sampling probe deployed beneath the machine. This probe is a lightweight flexible device carrying ground-sensing equipment that allows the pilot to fly at a terrain clearance similar to that used in crop-dusting operations. Vacuum equipment removes the surface microlayer of particulates in a semicontinuous sampling operation, at flight speeds varying between 40 and 80 km/h. Size sorting and cleaning of the particles is carried out by special inertial devices and, the sample is impacted upon high-purity adhesive tape that is advanced every few seconds. Marks on the tape are synchronized with fiducial marks placed on 35 mm film carried in a continuous-strip wide-angle camera. The navigation techniques that are used in conjunction with the wide-angle camera allow for pinpoint location of the sample. Microphone equipment in the sampling boom provides event marking of every contact with the ground on the flight chart recorder, thereby allowing a continuous record to be maintained of the locations of sampling contacts with the ground. Sample intervals can be adjusted and, typically, the incoming sample is accumulated on to the tape for period of 5 s, thereby providing a 75 m interval when operating at 54 km/h. Thus thousands of samples can be collected in a single day's operations. Survey grids can be flown at traverse intervals that are selected by criteria depending on the size of the target expected and the priority rating of the area being surveyed. The equipment is portrayed in Plate 6.1 (top left) and on the dust jacket.

A portable back pack system is also available. Samples are collected as before on adhesive sampling tape, which can be processed later. A vehicle-mounted system is also available. The reader is referred to Horler et al. [359] for a further discussion of aerial sampling and analysis.

20.5.   CASE HISTORIES OF USE OF AIRTRACE AND SURTRACE

20.5.1.   AIRTRACE

A case history of the use of AIRTRACE in prospecting for minerals under somewhat unfavorable conditions has been reported by Barringer [42] for a site in Quebec Province, Canada. The site at Brouillan township, in the northern part of the province, was reported to comprise three separate zones, the first with a combined tonnage of more than 50 million tonnes of sulphide ore averaging 0.39% copper, 2.30% zinc, 30 μg/g silver, and 0.3 μg/g gold, beneath an overburden of 10m of glacial drift. Ore grades in the other two zones were reported to be much higher in copper than in the first zone, and occurred at greater depth.

The anomaly map was derived by a multiple regression technique (see Chapter 19) carried out on a line-by-line basis for copper and zinc. In addition, spatial filtering was employed to smooth out noise and to generate a contoured map.

The area was flown twice, both times under far-from-ideal weather conditions, since the climate in this region is rather harsh and variable. Nevertheless, clear-cut delineation of an anomalous zone was obtained in both flights, in each case somewhat displaced but clearly relating to the geophysical anomaly determined by airborne EM conductimetric measurements. The data provided definite indications of a drilling target in a region which contains large number of barren EM conductors that normally have to be disproven by expensive drilling. The data are shown in Figure 20.3.

20.5.2. SURTRACE

A good example of the use of SURTRACE is provided by a case history from Arizona [43]. The Black Mountain prospect from Kingman, Arizona, is a large copper-molybdenum anomaly associated with porphyry copper-type mineralization. The anomalous area was characterized by 70 m of relief in which outcropping intrusives project through flat-lying alluvium. The entire area was covered with thin soil and mesquite that was sufficiently dense in growth to prevent the SURTRACE probe beneath the helicopter from touching the ground most of the time. Thus the SURTRACE samples consisted mainly of coarse particulate material of biological origin derived either from weathering of the mesquite surface or from local windblown surface soils entrained on the mesquite.

The traverse shown in Figure 20.4 represents less than 6 min of flight and 12 min of analytical time on the LASERTRACE equipment. This produced a 25-element set of data for approximately 5 km of traverse, and showed, with great clarity, the locations of the copper and molybdenum anomalies. A strong nickel-chromium anomaly that had not been reported previously was indicated by the SURTRACE survey, and required additional checking on the ground. Correlation of the flight traverse with the ground geochemical maps showed a very good match, and the flanking relationship of copper and molybdenum on the particular flight line chosen is illustrated very well in the SURTRACE data.

The above test demonstrates the speed with which the fairly rugged terrain in this region (40 km south of the Hoover Dam) could be covered with this type of fast, multielement survey.

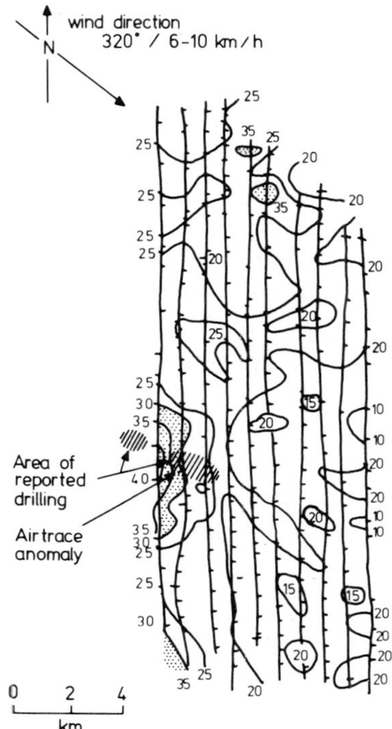

FIGURE 20.3. AIRTRACE activity contours computed from multiple regression analysis of copper + zinc concentrations in aerial particulates. The overflight was over an area of base metal mineralization in northern Quebec. *Source:* Barringer [42].

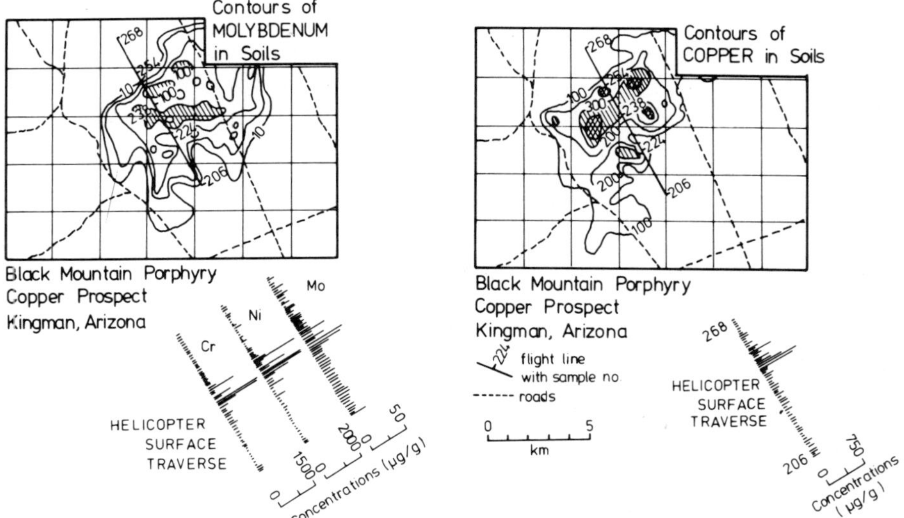

FIGURE 20.4. SURTRACE measurements over the Black Mountains porphyry copper deposit, Kingman, Arizona. *Source:* Barringer et al. [43].

20.6. AN ASSESSMENT OF AERIAL BIOGEOCHEMICAL PROSPECTING

An assessment of airborne biogeochemicl prospecting of necessity must be concerned almost exclusively with the Barringer AIRTRACE/SURTRACE systems. These are not exclusively biogeochemical in nature since excellent results can be obtained over desert terrain, but nevertheless a sound scientific basis has been established for perhaps one of the most effective potential applications of the biogeochemical method: sampling from the air. Its applications to terrain of difficult access are obvious. There are, nevertheless, several disadvantages of these aerial systems. The AIRTRACE method is heavily dependent on satisfactory meteorological conditions and the presence of convective plumes. No doubt as the technique is further refined, its use in the hands of experienced personnel will provide even more reliable data. The SURTRACE method is much less dependent on favorable weather conditions and will, in the long run, probably prove to be the more effective of the two techniques of aerial biogeochemical prospecting.

21

BIOGEOCHEMICAL PROSPECTING IN THE HERBARIUM

21.1. INTRODUCTION

The world's herbaria (plant museums) are a repository for about 200 million dried plant samples. There are nearly 2000 herbaria recorded in the *Index Herbariorum* [775]. Perhaps the oldest of these is the Paris Herbarium, which was founded in 1635 by Louis XIII and has about six million specimens. Similar numbers are held by such famous herbaria as those of Kew and the British Museum.

Herbarium specimens of flowering plants, ferns, and the larger algae are customarily dried between sheets of heavy blotting paper under pressure so that the dried specimen is flat. Most specimens will dry in about a week if the blotters are changed several times. The drying can be hastened by interspersing ventilators of corrugated cardboard among the blotters and by the gentle application of heat. Although the colors of flowers and leaves eventually fade, herbarium specimens last indefinitely, some being over 200 years old.

A herbarium is the most essential tool for research in plant taxonomy and serves as a permanent record of the distribution of each species, its range of variability, and the correlation of that variability with the geography and habitat. A botanist writing a flora of a region relies largely on herbarium specimens, which is studied against a background of his or her own observations on living plants. In more critical studies of smaller groups of species, cultural and breeding experiments, determination of chromosome number, and other detailed analyses are also customary. The herbarium, however, remains the essential foundation for the work, because no one individual can expect to see even a small group of species in all localities, and all habitats in which they occur and under the varying conditions of different years and seasons.

Although herbaria serve as samples of natural populations of plants, they must be interpreted with caution because they tend to be biased towards the unusual. When a region is first explored, collectors have their time cut out merely to make records of the common species. Thereafter, the less common species and unusual forms of common species are likely to be collected out of proportion to their abundance. Thus the usefulness of a specimen may be greatly enhanced by some indication on the label as to the frequency of occurrence of the species in the wild.

In 1948, Minguzzi and Vergnano [593] discovered hyperaccumulation of nickel by *Alyssum bertolonii* and had thought of analyzing herbarium material to see if this character was also possessed by other species of this genus. Their approach to the curator of the Florence Herbarium evoked such consternation (O. Vergnano Gambi, personal communication) that they desisted in making further requests. The reason for this reaction was not hard to find. At that time, using emission spectrography (ES) as an analytical method (see Chapter 18), virtually the entire specimen would have had to have been destroyed for a single analysis.

The first person to analyze herbarium material was probably E. M. Chenery [185–188] who determined aluminum in representatives of all of the 259 recognized families of Dicotyledons. His work was not oriented, however, towards mineral exploration, and consisted of semiquantitative colorimetric test of leaf samples of about 6 sq. cm in area.

Herbarium material has been used in geobotanical work (see Chapter 4). For example, Persson [650] noted the collection localities of Swedish herbarium specimens of a "copper moss" and discovered three localities with anomalous copper levels in the substrate, of which one was an existing copper mine. Later Cole [208], in the course of geobotanical investigations in southern Africa, identified plant indicators of copper and referred to an herbarium for other collection localities of these species.

In recent years, new techniques of chemical analysis (see Chapter 18) have resulted in the minimum sample size of vegetation being reduced to a few milligrams or even less for the determination of specific elements. Even the plant dust between herbarium sheets can be analyzed successfully without problems. Indeed the limiting factor is not the size of the sample needed for analysis, but rather the extent to which this small sample is representative of the entire plant specimen.

During the past few years, my research group has analyzed some 20,000 herbarium specimens (mainly leaf samples of area 1 sq. cm or less) in order to investigate the possibility of prospecting for minerals in countries that we have never visited. This proposition might at first sight appear to be facetious, but it will be shown below that analysis of herbarium material for this purpose is indeed a viable method for obtaining biogeochemical data of use for future field exploration.

21.2. ANALYSIS OF HERBARIUM MATERIAL

The first step in the analysis of herbarium material is to obtain the specimens. Provided that requests are modest and that only minute samples (1 sq. cm or less) are required, most herbarium curators are happy to donate small amounts of material. The following information is needed along with the sample.

1. *The binomial and authorities.* For example, *Rinorea bengalensis* (Wall.) O.K. The authorities are part of the name of the plant. In this example, the genus is *Rinorea* and the species is *bengalensis.* "(Wall.)" tells us that Wallich originally described the species but that it was later reclassified by O. Kuntze (O.K.)

2. *The collector's number or herbarium accession number.* The first is more useful because you can then check to see whether you have had a duplicate of this taxon from some other herbarium.

3. *The collection locality.* Full details should be sought. Early (nine-teenth-century) material often has a poor description of collection localities; for example, "near Vienna," whereas modern collectors usually give very detailed information from which it should be possible to pinpoint the locality to within a few hundred meters.

Herbaria are repositories for the sacrosanct **type material,** which are standards upon which names of species are based. A scientific name to be accepted must be properly published together with a description, but it is the type specimen that permanently determines the application of the name. In many herbaria, types are segregated from the rest of the collection in order to minimize the risk of damage. Curators normally will not supply type material for analysis, and it is ill-mannered to request it.

For chemical analysis of the material, weigh about two thirds of the sample (i.e., up to 50 mg) into a 5 ml borosilicate test tube using a five-figure balance. Place the tubes in squat borosilicate beakers and ignite them at 500°C in a muffle furnace for a period of 2–3 h. After cooling add a known volume (1–2 ml) of 2 M hydrochloric acid prepared from redistilled constant-boiling (6 M) reagent. Warm gently to dissolve the ash if it does not dissolve immediately. With care, four elements can be determined in a 1 ml sample using atomic absorption spectrometry (AAS; see Chapter 18). If ICP (see Chapter 18) is used for the analysis, about 3 ml of solution will be required. The data should be expressed on a dry weight basis. There are practical difficulties in use of an ash weight basis, since the weight of ash from 1 mg of dry leaf might be only 50 μg and would be impossible to weight accurately.

Replicate analyses of the same original sample should provide reasonably reproducible data (8–10%) but the question is often asked as to the inter-sample variability of different samples from the same specimen. This problem was examined by Wither [944], who obtained circular disc samples (6 mg) from four separate leaves of the same specimen of the nickel hyper-accumulator *Rinorea bengalensis*. Twenty separate samples (each containing one disc) gave 552 ± 12 µg/g nickel. Fifteen separate samples (each containing two discs) gave 536 ± 20 µg/g nickel. Ten separate samples of three discs each gave 518 ± 27 µg/g of this element. From these experiments it appeared that the sampling error for samples as small as 6 mg was only 2–5%.

21.3. SOME CASE HISTORIES OF HERBARIUM STUDIES

21.3.1. Nickel

One of the first herbarium studies was the determination of the nickel content of some 2000 specimens of *Hybanthus* and *Homalium* [110]. This survey was undertaken because a hyperaccumulator of nickel (i.e., > 1000 µg/g in dried leaves), *Hybanthus floribundus*, had been discovered in Western Australia [209, 740], and it seemed to be a reasonable assumption that others might be found in the same genus. Similarly, Jaffré and Schmid [397] had determined hyperaccumulators of nickel in *Homalium guillainii*. Fifty herbaria throughout the world were asked for plant material and 35 of these supplied samples. The specimens originally had been collected from all parts of the tropical and warm-temperate world, and represented a sampling density of about one per 2000 sq. km. The survey resulted in the reidentification of all previously known hyperaccumulators of nickel (5 species) and the discovery of five additional species (all from New Caledonia) with this same character. Fourteen previously unknown strong accumulators (100–1000 µg/g nickel in dry material) were also discovered. From the collection localities of the hyperaccumulators and strong accumulators, it was possible to pinpoint many of the world's major ultrabasic occurrences in tropical and warm-temperate regions. The principle was obviously applicable to other species.

Some mention has already been made of a nickel survey of all but one of the 168 species of *Alyssum* [115], which resulted in the discovery of over 40 previously unknown hyperaccumulators of nickel. The collection localities of these plants identified virtually every single major ultrabasic rock region in southern Europe and Asia Minor. Although this work was not of much value for mineral exploration because the area is already very well mapped geologically, it did point to the possibility of carrying out a similar survey in areas that had not been well mapped. An obvious place for this type of work was Southeast Asia.

Brooks and Wither [125] found hyperaccumulations of nickel in *Rinorea*

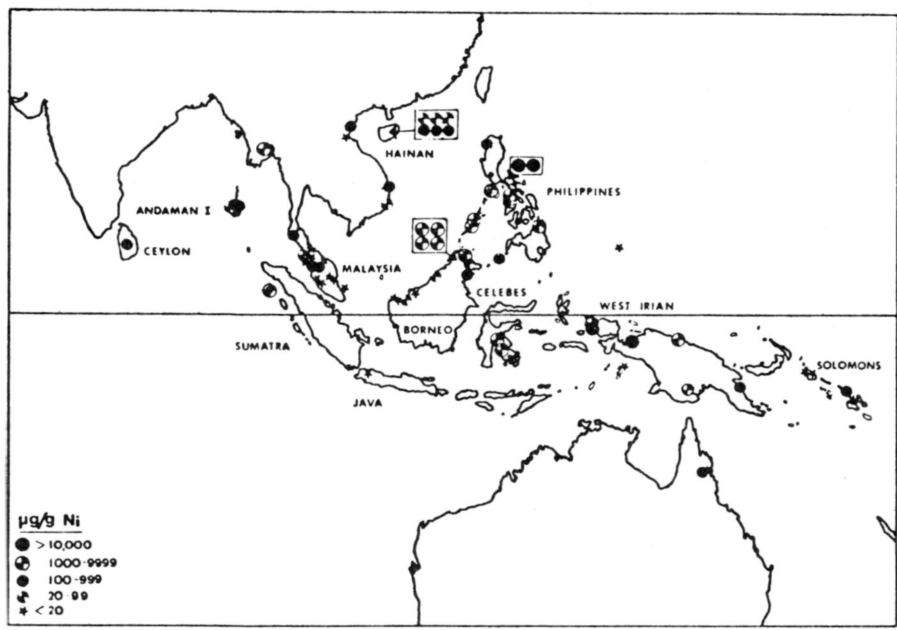

FIGURE 21.1. Map of Southeast Asia and Australasia showing collection locatities of herbarium specimens of *Rinorea bengalensis* and indicating nickel concentrations (µg/g dry weight) in leaf material. *Source:* Brooks and Wither [125].

bengalensis. A survey of the nickel and cobalt contents of 89 herbarium specimens of this species showed that many of the collection localities coincided with a number of important serpentine occurrences throughout the region. This is demonstrated in Figure 21.1. The data are also summarized in Table 21.1, where analytical data are arranged under the appropriate rock type (if known) of the collection locality.

TABLE 21.1. Nickel and Cobalt Concentrations (µg/g dry weight) in *Rinorea bengalensis*

Substrate	Number of Samples	Cobalt		Nickel		Co/Ni
		Mean	Range	Mean	Range	
Ultrabasic	21	87	6–545	6860	836–17,500	0.01
Limestone	12	12	1–33	113	2–560	0.11
Other sed.	14	75	3–290	674	3–3000	0.11
Basic rocks	11	27	1–217	103	1–550	0.26
Acid rocks	5	16	5–29	20	2–56	0.80
Unknown	26	38	1–300	177	1–2000	0.21
Overall	89	51	1–545	1810	1–17,500	0.03

Source: Brooks and Wither [125].

Rinorea bengalensis obviously grows over a wide range of substrates and is a hyperaccumulator of nickel when growing over ultrabasic rocks. Over such substrates, the Co/Ni ratio (*ca.* 0.01) is considerably lower than for plants growing over other rock types. This serves as an additional criterion for pinpointing ultramafic occurrences. It was invariably possible to identify ultrabasic substrates by nickel levels exceeding 3000 μg/g in the plant material. Values above 1000 μg/g usually indicated this type of substrate, but there was some overlap with the higher values corrresponding to sedimentary (excluding limestone) substrates. Whereas plants growing over ultrabasic substrates could almost always be differentiated from those growing over other rock types (i.e., by a combination of high nickel concentrations and low Co/Ni ratios), it was not possible to separate other substrates from each other by the nickel content of this species alone. However, this is relatively unimportant from the standpoint of mineral exploration because ultrabasic rocks are a favored target for exploration personnel, as they are hosts for many economic minerals such as nickel, chromium, cobalt, and the metals of the platinum group.

Figure 21.1 also shows the distribution of *R. bengalensis* throughout Southeast Asia and includes only those specimens whose collection localities could be established accurately. The range extends all the way from Sri Lanka to the Solomons, and the ultrabasic areas pinpointed (large circles) represent most of the important regions containing these rocks. In all but two cases, ultrabasic rocks were known to occur at the collection localities. The exceptions were the substrate of a plant from Dalman, Nabire, Irian Jaya (Indonesian New Guinea), which contained 1.75% nickel, and another from the Sorong area (1.2% nickel) also from Irian Jaya. Because of the very high nickel contents and low Co/Ni ratios (0.003 and 0.002, respectively) these plants were almost certainly growing over ultrabasic rocks. In the Nabire area, the presence of ultrabasic rocks can be inferred by extrapolation from partially surveyed, sporadic, undifferentiated mafic intrusives forming a long belt passing through Nabire. The location of the specimen from Sorong will probably have to be investigated *in situ* to establish the nature of the substrate.

The presence of previously unknown ultrabasic areas in Irian Jaya was revealed by a chain of events which began with the original collection of material by two Japanese botanists in 1940. The samples were stored for 37 years at the Arnold Arboretum (Harvard University) and then sent to New Zealand for analysis by persons who had never visited Irian Jaya.

R. bengalensis was not selected for analysis in a random manner. Since another hyperaccumulating genus (*Hybanthus*) belonged to the family Violaceae, it was logical that this be investigated for other such genera.

In another herbarium survey, an entire collection of plants from Obi Island, Indonesia, which had been collected by Dr. E. de Vogel (Leiden) in 1976, was analyzed for nickel. Most of these plants had been growing over a serpentine occurrence on the island. This survey resulted in the identifi-

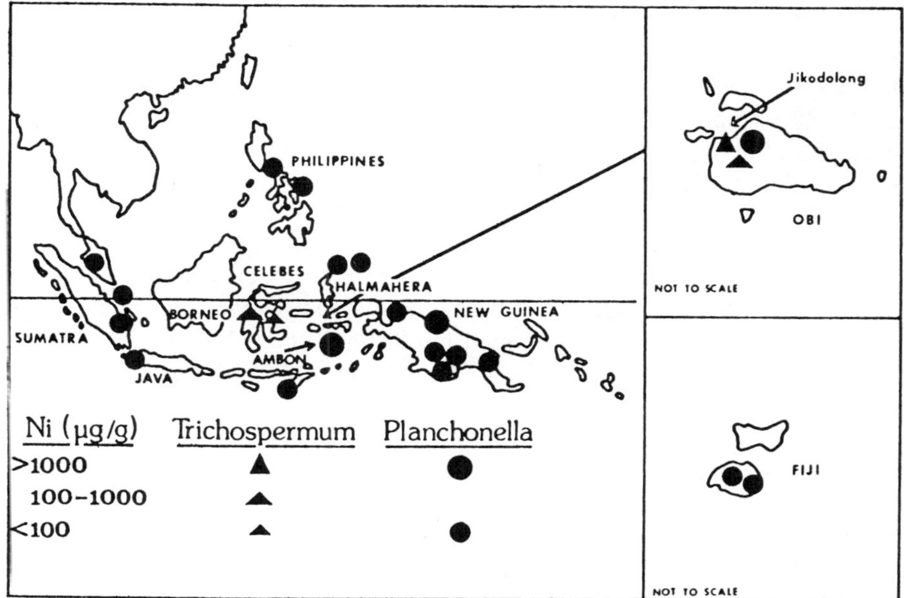

FIGURE 21.2. Map of Southeast Asia and Australasia showing collection locatities of herbarium specimens of *Planchonella oxyedra* and *Trichospermum kjellbergii*. Nickel concentrations (µg/g dry weight) in leaf material are also shown. *Source:* Wither and Brooks [945].

cation of two more hyperaccumulators of nickel, *Planchonella oxyedra* and *Trichospermum kjellbergii*. Once these species had been determined as having a hyperaccumulating capacity, further herbarium specimens of both species were obtained and analyzed for nickel [945]. The data are shown in Figure 21.2, and once again pinpointed ultrabasic areas throughout Southeast Asia. One of these areas (on the island of Ambon) had not been known previously to geologists.

Other herbarium surveys for nickel have involved the identification of seven hyperaccumulators of nickel in New Caledonian species of *Geissois* [393]. All of these species indicated nickeliferous rocks on the island. Similar studies were carried out by Jaffré et al. [394] on New Caledonian species of *Casearia, Lasiochlamys,* and *Xylosma*. As a result of this work, a further 18 "nickel plants" were discovered, all of which indicated ultrabasic rocks. A study of nickel levels in New Caledonian species of *Phyllanthus* [422] resulted in the identification of another ten hyperaccumulators of nickel including *Phyllanthus serpentinus,* which contains up to 3.8% nickel in dry leaves.

21.3.2. Cobalt

Hyperaccumulators of cobalt are far less numerous than those of nickel. This is not difficult to understand because, although nickeliferous (ultrabasic) rocks occupy a significant proportion of the earth's surface, cobalt-rich rocks (> 0.1%) are, apart from isolated ore occurrences, confined to the copper-cobalt province of Zaïre/Zambia in Central Africa, which is by far the largest source of cobalt in the world. These deposits support a highly unusual and diversified metal-tolerant vegetation which has been described by several workers [254, 255, 551, 553].

The first herbarium study on plants of Shaba Province, Zaïre, was by Brooks [101] who analyzed 19 species of *Haumaniastrum* and found up to 1.02% cobalt in the dried leaves of *H. robertii*. This species had been known as a "copper flower" (see also Chapter 4) for many years, but because of the coexistence of copper and cobalt in all the Central African deposits, it is not easy to decide which of these elements in really indicated. However, pot trials on seedlings of this species [605] appeared to indicate that *H. robertii* is more tolerant to cobalt than to copper, and therefore is probably a cobalt indicator. In all, 15 plant species (all from Zaïre) have been recognized as being hyperaccumulators of cobalt [119] as a result of herbarium studies.

The unusual accumulation of cobalt by the Nyssaceae has been reported by Brooks et al. [114]. Herbarium specimens (375) of species of *Nyssa, Camptotheca,* and *Davidia* were analyzed for cobalt and nickel. All species of *Nyssa* (four from the United States and two from Southeast Asia) possessed a marked ability to accumulate cobalt, not only in absolute terms but also relative to nickel. The mean cobalt contents of the species were sufficiently different to allow differentiation of many pairs of species. It was concluded that all species of *Nyssa* could be used for assessing the cobalt status of soils as had already been proposed by Kubota and his co-workers [490, 491] in the case of *N. sylvatica*. *Camptotheca acuminata* showed a much lower uptake of cobalt than *Nyssa* species, but this uptake was still considerably higher than for most other plants and strengthened the case for inclusion of *Camptotheca* in the Nyssaceae. *Davidia involucrata* showed no capability for accumulating cobalt to a level exceeding normal background for most vegetation, and strengthened powerful arguments that this monotypic genus does not belong to the Nyssaceae. This survey therefore, though of marginal significance for mineral prospecting, did have some importance in the field of plant taxonomy.

21.3.3. Copper

There are comparatively few examples of plants that hyperaccumulate copper. For example, although Brooks et al. [119] listed 15 species that could hyperaccumulate cobalt, only 12 were found (all from Zaïre) that could take up copper to the same extent. This is in spite of the fact that copper deposits are very much more widespread than are those of cobalt.

In an herbarium survey of *Aeolanthus,* the extraordinary accumulation of copper by *A. biformifolius* was observed [116, 552]. This species is confined to copper-rich substrates and contains the highest copper concentration ever recorded for phanerogams (flowering plants). The whole plant contained up to 1.37% copper on a dry weight basis.

Although there are very few hyperaccumulators of copper, a very much larger number of species is able to indicate copper mineralization without inordinate uptake of this element (see Chapter 4). One of these is the Australian "copper flower" *Polycarpaea spirostylis* which, like so many of this type of indicator, is an herb and belongs to the Caryophyllaceae (pink family). Brooks and Radford [117] analyzed 183 herbarium specimens of 11 out of 12 known species of *Polycarpaea,* and found statistically the threshold copper concentrations that apppeared to be anomalous. Fourteen specimens were in this category, and the collection localities are now being investigate further.

The well-known "kisplanten" (pyrite plants) of Fennoscandia (see Chapter 4) include *Lychnis* (= *Viscaria*) *alpina* and *Silene* (= *Melandrium*) *dioica.* Brooks et al. [122] analyzed over 700 herbarium specimens of these two species for copper, lead, and nickel. Anomalous values for copper (> 20 μg/g), nickel (> 20 μg/g), and lead (> 80 μg/g) revealed a number of biogeochemical anomalies, including virtually all of the major copper deposits of Fennoscandia, particularly those of the Røros area of Norway and at Bergslagen, Sweden. The highest value was 250 μg/g for a specimen growing over copper slag in Garpenberg at Bergslagen. Lead "anomalies" should be treated with caution in collection localities close to a main road where automotive emissions can easily pollute vegetation samples [888]. Nickel anomalies invariably coincided with ultrabasic rocks. An unusual case of prospecting for copper by use of herbarium material has been reported for Salajar Island in Indonesia [126].

The island of Salajar measures about 70 km in length and 10 km in width It is situated about 25 km south of the island of Sulawesi (Celebes) in Indonesia. A terraced mountain chain traverses the island, descending steeply to a rocky, coral-fringed coast on the east side and descending more gradually to the west.

There is no geological map of the entire island, although the northern third is included in a recently published 1:1,000,000 geological map of Sulawesi. From this and other information [245] it seems that the island is ringed with coral limestones around a central core of Miocene sandstones and volcanics.

Very little of the original vegetation was present in 1913 when Docters van Leeuwen [245] carried out a botanical expedition. The situation is much worse today after an interval of 70 years. A map of the island is shown in Figure 21.3.

The Docters van Leeuwen collection at Utrecht contains about 300 specimens from 73 families of Angiosperms. Eighty specimens from 19 families were analyzed by Brooks et al. [126]. Material was obtained not only from

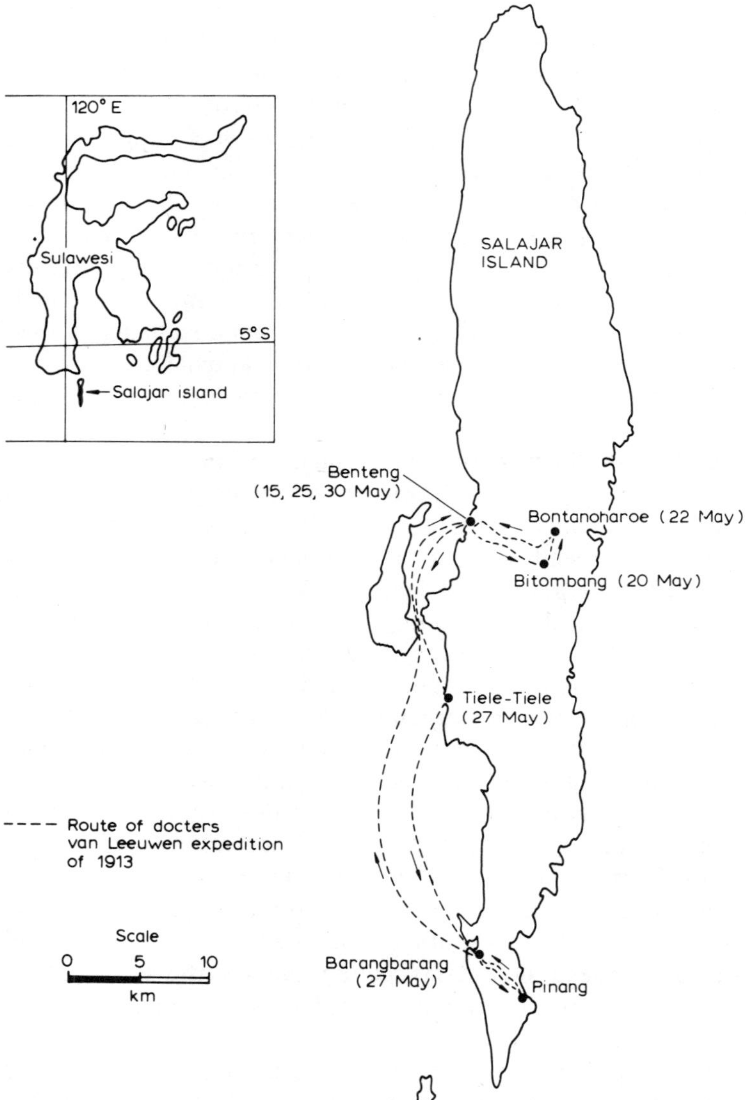

FIGURE 21.3.   Map of Salajar Island, Indonesia, showing the route of the Docters van Leeuwen expedition of 1913. *Source:* Brooks et al. [126].

Utrecht but also from the herbaria at Leiden, Kuala Lumpur, and Cambridge, Massachusetts.

A total of 20 species of plants from Salajar were found to have copper values exceeding 80 µg/g in their dried leaves. The highest value was 600 µg/g for *Laportea ruderalis*. The data are presented in Table 21.2. Value are also given for the identical species collected from other parts of South-

TABLE 21.2.   Mean Copper Concentrations in Plant Species with Anomalous Values in Salajar Island (Indonesia) Compared with Values from Elsewhere

Species	Family	Copper Concentrations (μg/g) Salajar	Outside Salajar
Abrus precatorius	Papilionaceae	167	12
Aerva scandens	Amaranthaceae	395	17
Allmania nodiflora	Amaranthaceae	175	24
Cassia sophora	Papilionaceae	333	12
Coleus scutellarioides	Labiatae	500	10
Cyathula prostrata	Amaranthaceae	553	12
Cymaria acuminata	Labiatae	250	13
Desmodium laxiflorum	Papilionaceae	167	17
Dolichos falcatus	Papilionaceae	276	7
Euphorbia plumerioides	Euphorbiaceae	185	6
Fatoua pilosa	Moraceae	180	13
Hyptis capitata	Labiatae	80	26
Laportea ruderalis	Urticaceae	600	21
Macaranga hispida	Euphorbiaceae	150	4
Moringa oleifera	Moringaceae	183	15
Peperomia pellucida	Piperaceae	300	19
Pisonia aculeata	Nyctaginaceae	258	26
Rubus rosaefolius	Rosaceae	170	10
Sesbania grandiflora	Papilionaceae	158	8
Vernonia actaea	Composites	300	11

Source: Brooks et al. [126].

east Asia outside of Salajar. It is clear that the Salajar samples had copper values greatly exceeding the highest values found outside the island. This is further demonstrated in Figure 21.4, which shows the data in histogram form. The histogram shows virtually two separate popualtions. The first of these (samples outside Salajar) hardly overlaps with the second (Salajar) population, which has a median value of about 250 μg/g copper.

The magnitude of the high copper values in the Salajar samples is such as to exceed in some cases the values for cuprophytes growing over copper deposits in Central Africa. For example, Duvigneaud and Denaeyer-de Smet [255] reported 3–356 μg/g copper in dried leaves of 43 plant species growing over a copper deposit in Zaïre. Only six of these had copper levels exceeding 100 μg/g.

To check that the high copper values in the Salajar samples were not due to contamination, pieces of the original herbarium labels and of the paper of the herbarium sheets were also analyzed for copper. In all cases negligible concentrations of copper were found and showed that the vegetation anomalies were genuine.

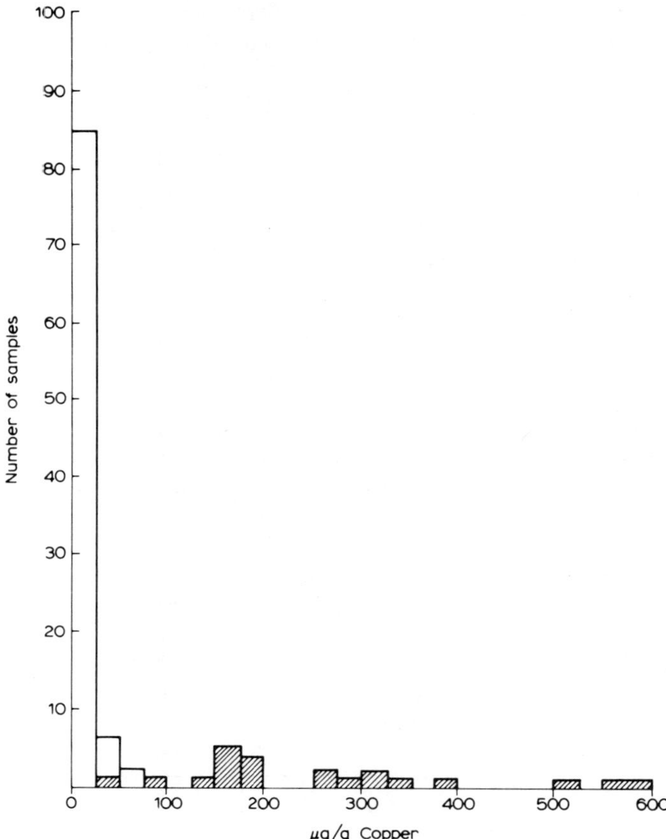

FIGURE 21.4. Histograms of anomalous copper concentrations (μg/g dry weight) in plants from Salajar Island, Indonesia, compared with levels in the same species from elsewhere. Salajar values are shaded. *Source:* Brooks et al. [126].

In common with many older collections, details on the labels of the Docters van Leeuwen collection were somewhat vague. However, from the dates and from the precise details of his expedition (see Figure 21.3) contained in his journal [245], it was possible to identify collection localities with more precision.

Most of the specimens with anomalous copper values had been collected between Barangbarang and Pinang in the south of the island, and it is suggested that this area is worth a follow-up investigation by geochemical procedures such as stream sediment and soil sampling. By inference from the known geology of the top third of the island, the south Salajar is probably a contact zone between volcanics and coral limestones. Geological work in the southern part of Sulawesi adjacent to Salajar Island has revealed the presence of tephrites with a relatively high copper content (ca. 800 μg/g; T. M. van Leeuwen, personal communication) and the Salajar biogeochemical anomalies may therefore be derived from a similar sort of substrate.

21.3.4. Lead

There has already been some discussion of lead values in *Lychnis alpina* and *Silene dioica*. Recently, specimens of *Alyssum wulfenianum* and *Thlaspi rotundifolium* s. sp. *cepaeifolium* from the Austria–Italy border area near Raibl have been analyzed for lead by R.D. Reeves and R.R. Brooks, *Environ. Pollut.* (in press). Both species are endemic plants confined to a base metal (zinc and lead) mining area. Values of up to 860 µg/g in dried leaves of *A. wulfenianum* and 7000 µg/g in the *Thlaspi* species were found. This work confirmed earlier records [515] of lead levels which had previously been thought to be impossibly high. No other plant has ever been found with lead levels even approaching the concentrations found in *Thlaspi rotundifolium* s. sp. *cepaeifolium*.

The lead content of 454 herbarium specimens of *Lychnis alpina* (see also section 21.3.3 above) from Greenland was determined by Brooks et al. [123]. Some of the anomalous values were associated with areas of known mineralization or at sites of active or abandoned lead mines, such as those at Mesters Vig near Scoresby Sound on the east coast. Other unexplained anomalies were recorded and are worthy of follow-up investigations by other geochemical techniques.

21.3.5. Manganese

Very little work has been done on herbarium specimens of manganese-accumulating plants. However, Jaffré [389, 390] used herbarium material to identify a number of manganese-accumulating species from New Caledonia. In a later study [124], 31 species of *Alyxia* from the same island were analyzed for their manganese content. Most of the species showed excessive uptake of manganese with a maximum of 1.15% in dried leaves of *A. rubricaulis*. The manganese content alone was sufficient to distinguish a number of species from each other. Correlation analysis and calculation of the slopes of the regression lines showed that manganese uptake was largely at the expense of calcium rather than magnesium. Similarly, magnesium was preferred to potassium, and potassium to sodium. It is possible that manganese may have some physiological role in *Alyxia,* compensating to some extent for reduced uptake of the nutrients calcium and potassium.

As was the case with nickel plants, the species with very high concentrations of manganese indicated localities with manganese-rich ultrabasic rocks.

21.3.6. Zinc

One of the best known of all zinc accumulators is *Thlaspi calaminare,* which is confined to the zinc-rich calamine (zinc carbonate) deposits of Western Europe. In a survey of European species of *Thlaspi,* Reeves and Brooks

[682] determined that at least nine other species contained over 1% zinc in their dried leaves. These high values were found not only in zinc-rich substrates but also in a wide range of other formations with presumably much lower zinc levels.

An interesting feature of *Thlaspi* species is that they also hyperaccumulate nickel (ten species) and accurately reflect the presence of ultrabasic occurrences even when these are quite localized.

21.4. HYPERACCUMULATING SPECIES IDENTIFIED BY HERBARIUM STUDIES

An abbreviated list of metallophytes identified by herbarium studies is given in Table 21.3. Some of these species are common to a list of indicator plants given in Chapter 4. It should be noted, however, that the metallophytes

TABLE 21.3. Hyperaccumulators of Various Elements Identified by Analysis of Herbarium Material

Family	Genus	No. of Species	Location	Element	Refs.
Monocotyledons					
Commelinaceae	*Commelina*	1	Zaïre	Copper	119
	Cyanotis	1	Zaïre	Cobalt	119
Cyperaceae	*Ascolepis*	1	Zaïre	Copper	119
	Bulbostylis	1	Zaïre	Copper/Cobalt	119
Gramineae	*Eragrostis*	1	Zaïre	Copper	119
Proteaceae	*Macadamia*	1	New Caledonia	Manganese	390
Dicotyledons					
Amaranthaceae	*Pandiaka*	4	Zaïre	Copper	119
Apocynaceae	*Alyxia*	18	New Caledonia	Manganese	124
Caryophyllaceae	*Silene*	1	Zaïre	Copper/Cobalt	119
Celastraceae	*Maytenus*	1	New Caledonia	Manganese	389
Compositae	*Anisopappus*	1	Zaïre	Cobalt	119
Convolvolaceae	*Ipomoea*	1	Zaïre	Copper	119
Crassulaceae	*Crassula*	1	Zaïre	Cobalt	119
Cruciferae	*Alyssum*	46	Europe/Asia Minor	Nickel	115, 121
	Bornmuellera	5	Greece/Anatolia	Nickel	683
	Noccaea	2	Greece/France	Nickel	682
	Peltaria	1	Greece	Nickel	686
	Streptanthus	1	United States	Nickel	684
	Thlaspi	1	Central Europe	Lead	685
		13	Europe	Nickel	682
		9	Europe	Zinc	682
	Thlaspiceras	10	Anatolia	Nickel	685

TABLE 21.3.   *(Cont.)*

Family	Genus	No. of Species	Location	Element	Refs.
Cunoniaceae	*Geissois*	6	New Caledonia	Nickel	393
	Pancheria	1	New Caledonia	Nickel	391
Escalloniaceae	*Argophyllum*	2	New Caledonia	Nickel	393
Euphorbiaceae	*Phyllanthus*	9	New Caledonia	Nickel	422
Flacourtiaceae	*Casearia*	1	New Caledonia	Nickel	394
	Homalium	6	New Caledonia	Nickel	110, 394
	Lasiochlamys	1	New Caledonia	Nickel	394
	Xylosma	9	New Caledonia	Nickel	394
Labiatae	*Aeolanthus*	2	Zaïre	Copper/Cobalt	119
	Haumaniastrum	3	Zaïre	Copper/Cobalt	101, 119
Leguminosae	*Vigna*	1	Zaïre	Copper	119
Myristicaceae	*Myristica*	1	Indonesia	Nickel	945
Oncothecaceae	*Oncotheca*	1	New Caledonia	Nickel	391
Rubiaceae	*Psychotria*	1	New Caledonia	Nickel	397
Sapotaceae	*Planchonella*	1	Indonesia	Nickel	945
	Sebertia	1	New Caledonia	Nickel	392
Scrophulariac.	*Alectra*	1	Zaïre	Cobalt	119
	Buchnera	1	Zaïre	Copper/cobalt	119
	Lindernia	2	Zaïre	Copper/cobalt	119
	Sopubia	1	Zaïre	Cobalt	119
Tiliaceae	*Trichospermum*	1	Indonesia	Nickel	945
	Triumfetta	1	Zaïre	Copper	119
Violaceae	*Agatea*	1	New Caledonia	Nickel	391
	Hybanthus	2	New Caledonia	Nickel	397
	Rinorea	2	Indonesia	Nickel	125, 127

thus identified are only those with an inordinate uptake (hyperaccumulation) of elements. Other indicators shown in Chapter 4, such as *Polycarpaea spirostylis,* do not accumulate high elemental levels even when growing directly over ore deposits.

21.5.   THE GEOGRAPHICAL DISTRIBUTION OF HYPERACCUMULATING SPECIES

The worldwide distribution of hyperaccumulators of various elements is shown in Figure 21.5. There are certain salient features in this figure. The first of these is the great preponderance of hyperaccumulators (nickel) in the single small island of New Caledonia. The second is the very large

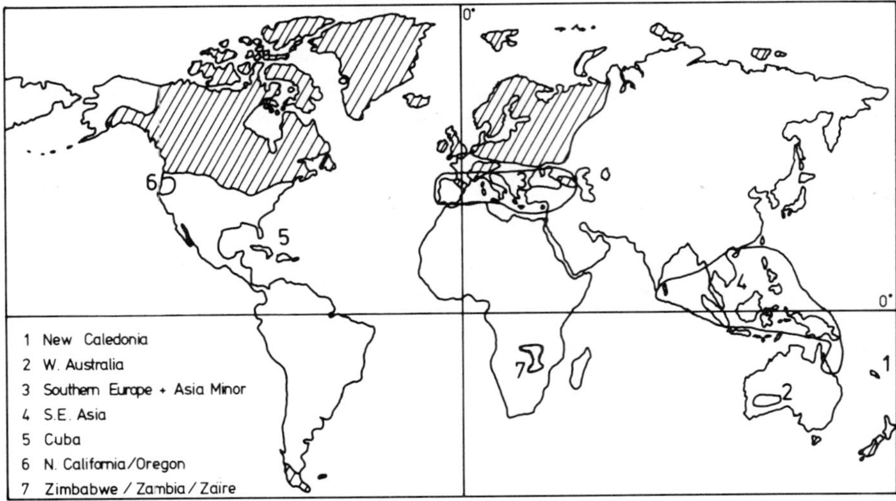

FIGURE 21.5.   The geographical distribution of plant hyperaccumulators of various elements. The shaded areas indicate the limits of the ice caps during the last glaciation.

number of metallophytes (nickel and zinc accumulators) found in southern and central Europe and in Asia Minor. The third feature is the island of "copper-cobalt plants" in Central Africa. Other smaller groups are found in the Western United States, Western Australia (one species), and Southeast Asia.

It will be observed from Figure 21.5 that hyperaccumulators are never found over previously glaciated areas. It may be concluded from this that hyperaccumulation of a particular element is a characteristic that takes a very long time to develop.

Although a great deal of work has been carried out on plants from southeast Europe, Asia Minor, and New Caledonia, it should not be assumed that this geographical distribution is apparent rather than real; that is, that most hyperaccumulators have been found in the above areas because these were where the most work was carried out. During the past few years we have analyzed very large numbers of plant from areas throughout the world ranging from Greenland to Patagonia, as well as covering every continent. So far, no hyperaccumulator has been discovered in South America or even in the greater part of the Asian land mass, despite the fact that these two continents have never been glaciated to any extent. To date we have discovered about 160 hyperaccumulators of which 138 are for nickel. We do not believe that there are many more to be found.

Apart from the factor of glaciation, it is not difficult to account for the geographical distribution of these metallophytes. Those that tolerate copper and cobalt depend on the existence of an area of mineralization sufficiently large to allow a diversity of metal-tolerant taxa to evolve. Only in Zaïre/Zambia is this requirement satisfied. Nickel plants evolve on extensive nonglaciated ultrabasic occurrences. This explains their multiplicity in southern Europe, Asia Minor, Southeast Asia, and New Caledonia, where extensive areas of serpentinic rocks are to be found.

21.6. AN ASSESSMENT OF BIOGEOCHEMICAL PROSPECTING IN THE HERBARIUM

Although some space in this book has been devoted to the topic of biogeochemical prospecting in the herbarium, this is partly because of its relative novelty and because there is no review on the subject that is available at present. Herbarium surveys will never, of course, replace *in situ* exploration work, but they can provide useful information at very low cost indeed. By far the single greatest cost in exploration geochemistry is that of collection of the sample itself. An herbarium sample has already been collected and the postal charges involved in sending a tiny leaf fragment for analysis are negligible. Against this advantage is the undeniable drawback that the original herbarium sample was usually not collected with mineral exploration in mind, although botanists are often attracted to such sites because of the unusual flora that might be expected there, as well as because of the ready access provided by mining roads and tracks. Very often botanists are not even aware of the geological nature of the substrate.

There is an interesting paradox inherent in herbarium surveys. Herbaria, like many other scientific institutions today, are increasingly under siege by budget-cutting zealots who can see no use in preserving a collection of dried leaves. The onus is increasingly on the herbarium curator to justify his existence. For this reason he is usually very happy to oblige in sending material to biogeochemists. It has been my own experience that 80–90% of all requests to herbaria are fulfilled speedily and cheerfully. If, however, herbarium surveys were ever to become a part of the usual armory of the exploration geochemist, it could be imagined that the attitude of curators might change. The herbarium sample, after all, is supposed to be preserved for posterity, and few curators would be happy to see a significant part of their collections literally go up in smoke.

In spite of the above reservations, there can be little doubt that a judicious and restrained use of herbarium material can furnish useful information to the exploration geochemist at very little cost.

22

BIOGEOCHEMICAL PROSPECTING IN RETROSPECT

22.1. SUCCESSFUL USE OF BIOGEOCHEMICAL PROSPECTING

It is difficult to assess the true effectiveness of biogeochemical prospecting because the method is seldom used by itself. It is usually used in combination with other methods such as soil surveys or stream sediment geochemistry. A number of surveys are also retrospective, that is, are used over anomalies previously determined by other methods. It is my personal experience that almost every subsurface anomaly that I have studied would have been indicated by the biogeochemical method if it had been used alone. Warren and Delavault [887] reached the same conclusion.

It is difficult to imagine that any mining company would undertake the expense of drilling any anomaly delineated by a single exploration method without confirmation by at least one other technique. Furthermore, the innate conservatism of human nature dictates that even if a single method were to be used, it would not be likely to be biogeochemistry, and would more likely be geophysics or stream sediment geochemistry, as these are "standard" methods. For this reason it is unlikely that many new discoveries will be made by biogeochemistry alone, except perhaps in the case of aerial sampling and analysis (see Chapter 20). However, there are in the literature many examples of ore discoveries in which biogeochemistry played a leading role. Some of these are listed in Table 22.1. In my opinion some of the more important contributions (all involving uranium) have

TABLE 22.1.   Some Examples of Successful Use of Biogeochemical Methods of Prospecting

Element	Location	Sample Type	Ore Depth (m)	References
Copper	Gaspé, Quebec	Balsam fir twigs	—	691
Copper	Caucasus, USSR	Birch leaves	2.4	813
Copper	Kazakhstan, USSR	Wormwood twigs	—	477
Copper/molybdenum	Uzbekistan, USSR	Cherry, almond	10–35	788
Copper/lead/zinc	Kazakhstan, USSR	Wormwood, juniper	1–5	863
Lead/zinc	Mississippi Valley, USA	Elms, maples, oaks	6–20	419
Uranium	La Ventura, USA	Pine, juniper, twigs	20	156
Uranium	Yellow Cat, USA	Astragalus, juniper	2–50	160
Uranium	Grants, USA	Juniper, pine	—	157
Uranium	Circle Cliffs, USA	Conifers	20	426
Uranium	Saskatchewan, Canada	Black spruce twigs	150	252

featured H. L. Cannon and her co-workers [156, 157, 160, 423] and more recently the work of Dunn [252], who showed convincingly that uranium anomalies in northern Saskatchewan could be detected beneath an overburden of 150 m of Athabasca sandstone by use of twigs of *Picea mariana* (black spruce). Up to 154 μg/g uranium was determined in the biological material, whereas the soil contained only 1–3 μg/g of this element.

The success achieved by the considerable volume of Russian work is perhaps more difficult to assess, though Malyuga [563] has reported at least six cases where ores were found by use of the biogeochemical method alone.

To sum up, it is a general rule that the number of mineral discoveries made by exploration personnel is linked to the amount of effort and money devoted to the search. Very little time or money has been devoted to biogeochemical research in the past. If this pattern were to change, there is no reason why biogeochemical prospecting should not be at least as successful as the other methods of exploration for minerals.

22.2.   THE ECONOMICS OF BIOGEOCHEMICAL PROSPECTING

In these days of hyperinflation, it is unwise to try to express the economics of any exploration technique in absolute monetary units. This was attempted, however, by Hawkes and Webb in 1962 [347], and by myself in 1972 [97], but today any figure assigned is liable to be out of date even before the book is published. In comparative terms however, biogeochemical methods can be compared with other techniques.

Kovalevsky [477], drawing on the Russian experience, has suggested that biogeochemical prospecting methods cost slightly more than the usual metallometric (soil sampling) programs, but considerably less when prospecting at depth is required. He suggested that one third of the cost of other methods is usual for depths of 2 m, one tenth for 4 m, and one hundredth at 30–35 m. Correct application of the biogeochemical technique should produce information very similar to that obtained from small mine workings.

A typical example of the favorable potential of biogeochemical methods was experienced by myself in the course of gold exploration in New Zealand. Soil samples were taken to a depth of about 1 m by means of an auger. Unfortunately, due to the stony nature of the substrate, only one attempt in four was successful, and the rate of sampling was only about four an hour. Biogeochemical samples (twigs of *Quintinia acutifolia*) could be taken at a rate many times faster.

Whether soil sampling or vegetation sampling is speedier depends not only on the physical state of the soil but also on the nature of the terrrain. In semidesert areas of gentle relief and sparse vegetation cover, soil samples can be taken at a very fast rate (several hundred per day). On the other hand, in areas of thick forest cover with a deep layer of overlying humus, the true soil is often difficult to find. This is especially true in some parts of New Zealand, where the humic layer can be several meters thick and covered with a dense mat of roots and forest litter. Under such conditions, plant sampling is appreciably quicker and less tiring.

Figure 22.1 compares biogeochemical prospecting with other exploration

Ore guides	Normal target diameter (2m · 20m · 200m · 1km · 10km · 100km · 1,000 km)	Property control			Reliability			Relative cost per square km (1 · 10 · 10² · 10³ · 10⁴ · 10⁵ · 10⁶)	Principal field of applicability		
		required	desirable	immaterial	excellent	good	fair		regional appraisal	reconnaissance exploration	detailed exploration
Favourable geological features											
Geochemical provinces				X					X	X	
Hydrothermal dispersion patterns		X	X							X	X
Drainage anomalies				X						X	X
Area soil anomalies			X	X						X	
Localized soil anomalies		X	X								X
Biogeochemical anomalies – on foot			X								X
– aerial			X						X	X	X
Geobotanical indicators – on foot				X							X
– aerial				X					X	X	
Botanical herbarium surveys				X					X		

FIGURE 22.1. Comparison of factors involved in various methods of prospecting.

techniques from the viewpoint of economics and other factors. It will be noted that apart from herbarium surveys, by far the cheapest botanical method is remote sensing using LANDSAT. It is difficult to assess the cost and reliability factors for airborne biogeochemistry, because so many factors (principally meteorological) affect the operation of the systems.

22.3. ADVANTAGES AND DISADVANTAGES OF BIOGEOCHEMICAL PROSPECTING

Some of the advantages and disadvantages of biogeochemical prospecting methods are summarized below.

Advantages

1. Large trees can penetrate through a thick overburden to give evidence of mineralization at depth. An effective depth of at least 30 m has been reported for the juniper in prospecting for uranium in the Colorado Plateau.
2. Plants with extensive root systems can effectively sample a large volume of soil.
3. Plant sampling is easier and quicker than soil sampling in forested areas.
4. Plant samples are much lighter than soil samples and this is useful in field work.
5. Chemical analysis of plant material is sometimes freer from interference problems than soil analysis because of the low concentrations of elements such as iron, which can cause problems in some spectroscopic methods.
6. In cases where accurate soil sampling depends on recognition of a particular horizon, plant sampling can be much easier, since it only depends on collection of an easily recognizable plant organ.
7. In cases where the soil is either nonexistent or complicated by the presence of such features as siliceous hardpan, biogeochemical prospecting can be more useful than soil surveys.
8. The method is particularly useful in areas of permafrost, and can always be used in winter where there is a thick snow cover.
9. The method is well adapted to aerial sampling and analysis.

Disadvantages

1. Plant sampling demands more skill from the field worker, particularly if there is a question of recognition of difficult species.
2. Unknown factors such as pH, drainage, aspect, age of the plant organ,

and so on can affect the reliabililty of the method, unless they are compensated for.

3. The method depends to some extent on an even distribution of species.
4. Usually some sort of orientation survey is needed in a new area. This is often more troublesome than a soil orientation survey.
5. Sometimes the method can only be used at certain times of the year, such as during the growing period if deciduous leaves are sampled.
6. Biogeochemical methods are not universally applicable, whereas soil surveys usually are.
7. Plant samples are much more subject to contamination than are soils.
8. Elements essential for plant nutrition can sometimes cause difficulty in some biogeochemical surveys.

The above listing of the advantages and disadvantages of biogeochemical methods of exploration is far from complete, but it may be concluded that the single greatest advantage of the technique is its penetrating power, and its greatest disadvantage is its variability caused by factors difficult to control.

22.4. CONSIDERATIONS GOVERNING THE USE OF BIOGEOCHEMICAL METHODS

Whether or not the biogeochemical method is to be employed in a particular area depends on a number of factors. In areas of permafrost or where the soil is shallow or virtually nonexistent, the usefulness of vegetation sampling is obvious. In other cases, however, the need for the method may not be as great as for a soil survey. Similarly, biogeochemical prospecting can be very useful in regions containing glacial till or transported soils where soil geochemistry is ineffective.

If biogeochemical methods of prospecting are to be used, there are a few conditions that must be satisfied before making an attempt. The first of these is that suitable, well-distributed species must be present in the exploration area. The easiest way of making this assessment is to use infrared color photography or satellite imagery. The latter technique does not really have sufficient resolution at present to identify individual species, though it can differentiate plant communities.

Another condition that should be satisfied is that the species selected should be suitable, that is, should have deep root systems and should not be so large that twigs, needles, or leaves cannot be reached from the ground.

In practice, few areas will be completely suitable or unsuitable for biogeochemical prospecting. The ultimate decision as to whether or not to use the technique will depend on the above factors as well as on economic considerations. A further favorable factor would be the existence of a pool of knowledge on background levels of elements in selected species. Such data are now available for all the common boreal species of the world, though not for the tropics or south-temperate regions.

Biogeochemical methods should be used preferably in conjunction with other procedures if economic considerations render this possible, since additional information can never be otherwise than beneficial. When biogeochemistry indicates anomalies confirmed by other methods this is useful, but when there is a conflict of data, investigation of the reasons for such discrepancies can often lead to fresh discoveries of direct use in prospecting. For example, an anomaly in plants not confirmed by soil geochemistry in an area of steep terrain [807] might indicate that soluble salts had been leached from an upslope zone of mineralization that might never have been discovered without the biogeochemical survey.

22.5.   THE FUTURE

It is notoriously difficult to forecast the future. In a previous work [97] written a decade ago, I predicted that significant advances would occur in the field of aerial biogeochemical sampling. This prediction did, in fact, come true with the development of the AIRTRACE and SURTRACE systems (see Chapter 20). It is not difficult to envisage that further refinements of this novel technique will again represent the direction in which very significant progress in biogochemical prospecting will occur.

In the earlier work, I also predicted increasing use of computer-assisted sophisticated statistical procedures. Several studies [807, 955] have employed such techniques with success, and there is no reason to suppose that future progress will not be made in this field also.

In the analytical field, the most exciting event during the past decade has probably been the development of inductively coupled plasma emission spectrometry (ICP; see Chapter 18) for the simultaneous analysis of up to 24 elements with a high degree of sensitivity and precision and with good accuracy. Future developments in this field will hopefully involve a lowering of relative costs of instrumentation and miniaturization of size, though even now the instrument can be carried and operated in a helicopter.

Biogeochemical exploration techniques have had much greater acceptance in recent years as a result perhaps of the multiplicity of new ideas and new instrumentation that have been applied to it. There is no reason why the method should not make progress in the future, greatly exceeding that which has gone before.

23

AN ELEMENT-BY-ELEMENT LISTING OF BIOLOGICAL PROSPECTING REFERENCES

The purpose of this chapter is to summarize, element by element, references on biological methods of prospecting. Data are also included for elemental abundances in igneous rocks [22, 327, 851], soils [79, 156], and vegetation (expressed on dry weight basis). To convert vegetation values to µg/g in ash, a factor of 20 should be employed. The references cited for vegetation are for wild plants only and do not include crop species for which there is an extensive literature.

The data are set out in Table 23.1. A few of these references concern analytical procedures for determining trace elements in plant material, but are only given in cases of special interest (e.g., gold in vegetation) and where the ultimate purpose of the work was to service samples used in biogeochemical prospecting.

TABLE 23.1. Abundance Data and References for Biological Prospecting Methods

Element	Abundance Data (μg/g)			Element Toxicity to Plants	References		
	Rocks	Soils	Plants		Biogeochem.	Geobot.	Geozool.
Antimony	0.3	0.5	0.05	Moderate	108, 259, 294, 621, 564, 792	294	
Arsenic	2	5	0.20	Severe	64, 106, 259, 315, 471, 547, 564, 592, 792, 894, 897, 934, 935	108, 937	
Barium	600	500	15	Moderate	14, 69, 80, 200, 227, 309, 318, 358, 462, 463, 467, 470, 638, 702, 711, 792, 868		
Beryllium	4	6	0.14	Severe	137, 141, 204, 358, 463, 465, 467, 473, 480, 493, 622, 631, 705, 792, 957, 963, 965		
Bismuth	0.2	0.8	0.04	Moderate	339, 478, 564, 764	38	
Boron	13	10	35	Slight	14, 17, 38, 139, 143, 166, 200, 227, 410, 438, 439, 638, 658, 697, 736, 737, 751, 767, 838, 844, 864	102, 143, 161	
Cadmium	0.13	0.20	0.005	Moderate	37, 103, 176, 227, 237, 339, 350, 354, 358, 466, 470, 500, 638, 711, 748		74, 75, 111
Caesium	10	5	10	Slight	963		
Chromium	100	150	0.05	Severe	14, 71, 166, 227, 318, 358, 388, 391, 411, 506, 509, 535, 536, 564, 625, 638, 640 660, 692, 792, 826, 839, 842, 864, 933	60, 799	
Cobalt	20	10	0.5	Severe	14, 58, 69, 101, 103, 113, 114, 116, 119, 127, 144, 227, 253, 254, 270, 363, 364, 365, 388, 391, 422, 438, 490, 491, 506, 547, 535, 536, 557, 568, 638, 640, 692, 722, 754, 765, 767, 792, 818, 826, 835, 839, 842, 846, 864, 893, 948	102, 161, 205, 241, 253, 254, 550, 553, 754, 779, 937	

(*Table continues on next page.*)

TABLE 23.1.   (Continued)

Element	Abundance Data (µg/g)			Element Toxicity to Plants	References		
	Rocks	Soils	Plants		Biogeochem.	Geobot.	Geozool.
Copper	70	20	20	Severe	2, 3, 11, 14, 23, 36–38, 47, 68, 69, 101, 103, 112, 113, 116, 117, 122, 126, 128, 130, 166, 173–178, 192–195, 200, 207, 215, 219, 224, 227, 231, 232, 226, 237, 256, 258, 268, 270, 286, 287, 294, 298, 299, 302, 310, 334, 342, 344, 345, 348, 350, 358, 361, 362, 366, 367, 370, 399, 414, 423, 438, 466, 468, 485, 494, 500, 524, 525, 527, 529, 546–548, 552, 558, 559, 561, 564–567, 573, 579, 591, 600, 601, 603, 612, 615, 619, 625–627, 637, 638, 650, 662, 674, 678, 687, 691, 692, 704, 711, 713, 719, 722, 724, 742, 745–747, 754, 755, 766, 767, 768, 776, 778, 785–788, 792, 793, 798, 805–809, 823, 830, 841, 846, 857, 860–862, 864, 884, 885, 890, 892, 895, 899, 903–905, 909, 924, 938, 947–949, 951, 955, 956	5, 13, 23, 36, 38, 68, 76, 102, 161, 177, 193, 205, 207, 208, 211, 214, 215, 231, 241, 253, 255, 287, 290, 342, 344, 348, 370–372, 527, 548, 553, 566, 615, 619, 627, 650, 704, 724, 742, 745–747, 754, 768, 823, 836, 837, 857, 930, 937, 938	74, 75, 95, 111, 247, 269, 408, 624, 624, 630, 802, 901
Fluorine	500	100	1	Moderate	462, 463, 467, 470, 479		
Gallium	15	15	0.5	Slight	141, 244, 259, 261, 294		
Germanium	2	5	0.2	Slight	261, 313, 643, 792		

Gold	0.001	0.002	0.001	Slight	2, 6, 28, 34, 38, 85, 103, 108, 167, 227, 229, 230, 240, 315, 316, 405, 416, 424, 462, 463, 467, 472, 477, 484, 495, 497, 519, 532, 613, 680, 712, 716, 725, 750, 778, 789, 790, 792, 793, 846, 878, 886, 908	38, 161, 635	24, 25, 34, 635, 911, 912, 921, 922
Hafnium	3	1	0.01	Slight	276		
Iron	4.6%	2.5%	0.01%	Slight	161, 166, 196, 197, 200, 205, 213, 227, 264, 286, 358, 391, 462, 463, 467, 470, 625, 638, 692, 711, 722, 778, 792, 812, 906	102, 136, 205, 213, 391, 404, 513, 598	111
Lead	15	10	10	Severe	14, 37, 38, 47, 69, 76, 103, 122, 123, 132, 141, 163, 166, 170, 181, 189, 191, 220, 227, 237, 256–259, 261, 268, 270, 273, 286, 287, 294, 297, 298, 310, 317, 334, 340, 344, 345, 350, 358, 361, 370, 399, 414, 419, 462, 463, 466–468, 470, 471, 485, 494, 500, 539, 545, 546, 564, 565, 567, 581, 612, 618, 619, 625, 626, 638, 639, 659, 674, 711, 719, 741, 755, 767, 776, 778, 785–788, 792, 793, 798, 805, 830, 846, 879, 895, 914, 959, 961	38, 76, 205, 207, 215, 287, 344, 619, 696, 937	74, 75, 95, 191, 269, 408, 624, 630, 802, 901
Lithium	50	30	0.1	Slight	77, 164, 622, 638, 792, 958		
Manganese	1000	850	100	Moderate	72, 77, 88, 124, 164, 179, 182, 197, 200, 227, 286, 310, 314, 318, 350, 358, 367, 391, 415, 437, 438, 462, 463, 467, 470, 511, 517, 555, 582, 598, 610, 625, 638, 640, 656, 658, 757, 778, 792, 815, 826, 833, 843, 844, 906, 958	102, 161, 833, 937	111

(Table continues on next page.)

Element	Abundance Data (μg/g)			Element Toxicity to Plants	References		
	Rocks	Soils	Plants		Biogeochem.	Geobot.	Geozool.
Mercury	0.1	0.01	0.01	Severe	38, 260, 412, 470, 621, 846, 898	38, 161	111
Molybdenum	2	3	0.5	Moderate	14, 28, 88 103, 112, 169, 174–176, 178, 227, 236, 249, 276, 279, 318, 360, 378, 406, 418, 434, 438, 442, 447, 457, 462–464, 467, 470, 517, 534, 547, 559–562, 564, 565, 579, 610, 622, 638, 658, 699, 764, 786–788, 792, 793, 896, 900, 907, 947, 948, 953, 954	161, 265, 378, 442, 537	877, 913
Nickel	100	40	1	Severe	13, 14, 71, 104, 109, 110, 114, 115, 118, 121, 122, 125, 127, 166, 209, 227, 244, 270, 297, 299, 318, 329, 337, 361, 387, 388, 391–397, 421, 422, 429, 504, 505, 509, 535, 536, 538, 554, 558, 564, 565, 579, 588, 593, 598, 602, 620, 625, 638, 640, 660, 682–684, 686, 692, 708, 711, 739, 765, 767, 772, 778, 792, 805, 826, 835, 839, 840, 842, 846, 864, 869, 889, 891, 931, 932, 938, 945, 948, 966	60, 102, 109, 161, 205, 207, 387, 392, 395–397, 538, 593, 601, 779, 931, 932, 937, 966	247
Niobium	20	15	0.01	Slight	276, 597, 622, 631, 828, 834		
Palladium	0.01	0.01	0.001	Slight	307, 435		
Platinum	0.01	0.01	0.001	Slight	307		
Radium (pg/g)	1	1	0.01	Severe	40, 332, 333, 336, 452–454, 462, 463, 467, 471, 492, 544, 810, 811, 845		
Rhenium	0.005	0.005	0.001	Slight	160, 585, 586		

Element						
Rubidium	280	80	2	Slight	638, 674, 963	
Scandium	22	10	1	Slight	141, 579, 722	
Selenium	0.01	0.05	0.05	Moderate	18, 152, 154, 155, 157, 158, 165, 168, 225, 295, 306, 338, 403, 425, 426, 432, 445, 510, 579, 580, 589, 821	102, 152, 154, 155, 157, 158, 160, 161, 165, 168, 306, 425, 426, 580, 589, 821
Silver	0.2	0.2	0.05	Severe	14, 38, 167, 175, 176, 227, 232, 244, 259, 294, 330, 358, 367, 462, 463, 467, 470, 471, 500, 547, 564, 579, 657, 658, 755, 764, 767, 792, 793, 886, 914, 948	36, 161, 232, 330, 348
Strontium	350	300	2	Slight	46, 200, 227, 310, 358, 438, 440, 441, 449, 638, 792	161, 441, 449
Sulfur	150	150	100	Slight	198, 199, 201	
Tantalum	2	2	0.01	Slight	631	
Thallium	0.5	0.1	0.01	Severe	259, 261	
Thorium	13	13	0.05	Slight	40, 86, 276, 291, 332, 336, 501, 518, 811, 845, 959, 961	
Tin	10	10	0.05	Moderate	14, 41, 224, 227, 244, 259, 318, 334, 382, 409, 466, 514, 547, 579, 591, 603, 640, 657, 792, 846, 963	161
Tungsten	2	1	0.05	Moderate	103, 227, 243, 276, 407, 409, 455, 462, 463, 476, 478, 481, 564, 657, 669, 670, 671, 672, 764, 792, 846	
Uranium	3	1	0.05	Moderate	19–21, 26, 29, 30–32, 39, 70, 73, 78, 82, 86, 92, 93, 102, 103, 107, 145, 151, 152, 154, 155, 157, 160, 161, 165, 168, 238, 239, 242, 250–252, 262, 263, 276–278, 280, 288, 291, 306, 321, 326, 332, 333, 355, 356, 373, 425, 426, 431, 433–436, 451, 452, 454, 456, 460–462, 467,	102, 151, 154, 155, 157, 160, 161, 165, 577, 666, 743, 744

(Table continues on next page.)

TABLE 23.1. (Continued)

Element	Abundance Data (μg/g)			Element Toxicity to Plants	References		
	Rocks	Soils	Plants		Biogeochem.	Geobot.	Geozool.
Vanadium	20	15	1	Slight	475, 481, 498, 501, 512, 516, 518, 522, 523, 569, 571, 572, 577, 599, 609, 611, 651, 666, 673, 693, 707, 726, 727, 734, 744, 771, 782, 783, 792, 801, 806, 810, 811, 832, 845, 850, 865, 869, 870, 919, 920, 925–928, 947, 952, 959, 961 65, 151, 157–160, 227, 244, 318, 463, 471, 514, 564, 638, 660, 792, 948	160, 161	
Yttrium	20	15	1	Slight	141, 959, 961		
Zinc	80	80	50	Moderate	14, 37, 47, 56, 69, 132, 166, 174–176, 181, 189, 194, 195, 215, 227, 237, 256–259, 268, 270, 286, 287, 294, 297, 317, 334, 340, 344, 350, 358, 361, 367, 401, 414, 419, 438, 450, 466, 468, 470, 471, 485, 500, 515, 539, 546, 547, 564, 565, 567, 579, 581, 612, 618, 625, 638, 619, 661, 662, 674, 691, 701, 711, 713, 719, 732, 733, 741, 755, 767, 776–778, 786–788, 793, 794, 798, 827, 846, 884, 885, 890, 892, 895, 899, 903, 905, 909, 914, 924, 947, 948, 951	102, 161, 207, 215, 287, 310, 344, 401, 515, 732, 937	74, 75, 95, 269, 408, 624, 630, 802, 901

pg/g = 10^{-6} μg/g.

REFERENCES

1. Åberg, B., 1948, *Ann. Kgl. Lantbr. Högsk.*, **15**, 37.
2. Adams, J., 1974, A Biogeochemical investigation of dispersed patterns of the Ste. Genevieve Co. Missouri copper deposits, MSc thesis, Univ. S. Illinois, Carbondale.
3. Adams, S. C., Hood, W. C., 1976, *J. Geochem. Explor.*, **6**, 163.
4. Adriani, M. J., 1958, in *Encyclopedia of Plant Physiology*, v. 4, Springer, Berlin.
5. Aery, N. C., 1977, *Geobios* (India), **4**, 225.
6. Aferov, Y. A., Zvygin, V. G., Roslyakova, M. V., Shabynin, L. L., Epov, L., 1968, *Vop. Geol. Pribaikal. Zabaikal.*, No. 3, 146.
7. Agricola, G., 1556, *De Re Metallica*, Eng. trans. 1912, *Mining Magazine*, London.
8. Agterberg, F. P., 1964, *Col. Sch. Mines Quart.*, **59**, 111.
9. Ahrens, L. H., 1954, *Geochim. Cosmochim. Acta*, **5**, 49.
10. Ahrens, L. H., Taylor, S. R., *Spectrochemical Methods of Analysis*, Addison-Wesley, Reading, Massachusetts.
11. Al'bov, M. N., Kostarev, I. I., 1968, *Sov. Geol.*, No. 2, 132.
12. Aldrich, D. G., Turrell, F. M., 1951, *Soil Sci.*, **70**, 83.
13. Aleskovsky, V. B., Mokhov, A. A., Spirov, V. A., 1959, *Geokhimiya*, No. 3, 266.
14. Allcott, G. H., 1970, *Saudi Arabia Min. Resourc. Res. Vol.*, 122.
15. Ally-Grégoire, R., 1976, Spectral reflectivity anomaly of *Populus canadensis* and *Salix pentandra* leaves grown in the laboratory on copper-contaminated soil (in Fr.), MSc thesis, École Polytéchnique, Montreal.
16. Alstrup, V., Hansen, E. S., 1977, *Oikos*, **29**, 290.
17. Ananichev, A. V., 1960, *Trudy Biogeokhim. Lab.*, No. 11, 226.
18. Anderson, M. S., Lakin, H. W., Beeson, K. C., Smith, F. F., Thacker, E., 1961, *U.S. Dept. Agr. Hdbk.*, **200**, 65 pp.
19. Anderson, R. Y., Kurtz, E. B., 1954, *Bull. Geol. Soc. Am.*, **65**, 1226.
20. Anderson, R. Y., Kurtz, E. B., 1955, *Econ. Geol.*, **50**, 227.
21. Anderson, R. Y., Kurtz, E. B., 1956, *Econ. Geol.*, **51**, 64.
22. Andrews-Jones, D. A., 1968, *Mineral Ind. Bull.*, **11**, 1.

23. Anon., 1959, *Horizon* (Zambia), **1**, 35.

24. Anon., 1969, *Chamber of Mines J.* (Zimbabwe), **11**, 16.

25. Anon., 1969, *Chamber of Mines J.* (Zimbabwe), **11**, 19.

26. Anon., 1974, *Min. Eng.*, **26**, 20.

27. Antonovics, J., Bradshaw, P. D., Turner, R. G., 1971, *Adv. Ecol. Res.*, **7**, 1.

28. Aripova, Kh., Talipov, R. M., 1966, *Uzbek. Geol. Zhur.*, **10**, 45.

29. Armands, G., 1967, in *Geochemical Prospecting in Fennoscandia*, Wiley, New York, 154.

30. Armands, G., Landergren, S., 1960, *Rep. 21st Int. Geol. Congr. Norden*, Part 15, 51.

31. Armstrong, C. W., Brewster, N. W., Ferguson, J., Goleby, A. B., 1979, Proc. *IAEA Ser. III*, STI/PUB/555, 733.

32. Arribas, A., Herrero, J., 1977, *Abs. 2nd Symp. Origin and Distribution of Elements*, Spain.

33. Aubrey, K. V., 1956, *Geochim. Cosmochim. Acta*, **9**, 83.

34. Babička, J., 1943, *Mikrochim. Verein Mikrochim. Acta*, **31**, 201.

35. Babička, J., Komarek, J. M., Nemec, B., 1945, *Bull. Int. Acad. Tcheque Sci.*, **45**, 1.

36. Bailey, F. M., 1899, *The Queensland Flora*, Diddams, Brisbane.

37. Baker, A. J. M., 1981, *J. Pl. Nutr.*, **3**, 643.

38. Ballew, G. I., 1975, *Proc. 10th Int. Symp. Rem. Sens. Environ.*, **2**, 1045.

39. Barakso, J. J., 1979, *Can. Min. Metall. Bull.*, April, 135.

40. Baranov, V. I., Kunasheva, K. G., 1954, *Trudy Biogeokhim. Lab.*, No. 10, 94.

41. Bardyuk, V. V., Ivashov, P. V., 1970, *Geol. Geofiz. Sib. Otd. Akad. Nauk SSSR*, No. 12, 125.

42. Barringer, A. R., 1977, *U. S. Geol. Surv. Prof. Pap.*, 1015, 231.

43. Barringer, A. R., Davies, J. H., Daubner, L., 1978, *Proc. 12th Int. Symp. Rem. Sens. Environ.*, Manila, 975.

44. Barshad, I., 1948, *Soil Sci.*, **66**, 187.

45. Barshad, I., 1951, *Soil Sci.*, **71**, 297.

46. Basitova, S. M., Zasorina, E. F., 1965, *Dokl. Akad. Nauk Tadzhik. SSR*, No. 8, 20.

47. Basitova, S. M., Zasorina, E. F., Khamidova, R. M., Kalashnikova, G. M., 1964, *Izv. Akad. Nauk SSSR Otd. Fiz.-Tekh. Khim. Nauk*, No. 1, 30.

48. Bateman, W. G., Wells, L. S., 1917, *J. Am. Chem. Soc.*, **39**, 811.

49. Baumeister, W., 1954, *Ber. Dt. Bot. Gesell.*, **67**, 205.

50. Bayer, E., Egeter, H., Fink, A., Nether, K., Wegman, K., 1966, *Angew. Chem. Int. Edn.*, **5**, 791.

51. Bazilevskaya, N. A., Sibireva, Z. A., 1950, *Byull. Glav. Bot. Sada Leningr.*, No. 6, 32.

52. Beath, O. A., Eppson, H. F., Gilbert, C. S., 1935, *Wyoming Agr. Res. Stn. Bull.*, **206**, 55 pp.

53. Beath, O. M., Gilbert, C. S., Eppson, H. F., 1939, *Am. J. Bot.*, **26**, 257.

54. Beauford, W., Barber, J., Barringer, A. R., 1975, *Nature*, **256**, 35.

55. Beauford, W., Barber, J., Barringer, A. R., 1977, *Science*, **195**, 571.

56. Beck, C. R., 1909, *Lehre von den Erzlagerstätten*, Borntraeger, Berlin.

57. Beerstecher, E., 1954, *Petroleum Microbiology*, Elsevier, New York.

58. Beeson, K. C., Lazar, V. A., Boyce, S. G., 1955, *Ecology*, **36**, 155.

59. Bélanger, J. R., 1980, *Geol. Surv. Can. Pap.*, **80-1B**, 287.

60. Bélanger, J. R., 1980, *Geos* (Ottawa), Summer, 10.

61. Bélanger, J. R., Rencz, A. N., Shilts, W. W., 1979, *Télédétection et Gestion des Ressources,* Association Québecoise de Télédétection, Ste-Foy, Québec.

62. Bell, J. M., Clarke, E. de C., Marshall, P., 1911, N. Z. Geol. Surv. Bull., **12,** 71 pp.

63. Benedict, R. C., 1937, in *Standard Cyclopedia of Horticulture,* Macmillan, New York.

64. Berbenni, P., 1959, *Ann. Chim.,* **49,** 614.

65. Bertrand, D., 1950, *Bull. Am. Mus. Nat. Hist.,* **94,** 409.

66. Beyer, W. H., 1968, *Handbook of Tables for Probability and Statistics,* 2nd ed., Chemical Rubber Co., Cleveland.

67. Billings, W. D., 1950, *Ecology,* **31,** 62.

68. Birnie, R. W., Francica, J. R. Jr., 1979, *Abs. Progr. Geol. Soc. Am.,* **11,** 389.

69. Björklund, A., 1971, *Bull. Geol. Soc. Finl.,* **251,** 42 pp.

70. Björklund, A., Tenhola, M., 1976, *J. Geochem. Explor.,* **5,** 244.

71. Blondiaux, G., Crousilles, M., Debrun, J. L., Dixsaut, C., 1978, *Bull. Bur. Fr. Rech. Géol. Min.,* Sec. 2, 323.

72. Bloss, D., Steiner, A. L., 1960, *Bull. Geol. Soc. Am.,* **71,** 1053.

73. Bogdanov, Y. V., Prakash, R., Mirkina, S. L., Onoshko, I. S., 1977, *Dokl. Akad. Nauk SSSR,* **224,** 30.

74. Bollingberg, H. J., 1975, *Geochim. Cosmochim. Acta,* **39,** 1567.

75. Bollingberg, H. J., Johansen, P., 1979, *J. Fish. Res. Bd. Can.,* **36,** 1023.

76. Bølviken, B., Honey, F., Levine, S. R., Lyon, R. J. P., Prelat, A., 1977, *J. Geochem. Explor.,* **8,** 457.

77. Borovik-Romanova, T. F., 1965, *Probl. Geokhim.,* 1965, 620.

78. Botova, M. N., Malyuga, D. P., Moiseenko, U. I., 1963, *Geochemistry,* No. 8, 379.

79. Bowen, H. J. M., 1966, *Trace Elements in Biochemistry,* Macmillan, London.

80. Bowen, H. J. M., Dymond, J. A., 1955, *Proc. Roy. Soc. Lond. Sec. B,* **144,** 355.

81. Bowen, N. L., 1928, *The Evolution of the Igneous Rocks,* Princeton University Press, Princeton.

82. Bowes, W. A., Boles, W. E., Hazelton, G. M., 1957, *U. S. At. Energ. Comm. Rep.,* RME-2063, Part 7.

83. Boyle, R. W., 1958, *Proc. 20th Int. Geol. Congr.,* Mexico City, 175.

84. Boyle, R. W., 1977, *J. Geochem Explor.,* **8,** 495.

85. Boyle, R. W., 1979, *Geol. Surv. Can. Bull.,* **280,** 576 pp.

86. Boyle, R. W., 1980, *At. Energ. Rev.,* **18,** 3.

87. Boyle, R. W., Cragg, C. B., 1957, *Bull. Mines Brch. Can.,* No. 39, 27 pp.

88. Bradis, A. V., Malinovsky, V. S., Sorokin, V. K., 1963, *Trudy Kalinin. Gos. Med. Inst.,* No. 10, 19.

89. Bradshaw, A. D., 1952, *Nature,* **169,** 1098.

90. Braun-Blanquet, J., 1932, *Plant Sociology,* McGraw-Hill, New York.

91. Bray, J. R., Curtis, J. T., 1957, *Ecol. Monogr.,* **27,** 325.

92. Breger, I. A., 1974, Proc. *Symp. Form. Uranium Ore Dep.,* Athens, IAEA-SM-183/29, 99.

93. Breger, I. A., Deul, M., 1956, *U. S. Geol. Surv. Prof. Pap.,* **300,** 505.

94. Brenchley, W. E., 1936, *Bot. Rev.,* **2,** 173.

95. Brock, J. S., 1972, *W. Miner,* **45,** 28.

96. Brookes, C. J., Betteley, I. G., Loxston, S. M., 1966, *Mathematics and Statistics for Chemists,* Wiley, New York.

97. Brooks, R. R., 1972, *Geobotany and Biogeochemistry in Mineral Exploration*, Harper & Row, New York.

98. Brooks, R. R., 1972, *Proc. Australas. Inst. Min. Metall.*, No. 241, 15.

99. Brooks, R. R., 1973, *J. Appl. Ecol.*, **10**, 825.

100. Brooks, R. R., 1977, in *Environmental Chemistry*, Plenum, New York.

101. Brooks, R. R., 1977, *Pl. Soil*, **48**, 541.

102. Brooks, R. R., 1979, *J. Geochem. Explor.*, **12**, 67.

103. Brooks, R. R., 1979, *Geol. Surv. Can. Econ. Geol. Rep.*, **31**, 397.

104. Brooks, R. R., 1980, in *Biogeochemistry of Nickel*, Wiley, New York.

105. Brooks, R. R., Crooks, H. M., 1980, *Pl. Soil*, **54**, 491.

106. Brooks, R. R., Fergusson, J. E., Holzbecher, J., Ryan, D. E., Zhang, H. F., Dale, J. M., Freedman, B., 1982, *Environ. Pollut.*, **13**, 109.

107. Brooks, R. R., Holzbecher, J., Robertson, D. J., Ryan, D. E., 1982, *J. Geochem. Explor.*, **16**, 189.

108. Brooks, R. R., Holzbecher, J., Ryan, D. E., 1981, *J. Geochem. Explor.* **16**, 24.

109. Brooks, R. R., Lee, J., Jaffré, T., 1974, *J. Ecol.*, **62**, 523.

110. Brooks, R. R., Lee, J., Reeves, R. D., Jaffré, T., 1977, J. Geochem. Explor., **7**, 49.

111. Brooks, R. R., Lewis, J. R., Reeves, R. D., 1976, *N. Z. J. Mar. Freshw. Res.*, **10**, 233.

112. Brooks, R. R., Lyon, G. L., 1966, *N. Z. J. Sci.*, **9**, 706.

113. Brooks, R. R., McCleave, J. A., Malaisse, F., 1977, *Proc. Roy. Soc. Lond. Ser. B*, **197**, 231.

114. Brooks, R. R., McCleave, J. A., Schofield, E. K., 1977, *Taxon*, **26**, 197.

115. Brooks, R. R., Morrison, R. S., Reeves, R. D., Dudley, T. R., Akman, Y., 1979, *Proc. Roy. Soc. Lond. Ser. B*, **203**, 387.

116. Brooks, R. R., Morrison, R. S., Reeves, R. D., Malaisse, F., 1978, *Pl. Soil*, **50**, 503.

117. Brooks, R. R., Radford, C. C., 1978, *Proc. Australas. Inst. Min. Metall.*, No. 268, 33.

118. Brooks, R. R., Radford, C. C., 1978, *Proc. Roy. Soc. Lond. Ser. B*, **200**, 217.

119. Brooks, R. R., Reeves, R. D., Morrison, R. S., Malaisse, F., 1980, *Bull. Soc. Roy. Bot. Belg.*, **113**, 166.

120. Brooks, R. R., Ryan, D. E., Zhang, H. F., 1981, *At. Spectros.*, **2**, 161.

121. Brooks, R. R., Shaw, S., Arsensi Marfil, A., 1981, *Vegetatio*, **45**, 183.

122. Brooks, R. R., Trow, J. M., Bølviken, B., 1979, *J. Geochem. Explor.*, **11**, 73.

123. Brooks, R. R., Trow, J. M., Fredskild, B., 1980, *Oikos*, **35**, 425.

124. Brooks, R. R., Trow, J. M., Veillon, J.-M., Jaffre, T., 1981, *Taxon*, **30**, 420.

125. Brooks, R. R., Wither, E. D., 1977, *J. Geochem. Explor.*, **7**, 295.

126. Brooks, R. R., Wither, E. D., Westra, L. Y., 1978, *J. Geochem. Explor.*, **10**, 181.

127. Brooks, R. R., Wither, E. D., Zepernick, B., 1977, *Pl. Soil*, **47**, 707.

128. Brooks, R. R., Yates, T. E., Ogden, J., 1973, *N. Z. J. Bot.*, **11**, 443.

129. Brown, A. C., 1964, *Geol. Surv. Can. Pap.*, **63-42**, 26 pp.

130. Brown, A. G., 1971, *Trans. Inst. Min. Metall.* Sec. B, **80**, 138.

131. Brown, D. H., 1973, *Proc. Bristol Naturalist Soc.*, **32**, 267.

132. Brown, J. S., Meyer, P. A., 1960, *Proc. Geochem. Explor. Symp.*, 20th Int. Geol. Congr., Mexico City, 623.

133. Brundin, N. H., Nairis, B., 1972, *J. Geochem. Explor.*, **1**, 7.

134. Brundin, N. J., 1939, U. S. Pat. 2158980.

135. Brussell, D., 1978, *Phytologia*, **38**, 469.

136. Buck, L. J., 1951, *J. N. Y. Bot. Gdn.,* Jan.-Feb., 22.

137. Buehling, A., Carl, C., Herr, W., Ney, P., Holzer, H. F., Stumpfl, E. F., 1978, *Verh. Öst. Bundesanst.,* No. 3, 267.

138. Buks, I. I., 1965, in *Plant Indicators of Soils, Rocks, and Subsurface Waters,* Consultants Bureau, New York.

139. Burenkov, E. K., Kuzina, 1968, *Int. Geol. Rev.,* **10,** 421.

140. Burghardt, H., 1956, *Flora,* **143,** 1.

141. Burkser, E. S., Mitskevich, B. F., 1960, *Dopovidi Akad. Nauk Ukr. RSR,* No. 3, 349.

142. Butterworth, J., Lester, P., Nickless, G., 1972, *Mar. Pollut. Bull.,* **3,** 72.

143. Buyalov, N. I., Shvyryayeva, A. M., 1961, *Int. Geol. Rev.,* **3,** 619.

144. Cabrol, J., Cabrol, P., Magny, J., 1961, *Bull. Soc. Chim. Biol.,* **43,** 1111.

145. Calvo, M. M., 1974, *Proc. Symp. Form. Uranium Ore Dep.,* Athens, IAEA-SM-183/33, 125.

146. Cameron, E. M., 1968, *Can. J. Earth Sci.,* **5,** 287.

147. Cameron, E. M., 1969, *Can. J. Earth Sci.,* **6,** 247.

148. Canney, F. C., 1971, *4th Ann. Earth Res. Progress Rev.* (NASA), Houston.

149. Canney, F. C., Cannon, H. L., Cathrall, J. B., Robinson, K., 1979, *Econ. Geol.,* **74,** 1673.

150. Canney, F. C., Wenderoth, S., Yost, E., 1970, *3rd Ann. Earth Res. Progress Rev.* (NASA), Houston, **1,** 18.

151. Cannon, H. L., 1952, *Am. J. Sci.,* **250,** 735.

152. Cannon, H. L., 1953, *U. S. Geol. Surv. Circ.,* **264,** 8 pp.

153. Cannon, H. L., 1955, *U. S. Geol. Surv. Bull.,* **1000D,** 119.

154. Cannon, H. L., 1957, *U. S. Geol. Surv. Bull.,* **1030M,** 399.

155. Cannon, H. L., 1959, *Proc. Geochem. Symp., 20th Int. Geol. Congr.,* Mexico City, 235.

156. Cannon, H. L., 1960, *Science,* **132,** 591.

157. Cannon, H. L., 1960, *U. S. Geol. Surv. Bull.,* **1085A,** 50 pp.

158. Cannon, H. L., 1960, *U. S. Geol. Surv. Prof. Pap.,* **400B,** 96.

159. Cannon, H. L., 1963, *Soil Sci.,* **96,** 196.

160. Cannon, H. L., 1964, *U. S. Geol. Surv. Bull.,* **1176,** 127 pp.

161. Cannon, H. L., 1971, *Taxon,* **20,** 227.

162. Cannon, H. L., 1979, *Geol. Surv. Can. Econ. Geol. Rep.,* **31,** 385.

163. Cannon, H. L., Bowles, J. M., 1962, *Science,* **137,** 765.

164. Cannon, H. L., Harms, T. F., Hamilton, J. C., 1975, *U. S. Geol. Surv. Prof. Pap.,* **918,** 23 pp.

165. Cannon, H. L., Kleinhampl, F. J., 1956, *U. S. Geol. Surv. Prof. Pap.,* **300,** 681.

166. Cannon, H. L., Papp, C. S. E., Anderson, B. M., 1972, *Ann. N. Y. Acad. Sci.,* **199,** 124.

167. Cannon, H. L., Shacklette, H. T., Bastron, H., 1968, *U. S. Geol. Surv. Bull.,* **1278A,** 21 pp.

168. Cannon, H. L., Starrett, W. H., 1956, *U. S. Geol. Survey Bull.,* **1009M,** 391.

169. Carlisle, D., Cleveland, G. B., 1958, *Calif. Div. Mines Spec. Rep.,* No. 50, 31 pp.

170. Carter, M. W., 1966, A Preconcentration spectrographic method for the determination of trace elements in plant materials and the application of the biogeochemical method at the Silver Mine lead deposits, Cape Breton, Nova Scotia, MSc thesis, Carleton Univ. Ottawa.

171. Cassie, R. M., 1954, *Austral. J. Mar. Freshw. Res.,* **5,** 513.

172. Cattell, R. B., 1965, *Biometrics,* **21,** 190, 405.

173. Chaffee, M. A., 1975, *U. S. Geol. Surv. Prof. Pap.,* **907B,** 26 pp.

174. Chaffee, M. A., 1976, *U. S. Geol. Surv. Bull.,* **1278D,** 55 pp.

175. Chaffee, M. A., 1976, *U. S. Geol. Surv. Open File Rep.,* **76-559,** 34 pp.

176. Chaffee, M. A., 1977, *U. S. Geol. Surv. Bull.,* **1278E,** 78 pp.

177. Chaffee, M. A., Gale, C. W. III, 1976, *J. Geochem. Explor.,* **5,** 59.

178. Chaffee, M. A., Hessin, J. D., 1971, *Can. Inst. Min. Metall. Spec. Vol.,* 11, 401.

179. Chamberlain, G. T., Searle, A. J., 1963, *E. Afr. Agric. For. J.,* **29,** 114.

180. Chandler, R. F., 1942, *Proc. Soil. Sci. Am.,* **7,** 454.

181. Chapman, R. A., Shacklette, H. T., 1960, *U. S. Geol. Surv. Prof. Pap.,* **400B,** 104.

182. Chartko, M. K., *Vesti Akad. Navuk BSSR Ser. Sel'sk. Navuk,* No. 4, 42.

183. Chayes, F., 1954, *Geochim. Cosmochim. Acta,* **6,** 119.

184. Chebaevskaya, V. S., 1960, *Izv. Sel's. Akad. K. A. Timiryazeva,* No. 5, 123.

185. Chenery, E. M., 1948, *Kew Bull.,* 1948, 173.

186. Chenery, E. M., 1951, *Col. Office Pap.,* 1529.

187. Chenery, E. M., 1951, *J. Soil Sci.,* **2,** 97.

188. Chenery, E. M., Sporne, K. R., 1976, *New Phytol.,* **76,** 551.

189. Cheuong, Ho Kun, 1973, *J. Geol. Soc. Korea,* **9,** 244.

190. Chikishev, A. G., 1965, *Plant Indicators of Soil, Rocks, and Subsurface Waters,* Consultants Bureau, New York.

191. Chisnall, K. T., Markland, J., 1971, *J. Ass. Pub. Analysts,* **9,** 116.

192. Chiu, Jung-Chin, Wang, T. C., 1964, *Acta Microbiol. Sin.,* **10,** 157.

193. Chou-Chin-Han, 1960, *Int. Geol. Rev.,* **2,** 361.

194. Chukhrov, F. V., Churikov, V. S., Yermilova, L. P., Kalentchuk, G. E., 1979, *J. Geochem. Explor.,* **12,** 79.

195. Chukhrov, F. V., Churikov, V. S., Yermilova, L. P., Kalentchuk, G. Ye., 1978, *Izv. Akad. Nauk SSSR Ser. Geol.,* No. 5, 5.

196. Chukhrov, F. V., Gorshkov, A. I., Berezovskaya, V. V., Sivtsov, A. V., 1980, *Izv. Akad. Nauk SSSR Ser. Geol.,* No. 11, 81.

197. Chukhrov, F. V., Gorshkov, A. I., Tyuryukanov, A. N., Berezovskaya, V. V., Sivtsov, A. V., 1980, *Izv. Akad. Nauk SSSR, Ser. Geol.,* No. 7, 5.

198. Chukhrov, F. V., Yermilova, L. P., Churikov, V. S., Kalentchuk, G. Ye., 1978, *Izv. Akad. Nauk SSSR Ser. Geol.,* No. 3, 5.

199. Chukhrov, F. V., Yermilova, L. P., Churikov, V. S., Nosik, L. P., 1978, *Geochem. Int.,* **15,** 25.

200. Chukhrov, F. V., Yermilova, L. P., Kalentchuk, G. E., Nikitina, I. B., 1981, *Izv. Akad. Nauk SSSR Ser. Geol.,* No. 7, 86.

201. Chukhrov, F. V., Yermilova, L. P., et al., 1976, *Proc. 25th Sess. Int. Geol. Congr., Moscow,* 55.

202. Clarke, F. W., 1924, *U. S. Geol. Surv. Bull.,* **770,** 841 pp.

203. Clymo, R. S., 1963, *Ann. Bot. Lond.,* **27,** 309.

204. Cohen, N. C., Brooks, R. R., Reeves, R. D., 1967, *N. Z. J. Geol. Geophys.,* **10,** 732.

205. Cole, M. M., 1965, *Biogeography in the Service of Man,* Inaug. Lect., Bedford College, London, 59 pp.

206. Cole, M. M., 1968, *Proc. Roy. Soc. Lond. Ser. A,* **308,** 173.

207. Cole, M. M., 1971, *Can. Inst. Min. Metall. Spec. Vol.,* **11,** 414.

208. Cole, M. M., 1971, *Geol. Mijnb.,* **50,** 771.

209. Cole, M. M., 1973, *J. Appl. Ecol.,* **10,** 269.

210. Cole, M. M., 1977, *Trans. Inst. Min. Metall. Sec. B,* **86,** 195.

211. Cole, M. M., 1980, *Trans. Inst. Min. Metall. Sec. B,* **89,** 73.

212. Cole, M. M., 1980, *Proc. 6th Ann. Pecora Symp. Rem. Sens. Environ.,* Sioux Falls.

213. Cole, M. M., Brown, R. C., 1976, *J. Biogeogr.,* **3,** 169.

214. Cole, M. M., Le Roex, H. D., 1978, *Trans. Geol. Soc. S. Afr.,* **81,** 277.

215. Cole, M. M., Provan, D. M. J., Tooms, J. S., 1968, *Trans. Inst. Min. Metall. Sec. B,* **77,** 81.

216. Collins, W. E., Raines, G. L., Canney, F. C., 1977, *Abs. Progr. Geol. Soc. Am.,* **9,** 932.

217. Colwell, R. N., 1968, *Sci. Am.,* **218,** 54.

218. Connor, J. J., Miesch, A. T., 1964, *Stanford Univ. Pub. Geol. Sci.,* **9,** 110.

219. Connor, J. J., Shacklette, H. T., Erdman, J. A., 1971, *U. S. Geol. Surv. Prof. Pap.,* **750B,** 151.

220. Cooke, H. R., 1978, *Greenl. Geol. Unders. Rapp.,* No. 90, 27.

221. Cooley, W. W., Lohnes, P. R., 1971, *Multivariate Data Analysis,* Wiley, New York.

222. Corbett, J. R., 1969, *The Living Soil,* Martindale Press, Sydney.

223. Core, E. L., 1955, *Plant Taxonomy,* Prentice-Hall, Englewood Cliffs, New Jersey.

224. Correll, R. L., Taylor, R. G., 1975, *Proc. Australas. Inst. Min. Metall.,* No. 255, 7.

225. Cowgill, U. M., 1979, *J. Pl. Nut.,* **1,** 73.

226. Craigie, J. S., Maas, W. S. G., 1966, *Ann. Bot. Lond.,* **30,** 153.

227. Curtin, G. C., Day, G. W., Tripp, R. B., 1980, *U. S. Geol. Surv. Open File Rep.,* **80-156,** 22 pp.

228. Curtin, G. C., King, H. D., Mosier, E. L., 1974, *J. Geochem. Explor.,* **3,** 245.

229. Curtin, G. C., Lakin, H. W., Hubert, W., Hubert, A. F., Mosier, E. C., Watts, K. C., 1971, *U. S. Geol. Surv. Bull.,* **1278B,** 16 pp.

230. Curtin, G. C., Lakin, H. W., Neuerburg, G. J., Hubert, A. E., 1968, *U. S. Geol. Surv. Circ.,* **562,** 11 pp.

231. Czehura, S. J., 1977, *Econ. Geol.,* **72,** 796.

232. Czehura, S. J., 1978, *Abs. Progr. Geol. Soc. Am.,* **10,** 214.

233. Dahlberg, E. C., 1968, *Econ. Geol.,* **63,** 409.

234. Davis, J. C., 1973, *Statistics and Data Analysis in Geology,* Wiley, New York.

235. Dawson, K. M., Sinclair, A. J., 1974, *Econ. Geol.,* **69,** 404.

236. Day, G. W., Curtin, G. C., Tripp, R. B., 1979, *U. S. Geol. Surv. Open File Rep.,* **79-1320.**

237. Day, G. W., Curtin, G. C., Tripp, R. B., 1979, *U. S. Geol. Surv. Open File Rep.,* **79-1319.**

238. Dean, M. H., 1966, *J. Ecol.,* **54,** 589.

239. Debnam, A. H., 1955, *Rec. Commonw. Australia Dep. Natl. Dev. Bur. Min. Res. Geol. Geophys.,* 1955/48, 24 pp.

240. Dekate, Y. G., 1971, *J. Ind. Acad. Geosci.,* **13,** 74.

241. De Plaen, G., Malaisse, F., Brooks, R. R., 1982, *Endeavour,* **6,** 72.

242. Dilabio, R. N. W., Rencz, A. N., 1980, *Can. Bot.,* **58,** 18.

243. Dobrolyubsky, O. K., Vozdeistvy, O., 1964, *Dokl. Nauchn. Uyssh. Shk. Ser. Biol. Nauk,* No. 1, 142.

244. Dobrovol'sky, V. V., 1963, *Dokl. Nauchno. Uyyssh. Shk. Ser. Biol. Nauk,* No. 3, 193.

245. Docters van Leeuwen, W. M., 1937, *Blumea,* 2, 239.

246. Dokuchaev, V. V., 1883, *The Russion Chernozem* (in Russ.), St. Petersburg.

247. D'Orey, F. L. C., 1973, *Trans. Inst. Min. Metall.*, **84**, 150.

248. Dorn, P., 1937, *Biologe*, **6**, 11.

249. Doyle, P., Fletcher, W. K., Brink, V. C., 1973, *Can. J. Bot.*, **51**, 421.

250. Dunn, C. E., 1979, *Sask. Dep. Min. Res. Misc. Rep.*, **79-10**, 166.

251. Dunn, C. E., 1980, *Sask. Dep. Min. Res. Misc. Rep.*, **80-4**, 60.

252. Dunn, C. E., 1981, *J. Geochem. Explor.*, **15**, 437.

253. Duvigneaud, P., 1958, *Bull. Soc. Roy. Bot. Belg.*, **90**, 1278.

254. Duvigneaud, P., 1959, *Bull. Soc. Roy. Bot. Belg.*, **91**, 111.

255. Duvigneaud, P., Denaeyer-De Smet, S., 1963, *Bull. Soc. Roy. Bot. Belg.*, **96**, 93.

256. Dvornikov, A. G., Ovsyannikova, L. B., 1970, *Dokl. Akad. Nauk SSSR*, **195**, 187.

257. Dvornikov, A. G., Ovsyannikova, L. B., 1972, *Geochem. Int.*, No. 9, 591.

258. Dvornikov, A. G., Ovsyannikova, L. B., 1972, *Geokhimiya*, No. 7, 873.

259. Dvornikov, A. G., Ovsyannikova, L. B., 1974, *Geol. Zhur.*, **34**, 49.

260. Dvornikov, A. G., Ovsyannikova, L. B., Sidenko, O. G., 1976, *Geochem. Int.*, **13**, 189.

261. Dvornikov, A. G., Ovsyannikova, L. B., Sidenko, O. G., 1973, *Dopovidi Akad. Nauk Ukr. RSR.*, No. 6, 490.

262. Dyck, W., Boyle, R. W., 1980, *Bull. Can. Inst. Min. Metall.*, June, 77.

263. Eakins, G. R., 1970, *Alaska Div. Mines Geol. Rep.*, **41**, 52 pp.

264. Easton, R. M., 1976, Geobotanical studies in the Back River volcanic complex, Northwest Territories, BSc thesis, Univ. W. Ontario, London, Ontario.

265. Eaton, M. K., 1979, Study of vegetation anomalies determined by using LANDSAT digital data on the Pine Nut Mountains of western Nevada, MSc thesis, Stanford Univ., Palo Alto.

266. Eckel, E. B., Williams, J. S., Galbraith, F. W., et al., *U. S. Geol. Surv. Prof. Pap.*, **219**, 179 pp.

267. Efroymson, M. A., 1960, in *Mathematical Methods for Digital Computers*, Wiley, New York.

268. Ek, J., 1974, *Sverig. Geol. Undersökn. Avh. Ser. C*, No. 698, 1.

269. Ekdahl, E., 1976, *J. Geochem. Explor.*, **5**, 296.

270. El Shazly, E. M., Barakat, N., Eissa, E. A., Emara, H. H., Ali, I. S., Shaltout, S., Sharaf, F. S., 1971, *Can. Inst. Min. Metall., Spec. Vol.*, **11**, 426.

271. Ellenberg, H., 1958, in *Encyclopedia of Plant Physiology* v. 4, Springer, Berlin.

272. Ellis, P. J., Thomas, I. L., McDonnell, M. J., 1978, *LANDSAT II Over New Zealand*, DSIR, Wellington.

273. Emmerling, O. E., Kolkwitz, R. K., 1914, *Mitt. Kgl. Landesamt. Wasserhyg.*, **19**, 167.

274. Erämetsä, O., Lounamaa, K. J., Haukka, M., 1969, *Suomen Kemistilehti*, **42**, 363.

275. Erämetsä, O., Yliruokanen, I., 1971, *Suomen Kemistlehti*, **B44**, 121.

276. Erämetsä, O., Yliruokanen, I., 1971, *Suomen Kemistlehti*, **B44**, 372.

277. Erdman, J. A., Harrach, G. H., 1980, *U. S. Geol. Surv. Open File Rep.*, **80-370**, 20 pp.

278. Erdman, J. A., Harrach, G. H., 1981, *J. Geochem. Explor.*, **14**, 83.

279. Erdman, J. A., McNeal, J. M., Pierson, C. T., 1979, *U. S. Geol. Surv. Prof. Pap.*, **1150**, 51.

280. Erdman, J. A., McNeal, J. M., Pierson, C. T., Harms, T. F., 1979, *Abs. Symp. Ass. Explor. Geochemists*, Tucson, 16.

281. Ermengen, S. W., 1957., *Can. Min. J.*, **78**, 99.

282. Ernst, W., 1965, *Ber. Dt. Bot. Ges.*, **78**, 205.

283. Ernst, W., 1967, *Excerpta Bot. Sec. B*, **8**, 50.

284. Ernst, W., 1968, *Mitt. Flor.-Soziol. Arbeitsgemeinschaft,* **13,** 263.

285. Ernst, W., 1969, *Vegetatio,* **18,** 393.

286. Ernst, W., 1972, *Kirkia,* **8,** 125.

287. Ernst, W., 1974, *Schwermetallvegetation der Erde,* Fischer, Stuttgart.

288. Eupene, G. S., William, B. T., Butt, C. R., Smith, R. E., 1980, *J. Geochem. Explor.,* **12,** 230.

289. Faber, F. C. von, 1925, *Flora,* **118,** 89.

290. Farago, M. E., Mullen, W. A., Cole, M. M., Smith, R. F., 1980, *Environ. Pollut.,* **21,** 225.

291. Faust, R. A., Bondietti, E. A., 1976, *Rep. 920,* ORNL/EIS-99, 142 pp.

292. Ferres, H. M., 1951, *Proc. Spec. Conf. Animal Nutr.* (Australia), 240.

293. Fisher, R. A., Yates, F., 1974, *Statistical Tables for Biological Agricultural and Medical Research* 6th ed., Longmans, London.

294. Flerova, T. P., Flerov, V. E., 1964, *Sbor. Mat. Geol. Polez. Iskop. Yuzh. Kazakh.,* No. 2, 144.

295. Fletcher, K., Doyle, P., Brink, V. C., 1973, *Can. J. Pl. Sci.,* **53,** 701.

296. Forrester, J. D., 1942, *Econ. Geol.,* **37,** 126.

297. Fortescue, J. A. C., Hornbrook, E. H. W., 1969, *Geol. Surv. Can. Pap.,* **67-23,** 39.

298. Fortescue, J. A. C., Hornbrook, E. H. W., 1967, *Geol. Surv. Can. Pap.,* **67-23,** 143 pp.

299. Fortescue, J. A. C., Usik, L., 1969, *Geol. Surv. Can. Pap.,* **67-23,** 10.

300. Fraser, D. C., 1961, *Econ. Geol.,* **56,** 1063.

301. Fraser, D. C., 1961, *Econ. Geol.,* **56,** 951.

302. Frederiksson, K., Lindgren, I., 1967, in *Geochemical Prospecting in Fennoscandia,* Interscience, New York, 193.

303. Freise, W., cited by Dorn, P., 1937, *Biologe,* **6,** 11.

304. Frey-Wyssling, A., 1935, *Naturwissenschaften,* **23,** 767.

305. Fritz, N. L., 1967, *Photogramm. Eng.,* **33,** 1128.

306. Froehlich, A. J., Kleinhampl, F. J., 1960, *U. S. Geol. Surv. Bull.,* **1085B,** 51.

307. Fuchs, W. A., Rose, A. W., 1974, *Econ. Geol.,* **69,** 332.

308. Fujimoto, C. K., Sherman, D. G., 1948, *Soil Sci.,* **66,** 131.

309. Gamble, J. F., 1964, Ann Arbor Univ. Microfilm, 64-10914, 116 pp.

310. Gandhi, S. M., Aswathanarayana, U., 1975, *J. Geochem. Explor.,* **4,** 247.

311. Garrels, R. M., 1960, *Mineral Equilibria at Low Temperature and Pressure,* McGraw-Hill, New York.

312. Garrett, R. G., Nichol, I., 1969, *Col. Sch. Mines Quart.,* **64,** 245.

313. Geilmann, W., Brünger, K., 1935, *Biochem. Z.,* **275,** 387.

314. Gerloff, G. C., Moore, D. D., Curtis, J. T., 1964, *Wis. Univ. Agr. Exp. Stn. Res. Rep.,* **14,** 26 pp.

315. Girling, C. A., Peterson, P. J., Minski, M. J., 1978, *Sci. Tot. Environ.,* **10,** 79.

316. Girling, C. A., Peterson, P. J., Warren, H. V., 1979, *Econ. Geol.,* **74,** 902.

317. Giuliani, C. A., Angelelli, V., 1975, *Prim. Symp. Nac. Geol. Econ.,* Buenos Aires, **1,** 25.

318. Glazovskaya, M. A., 1964, *Proc. 8th Int. Soil Sci. Congr.* Bucharest, 148.

319. Goldich, S. S., 1938, *J. Geol.,* **46,** 17.

320. Goldschmidt, V. M., 1937, *J. Chem. Soc.,* 655.

321. Goldzstein, M., 1957, *Bull. Soc. Fr. Minér. Cristallogr.,* **80,** 318.

322. Gonzalez, G. L., 1975, *Acta Bot. Malacitana,* **1,** 81.

323. Goodall, D. W., 1952, *Biol. Rev.*, **27**, 194.

324. Gorsuch, T. T., 1970, *The Destruction of Organic Matter,* Pergamon, Oxford.

325. Gower, J. C., 1966, *Biometrics,* **55**, 588.

326. Grabovskaya, L. I., 1965, *Biogeochemical Methods in Mineral Exploration,* Izd. Gosgeolkoma SSSR, Moscow.

327. Green, J., 1959, *Bull. Geol. Soc. Am.,* **70**, 1127.

328. Greig-Smith, P., 1964, *Guantitative Plant Ecology,* 2nd Ed., Butterworths, London.

329. Grigoryan, G. B., 1966, *Izv. Sel'sk. Nauk Minist. Sel'sk. Arm. SSR,* No. 7, 57.

330. Grimes, D. J., Earhart, R. L., 1975, *U. S. Geol. Surv. Open File Rep.,* 75-72, 25 pp.

331. Grimes, D. J., Whipple, J. W., 1977, *Abs. Progr. Geol. Soc. Am.,* **9**, 727.

332. Grodzinsky, A. M., 1959, *Ukr. Biokhim. Zhur.,* **16**, 30.

333. Grodzinsky, D. M., Golubkova, M. G., 1964, *Dokl. Akad. Nauk Ukr. SSR Inst. Fiz. Tast.,* 1964, 135.

334. Groves, R. W., Steveson, B. G., Steveson, E. A., Taylor, R. G., 1972, *Trans. Inst. Min. Metall. Sec. B,* **81**, 127.

335. Guha, M., 1961, A Study of the trace element uptake of deciduous trees, PhD thesis, University of Aberdeen.

336. Gvozdanovic, S. M., Overton, T. R., Spiers, F. W., 1967, *UKAEA Rep.,* AERE-R5474, 175.

337. Hall, J. S., Both, R. A., Smith, F. A., 1973, *Proc. Australas. Inst. Min. Metall.,* No. 247, 11.

338. Hamilton, J. W., 1975, *Rep. Agr. Exptl. Stn. Laramie, Wyom.,* **RJ94**, 5 pp.

339. Hanna, W. J., Grant, C. L., 1962, *Bull. Torrey Bot. Club,* **89**, 293.

340. Harbaugh, J. W., 1950, *Econ. Geol.,* **45**, 548.

341. Harman, W., 1967, *Modern Factor Analysis,* University of Chicago Press, Chicago.

342. Hartman, E. L., 1969, *Bryologist,* **72**, 56.

343. Hartsock, L., Pierce, R. P., 1952, *U. S. Geol. Surv. Circ.,* **127**, 37 pp.

344. Hattensaur, G., 1890, *Just Bot. Jahresber.,* **1**, 48.

345. Hattensaur, G., 1980, *Monatschr. Chem.,* **11**, 19.

346. Hawkes, H. E., Salmi, M. L., 1960, *U. S. Geol. Surv. Circ.,* **127**, 37 pp.

347. Hawkes, H. E., Webb, J. S., 1962, *Geochemistry in Mineral Exploration,* Harper & Row, New York.

348. Henwood, W. J., 1857, *Edinb. New Phil. J.,* **5**, 61.

349. Herodotus, 450 B.C., Eng. trans. by A. de Selincourt, Penguin Classics, London.

350. Herrero Payo, F. J., 1976, *Tecniterrae,* **2**, 27.

351. Hewitt, E. J., 1953, *J. Exp. Bot.,* **4**, 59.

352. Hewitt, E. J., Nicholas, D. J. D., 1963, in *Metabolic Inhibitors,* Academic Press, New York.

353. Hinchen, 1969, *Practical Statistics for Chemical Research,* Methuen, London.

354. Hinesley, T. D., Alexander, D. E., Ziegler, E. L., Barrett, G. C., 1978, *Agronom. J.,* **70**, 425.

355. Hoffman, J., 1941, *Naturwissenschaften,* **29**, 403.

356. Hoffman, J., 1942, *Bodenk. Pflanzennähr.,* **26**, 318.

357. Hopps, H. C., 1971, *Mem. Geol. Soc. Am.,* **123**, 1.

358. Horler, D. N. H., Barber, J., Barringer, A. R., 1980, *J. Geochem. Explor.,* **13**, 41.

359. Horler, D. N. H., Barber, J., Barringer, A. R., 1981, in *Remote Sensing in Geological and Terrain Studies,* Remote Sensing Society, London.

360. Hornbrook, E. H. W., 1969, *Geol. Surv. Can. Bull.*, **68-56**, 41 pp.

361. Hornbrook, E. H. W., 1969, *Geol. Surv. Can. Pap.*, **67-23**, 65.

362. Hornbrook, E. H. W., 1969, *Geol. Surv. Can. Rep.*, **69-49**, 39 pp.

363. Hornbrook, E. H. W., 1970, *Abs. 3rd Int. Geochem. Explor. Symp.*, Toronto, 38.

364. Hornbrook, E. H. W., 1971, *Can. Inst. Min. Metall. Spec. Vol.*, 435.

365. Hornbrook, E. H. W., 1972, *Geol. Surv. Can. Pap.*, **71-32**, 36 pp.

366. Hornbrook, E. H. W., 1972, *Geol. Surv. Can. Pap.*, **72-1**, 73.

367. Hornbrook, E. H. W., Allan, R. J., 1970, *Geol. Surv. Can. Pap.*, **70-36**, 35 pp.

368. Horvath, E., 1960, *Atomki Közl.*, **2**, 177.

369. Howard, J., 1971, *Proc. 7th Int. Symp. Rem. Sens. Environ.*, Ann Arbor, 285.

370. Howard-Williams, C., 1970, *J. Ecol.*, **58**, 745.

371. Howard-Williams, C., 1971, *Vegetatio*, **23**, 141.

372. Howard-Williams, C., 1972, *Nature*, **237**, 171.

373. Huffman, C. Jr., Riley, L. B., 1970, *U. S. Geol. Surv. Prof. Pap.*, **700B**, 181.

374. Hutchinson, G. E., 1943, *Quart. Rev. Biol.*, **18**, 1, 129, 242, 331.

375. Hvatum, O., 1964, *Abs., 6th Nordic Geol. Winter Mtg.*, Trondheim.

376. Ignatieff, V., 1952, *Efficient Use of Fertilizers*, Hill, London.

377. Igoshina, K. N., 1966, *Bot. Zhur. SSSR*, **51**, 322.

378. Il'in, V. B., 1966, *Izv. Sib. Otd. Akad. Nauk SSSR Ser. Biol.-Med. Nauk*, No. 3, 22.

379. Imbrie, J., 1956, *Bull. Am. Mus. Nat. Hist.*, **108**, 211.

380. Irving, H. M. N. H., Williams, R. P. J., 1953, *J. Chem. Soc.*, 3192.

381. Isohanni, M., Eeronheimo, J., Ilvonen, E., Kokkola, M., Kurki, J., Nurmi, P., Peuran-iemi, V., 1981, *Geologi* (Finland), **33**, 101.

382. Ivashov, P. V., Bardyuk, V. V., 1967, *Geokhimiya*, No. 2, 228.

383. Ives, R. L., 1939, *Ecology*, **20**, 433.

384. Jacobsen, W. B. G., 1967, *Kirkia*, **6**, 63.

385. Jacobsen, W. B. G., 1968, *Kirkia*, **6**, 259.

386. Jacobsen, W. B. G., 1970, *Kirkia*, **77**, 285.

387. Jaffré, T., 1974, *Candollea*, **29**, 427.

388. Jaffré, T., 1977, *Cah. ORSTOM, Sér. Biol.*, **12**, 323.

389. Jaffré, T., 1977, *Compt. Rend. Acad. Sci. Paris Sér. D*, **284**, 1573.

390. Jaffré, T., 1979, *Compt. Rend. Acad. Sci. Paris Sér. D*, **289**, 425.

391. Jaffré, T., 1980, *Végétation des Roches Ultrabasiques en Nouvelle Calédonie*, OR-STOM, Paris.

392. Jaffré, T., Brooks, R. R., Lee, J., Reeves, R. D., 1976, *Science*, **193**, 579.

393. Jaffré, T., Brooks, R. R., Trow, J. M., 1979, *Pl. Soil*, **51**, 157.

394. Jaffré, T., Kersten, W., Brooks, R. R., Reeves, R. D., 1979, *Proc. Roy. Soc. Lond. Ser. B*, **205**, 385.

395. Jaffré, T., Latham, M., 1974, *Adansonia*, **14**, 311.

396. Jaffré, T., Latham, M., Quantin, P., 1970, *Rapp. ORSTOM*, Nouméa, 20 pp.

397. Jaffré, T., Schmid, M., 1974, *Compt. Rend. Acad. Sci. Paris Sér. D*, **278**, 1727.

398. Jamieson, V. C., 1942, *Soil Sci.*, **53**, 287.

399. Jenni, J. P., 1973, *Beitr. Geol. Schweiz. Geotech. Ser.*, No. 53, 144 pp.

400. Jenny, H., 1941, *Factors of Soil Formation*, McGraw-Hill, New York.

401. Jensch, E., 1894, *Angew. Chem.*, **7**, 14.

402. Johnson, A. J., 1982, personal communication.

403. Johnson, C. M., Asher, C. J., Broyer, T. C., 1966, *Proc. 1st Int. Se Biomed. Symp.*, Oregon State Univ., 57.

404. Jolis, A. le, 1860, *Procès-Verb. 21-ième Séanc. Congr. Fr. Sci.*, Cherbourg, 38 pp.

405. Jones, R. S., 1970, *U. S. Geol. Surv. Circ.*, **625**, 15 pp.

406. Kabata-Pendias, A., Bolibrzuch, E., 1964, *Rocz. Naukro Inst. Ser. A*, **88**, 605.

407. Kabiashvili, V. I., 1964, *Soobsch. Akad. Nauk Gruz. SSR*, **35**, 83.

408. Kahma, A., Nurmi, A., Mattson, P., 1975, *Geol. Surv. Finl. Rep.*, No. 6.

409. Kakirov, K. Z., Rish, M. A., Yezdakov, V. I., 1959, *Uzbek. Biol. Zhur.*, No. 1, 15.

410. Kantor, M. Z., 1959, *Zap. Tadzhik. Otd. Vses. Mineral Obsch.*, No. 1, 149.

411. Karamata, S., 1967, *Yougoslav. Sci. Cons. Acads. RPF Bull. Sec. A*, **12**, 244.

412. Karasik, M. A., 1962, *Bull. Nauchno.-Tekh. Inf. Minist. Geol. Okhr. Nedr SSSR*, No. 1, 60.

413. Karpinsky, A. M., 1841, *Zhur. Sadovodstva*, Nos. 3 and 4.

414. Karunakaran, C., Krishnamurty, J. G., Bandyopadhyaya, S., 1969, *Geol. Surv. India Misc. Pub.*, **17**, 33.

415. Kashlev, V. F., Kurbatov, A. I., Kuznetsov, A. V., 1966, *Dokl. Sel'sk. Akad. K. A. Timiryazeva*, **119**, 269.

416. Kaspar, J., Hudec, I, Schiller, P., Cook, G. B., Kitzinger, A., Wölfl, E., *Chem. Geol.*, **10**, 299.

417. Kasynova, M. S., 1961, *Int. Geol. Rev.*, **3**, 626.

418. Kazitsyn, Y. V., Aleksandrov, G. V., 1960, *Mater. Vses. Nauchno.-Issled. Geol. Inst.*, **32**, 127.

419. Keith, J. R., 1968, *Prepr. Soc. Min. Eng. AME*, No. 68-L-320, 12 pp.

420. Keller, W. D., Frederickson, A. F., 1952, *Am. J. Sci.*, **250**, 594.

421. Kelley, P. C., Brooks, R. R., Dilli, S., Jaffré, T., 1975, *Proc. Roy. Soc. Lond. Sec. B*, **189**, 69.

422. Kersten, W., Brooks, R. R., Reeves, R. D., Jaffré, T., 1979, *Taxon*, **28**, 529.

423. Khamrabaev, I. K., Talipov, R. M., 1960, *Uzbek. Geol. Zhur.*, No. 5, 3.

424. Khatamov, Sh., Lobanov, E. M., Kist, A. A., 1966, *Dokl. Akad. Nauk. Tadzhik. SSR*, No. 9, 227.

425. Kleinhampl, F. J., 1962, *U. S. Geol. Bull.*, **1085D**, 105.

426. Kleinhampl, F. J., Koteff, C., 1960, *U. S. Geol. Surv.*, **1085C**, 85.

427. Klovan, J. E., 1966, *J. Sed. Petrol.*, **36**, 115.

428. Knight, A. H., Crooke, W. M., Inkson, R. H. E., 1961, *Nature*, **192**, 142.

429. Knypl, J. S., 1980, *Wiadomosci Bot.* (Poland), **24**, 17.

430. Koch, G. S., Link, R. F., 1971, *Statistical Analysis of Geological Data* Vol. 2, Wiley, New York.

431. Kochenov, A. V., Zineveev, V. V., Lavaleva, S. S., 1965, *Geochem. Int.*, **2**, 65.

432. Koljonen, T., 1974, *Oikos*, **25**, 353.

433. Konstantinov, V. M., 1963, *Sov. Geol.*, **3**, 151.

434. Kontas, E., 1975, *J. Geochem. Explor.*, **5**, 261, 311.

435. Kothny, E. L., 1979, *Pl. Soil*, **53**, 547.

436. Kotolov, B. A., Kiseleva, Ye. A., Rubeykin, V. Z., 1965, *Geochem. Int.*, **2**, 675.

437. Kovacs, L., 1966, *Bot. Közl.*, **53**, 175.

438. Koval'sky, V. V., 1969, *Dokl. Vses. Leningr. Akad. Sel'sk.*, No. 8, 2.

439. Koval'sky, V. V., Ananichev, A. V., Shakhova, I. K., 1965, *Agrokhimiya*, No. 11, 153.

440. Koval'sky, V. V., Basitova, S. M., Zasorina, E. F., 1965, in *Microchemistry in Agriculture* (in Russ.), Uzbek. Akad. Nauk, Tashkent.

441. Koval'sky, V. V., Blokhina, R. I., Zasorina, E. F., Nikitina, I., 1968, *Trudy Biogeokhim. Lab.*, No. 12, 123.

442. Koval'sky, V. V., Petrunina, N. S., 1964, *Dokl. Akad. Nauk SSSR*, **159**, 1175.

443. Koval'sky, V. V., Voronitskaya, I. E., 1966, *Ukr. Biokhim. Zhur.*, **38**, 119.

444. Koval'sky, V. V., Voronitskaya, I. E., 1966, *Ukr. Biokhim. Zhur.*, No. 4, 419.

445. Koval'sky, V. V., Voronitskaya, I. E., Lekarer, V. S., 1966, *Proc. Int. Symp. Radioecol. Concn. Processes,* Pergamon Press, New York, 329.

446. Koval'sky, V. V., Voronitskaya, I. E., Lekarer, V. S., 1973, *Mater. Vses. Simp. Teor. Prakt. Aspekti Deistuiya Malykh. Doz. Ioniz. Radiats.*, 83.

447. Koval'sky, V. V., Yarovaya, G. A., 1966, *Agrokhimiya*, No. 4, 78.

448. Koval'sky, V. V., Yermakov, V. V., 1967, *Geokhimiya*, No. 1, 86.

449. Koval'sky, V. V., Zasorina, E. F., 1965, *Agrokhimiya*, No. 4, 78.

450. Kovalevsky, A. L., Tomsky, I. V., 1981, *Dokl. Akad. Nauk SSSR*, **251**, 173.

451. Kovalevsky, A. L., 1962, *Izv. Sib. Otd. Akad. Nauk SSSR*, No. 4,108.

452. Kovalevsky, A. L., 1962, *Izv. Sib. Otd. Akad. Nauk SSSR*, No. 9, 112.

453. Kovalevsky, A. L., *Trudy Konf. Pochv. Sib. Dal'ne Vost.*, Novosibirski, 53.

454. Kovalevsky, A. L., 1965, *Izv. Sib. Otd. Akad. Nauk SSSR Ser. Biol.-Med. Nauk*, No. 1, 43.

455. Kovalevsky, A. L., 1966, *Geokhimiya*, No. 6, 737.

456. Kovalevsky, A. L., 1966, *Natural Radioactive Elements of Plants of Siberia* (in Russ.), Buriat Kn. Izd., Ulan Ude.

457. Kovalevsky, A. L., 1969, *Geol. Geofiz.* No. 9, 124.

458. Kovalevsky, A. L., 1969, *Trudy Buryat Inst. Yestvestvenn. Nauk*, No. 2, 195.

459. Kovalevsky, A. L., 1969, *Trudy Buryat Inst. Yestvestvenn. Nauk*, No. 2, 6.

460. Kovalevsky, A. L., 1972, *At. Energiya*, **33**, 557.

461. Kovalevsky, A. L., 1973, *Mater. Vses. Simp. Teor. Prakt. Aspekti Deistuiya Malykh. Doz. Ioniz. Radiats.*, 90.

462. Kovalevsky, A. L., 1974, *Dokl. Akad. Nauk SSSR*, **218**, 183.

463. Kovalevsky, A. L., 1974, *Dokl. Akad. Nauk SSSR*, **218**, 199.

464. Kovalevsky, A. L., 1974, *Izv. Akad. Nauk SSSR, Ser. Geol.*, No. 12, 105.

465. Kovalevsky, A. L., 1975, *Geochem. Int.*, No. 11, 1106.

466. Kovalevsky, A. L., 1975, *Sov. Geol.*, No. 2, 108.

467. Kovalevsky, A. L., 1976, *Geol. Rudnikh. Mest.*, No. 2, 89.

468. Kovalevsky, A. L., 1976, *Int. Geol. Rev.*, **18**, 1000.

469. Kovalevsky, A. L., 1976, *Int. Geol. Rev.*, **20**, 548.

470. Kovalevsky, A. L., 1976, *Obzor., Geol. Metod. Poisk. Razv. Metal. Polez. Iskop.*, 59 pp.

471. Kovalevsky, A. L., 1977, *Sov. Geol.*, No. 9, 37.

472. Kovalevsky, A. L., 1978, *Dokl. Akad. Nauk SSSR*, **242**, 430.

473. Kovalevsky, A. L., 1978, *Geochem. Int.*, **15**, 164.

474. Kovalevsky, A. L., 1978, *Geokhimiya*, No. 8, 1239.

475. Kovalevsky, A. L., 1978, *Int. Geol. Rev.*, **20**, 548.

476. Kovalevsky, A. L., 1978, *Izv. Akad. Nauk SSSR Ser. Geol.*, No. 4, 113.

477. Kovalevsky, A. L., 1979, *Biogeochemical Exploration for Mineral Deposits*, Amerind, New Delhi.

478. Kovalevsky, A. L., Kovalevskaya, O. M., 1980, *Geol. Geofiz. Sib. Otd. Akad. Nauk SSSR*, No. 8, 138.

479. Kovalevsky, A. L., Kovalevskaya, O. M., 1976, *Geol. Rudnikh. Mest.*, **43**, 97.

480. Kovalevsky, A. L., Kovalevskaya, O. M., 1979, *Trudy Otd. Geol. Bur. Fil. Akad. Nauk SSSR*, No. 6, 187.

481. Kovalevsky, A. L., Tauson, L. V., 1978, *Peculiarities of Biogeochemical Prospecting for Tungsten Ores* (in Russ.), Izd. Nauk, Novosibirsk.

482. Kranz, R. L., 1968, *Trans. Inst. Min. Metall. Sec. B*, **77**, 26.

483. Krause, W., 1958, in *Encyclopedia of Plant Physiology* v. 4., Springer, Berlin.

484. Krendelev, F. P., Pogrevniak, Yu. F., 1977, *Dokl. Akad. Nauk SSSR*, **234**, 184.

485. Krendelev, F. P., Pogrevniak, Yu. F., Tsyrenova, A. A., 1978, *Sov. Geol. Geophys.* (USA), **19**, 106.

486. Kruckeberg, A. R., 1954, *Ecology*, **35**, 267.

487. Kruckeberg, A. R., 1969, *Madroño*, **20**, 129.

488. Krumbein, W. C., 1959, *J. Geophys. Res.*, **64**, 823.

489. Krumbein, W. C., Graybill, F. A., 1965, *Introduction to Statistical Models in Geology*, McGraw-Hill, New York.

490. Kubota, J., Lazar, V. A., 1971, in *Instrumental Methods for Analysis of Soil and Plant Tissue*, Soil Science Society of America, Madison.

491. Kubota, J., Lazar, V. A., Beeson, K. C., 1960, *Soil Sci. Proc.*, **24**, 527.

492. Kunasheva, K. G., 1944, *Trudy Biogeokhim. Lab.*, No. 7, 98.

493. Kuzin, M. F., 1959, *Razv. Okhr. Nedr*, **11**, 16.

494. Kyuregyan, E. A., Burnutyan, R. A., 1973, *Dokl. Akad. Nauk SSSR*, **57**, 103.

495. Kyuregyan, E. A., Burnutyan, R. A., 1972, *Izv. Nauki Zemle Akad. Nauk Arm. SSR*, **25**, 83.

496. L'vov, B. V., 1961, *Spectrochim. Acta*, **17**, 761.

497. Lakin, H. W., Curtin, G. C., Hubert, A. E., 1971, *Can. Inst. Min. Metall. Spec. Vol.*, **11**, 196.

498. Larsson, J.-O., 1976, *J. Geochem. Explor.*, **6**, 233.

499. Lawley, D. N., Maxwell, A. E., 1963, *Factor Analysis as a Statistical Method*, Butterworth, London.

500. Leavitt, S. W., Goodall, H. G., 1979, *J. Geochem. Explor.*, **11**, 89.

501. Lecoq, J. J., Bigotte, G., Hinault, J., Leconte, J. R., 1958, *Proc. 2nd Int. Conf. Peacef. Uses At. Energ.*, Geneva, **2**, 744.

502. Lee, J., 1974, Biogeochemical studies on some nickel-accumulating plants from New Zealand and New Caledonia. MSc thesis, Massey University, New Zealand.

503. Lee, J., 1977, Phytochemical and biogeochemical studies on nickel accumulation by some New Caledonian plants, PhD thesis, Massey University, New Zealand.

504. Lee, J., Brooks, R. R., Reeves, R. D., Boswell, C. R., 1975, *Pl. Soil*, **42**, 153.

505. Lee, J., Brooks, R. R., Reeves, R. D., Boswell, C. R., Jaffré, T., 1977, *Pl. Soil*, **46**, 675.

506. Lee, J., Brooks, R. R., Reeves, R. D., Jaffré, T., 1977, *Bryologist*, **80**, 203.

507. Lehmann, K. B., 1896, *Arch. Hyg.*, **27**, 1.

508. Leroy, L. W., Koksoy, M., 1962, *Econ. Geol.*, **57**, 107.

509. Leutwein, F., Pfeiffer, L., 1954, *Geologie*, **3**, 950.

510. Leutwein, F., Starke, R., 1957, *Geologie*, **6, 349.**

511. Levanidov, L. Y., Khilyukova, M. I., 1953, *Metod. Sbor. Chelyabinsk. Gos. Ped. Inst.*, 26 pp.

512. Levy, Y., Miller, D. S., Friedman, G. M., 1977, *Rep. C. O. O.*, 3462-14, 45 pp.

513. Lidgey, E., 1897, *Trans. Australas. Inst. Min. Eng.*, **4,** 110.

514. Lindner, M., Sarosiek, J., 1963, *Przegl. Geol.*, **11,** 448.

515. Linstow, O., von, 1924, *Fedd. Rep.*, **31,** 1.

516. Lisitsin, A. K., Kruglov, A. I., Panteleev, V. M., Sidel'nikova, V. D., 1967, *Litol. Polez. Iskop.*, **3,** 103.

517. Liwski, S., 1961, *Roczn. Naukro Inst. Ser. F,* **75,** 7.

518. Lobanov, E. M., Khatamov, Sh., 1968, *Izv. Akad. Nauk Tadzhik SSR Otd. Fiz.-Mat.*, No. 2, 3.

519. Lobanov, E. M., Khatamov, Sh., Talipov, R. M., 1966, *Uzbek. Geol. Zhur.*, No. 6, 49.

520. Löhnis, M. P., 1950, *Lotsya,* **3,** 63.

521. Löhnis, M. P., 1951, *Pl. Soil.*, **3,** 193.

522. Lopatkina, A. P., *Geochem. Int.*, **4,** 577.

523. Lopatkina, A. P., Komarov, V. S., Sergeev, A. N., Andreev, A. G., 1970, *Geochem. Int.*, **7,** 277.

524. Lounamaa, J., 1956, *Ann. Bot. Soc. Zoo. Bot. Fenn. "Vanamo,"* **9,** Suppl. 170, 1.

525. Lounamaa, J., 1967, in *Geochemical Prospecting in Fennoscandia,* Interscience, New York, 287.

526. Lounamaa, K. J., 1965, *Ann. Bot. Fenn.*, **2,** 127.

527. Lovering, T. S., Huff, L. C., Almond, H., 1950, *Econ. Geol.*, **45,** 239.

528. Lovering, T. S., 1927, *U. S. Geol. Surv. Bull.*, **795C,** 45.

529. Lovstrom, K. A., Horsnail, R. E., Lovering, T. G., McCarthy, J. H. Jr., 1978, *J. Geochem. Explor.*, **9,** 222.

530. Lundberg, H. T. F., 1941, *Min. Metall.*, **22,** 256.

531. Lundegårdh, H., 1931, *Environment and Plant Development,* Arnold, London.

532. Lungwitz, E. E., 1900, *Eng. Min. J.*, **69,** 500.

533. Lyon, G. L., 1969, Trace elements in New Zealand plants, PhD thesis, Massey University, New Zealand.

534. Lyon, G. L., Brooks, R. R., 1969, *N. Z. J. Sci.*, **12,** 200.

535. Lyon, G. L., Brooks, R. R., Peterson, P. J., Butler, G. W., 1968, *Pl. Soil,* **29,** 225.

536. Lyon, G. L., Brooks, R. R., Peterson, P. J., Butler, G. W., 1970, *N. Z. J. Sci.*, **13,** 133.

537. Lyon, R. J. P., 1975, *Proc. 10th Int. Symp. Rem. Sens. Environ.*, Sioux Falls, **2,** 1031.

538. Lyons, M. T., Brooks, R. R., Craig, D. C., 1974, *Proc. Roy. Soc. N. S. W.*, **107,** 67.

539. Lyubofeev, V. N., Balisky, V. S., Cherkasov, M. I., 1962, *Trudy Geol. Polez. Iskop. Ser. Kavk.*, No. 2, 281.

540. Mahalanobis, P. C., 1936, *Proc. Nat. Inst. Sci.* (India), **12,** 49.

541. Makarochkin, B. A., 1974, *Apiculture Abs.*, **25,** 104.

542. Makarochkin, B. A., Yudenich, D. M., 1960, *Pchelovodstvo,* **37,** 34.

543. Makarochkin, B. A., Yudenich, D. M., 1962, *Priroda,* **51,** 67.

544. Makarov, M. S., 1965, *Vop. Rud. Geofiz. Minist. Geol. Okhr. Nedr SSSR*, No. 5, 33.

545. Makarova, A. I., 1960, *Geokhimiya,* No. 7, 624.

546. Makunina, G. S., 1972, *Geochem. Int.*, **9,** 883.

547. Makunina, G. S., 1972, *Geokhimiya,* **10,** 1302.

548. Malaisse, F., Brooks, R. R., 1982, *Pl. Soil,* **64,** 289.

549. Malaisse, F., Brooks, R. R., 1982, *Pl. Syst. Evol.* (in press).

550. Malaisse, F., Colonval-Elenkov, E., Brooks, R. R., 1982, *Pl. Syst. Evol.,* (in press).

551. Malaisse, F., Grégoire, J., 1978, *Bull. Soc. Bot. Bot. Belg.,* **111,** 252.

552. Malaisse, F., Grégoire, J., Brooks, R. R., Morrison, R. S., Reeves, R. D., 1978, *Science,* **199,** 887.

553. Malaisse, F., Grégoire, J., Brooks, R. R., Morrison, R. S., Reeves, R. D., *Oikos,* **33,** 472.

554. Malyuga, D. P., 1939, *Trudy Biogeokhim. Lab.,* No. 5, 91.

555. Malyuga, D. P., 1947, *Izv. Akad. Nauk SSSR Ser. Geograf. Geofiz.,* No. 11, 135.

556. Malyuga, D. P., 1950, *Dokl. Akad. Nauk SSSR,* **70,** 257.

557. Malyuga, D. P., 1951, *Dokl. Akad. Nauk SSSR,* **76,** 231.

558. Malyuga, D. P., 1954, *Trudy Biogeokhim. Lab.,* **10,** 28.

559. Malyuga, D. P., 1958, *Geochemistry,* No. 3, 314.

560. Malyuga, D. P., 1958, in *Trace Elements in Agriculture* (in Russ.), Moscow, 105.

561. Malyuga, D. P., 1959, *Razv. Okhr. Nedr,* **25,** 19.

562. Malyuga, D. P., 1960, *Trudy Biogeokhim. Lab.,* No. 11, 197.

563. Malyuga, D. P., 1964, *Biogeochemical Methods of Prospecting,* Consultants Bureau, New York.

564. Malyuga, D. P., Ayvasyan, A. D., 1970, *Geochem. Int.,* No. 7, 269.

565. Malyuga, D. P., Makarova, A. I., 1955, *Trudy Vses. Soveshch. Riga,* 1955, 485.

566. Malyuga, D. P., Malashkina, N. S., Makarova, A. I., 1959, *Geokhimiya,* No. 5, 423.

567. Malyuga, D. P., Nadiradze, V. R., Chargeishvili, Y. M., Makarova, A. I., 1960, *Geokhimiya,* No. 4, 330.

568. Malyuga, D. P., Petrunina, N. S., 1961, *Geokhimiya,* No. 3, 258.

569. Mamulea, O., Buracu, O., 1967, *Rom. Com. Geol. Dari Seama Sedin.,* **52,** 237.

570. Manskaya, S. M., Drozdeva, T. V., Emilianova, M. P., 1960, *Geokhimiya,* No. 6, 630.

571. Manskaya, S. M., Drozdova, T. V., 1968, *Geochemistry of Organic Substances,* Pergamon, New York.

572. Manskaya, S. M., Drozdova, T. V., Emelianova, M. P., 1956, *Geokhimiya,* No. 4, 10.

573. Marmo, V., 1953, *Econ. Geol.,* **48,** 211.

574. Marshall, N. J., 1970, *Proc. 2nd UNESCO Sem. Prospect. Meth. Tech.,* U. N., New York.

575. Mårtensson, O., 1956, *Kgl. Svensk. Vetens. Akad. Avh. Naturskydd.,* No. 14, 321 pp.

576. Mason, B., 1966, *Principles of Geochemistry* 3rd ed., Wiley, New York.

577. Massingill, G. L., 1979, *New Mex. Geol.* (Socorro), **1,** 49.

578. Mattson, W. J., Addy, N. D., 1975, *Science,* **190,** 515.

579. Matula, I., 1973, *Mineralia Slovaca,* **5,** 21.

580. McCray, C. W. R., Hurwood, I. S., 1963, *Qld. J. Agr. Sci.,* **20,** 475.

581. McNeill, R. J., 1978, A Biogeochemical investigation of the lead and zinc contents in eastern hemlock, *Tsuga Canadensis* (L.) Carr in northeastern United States, MSc thesis, Queens College, Flushing, New York.

582. Mehta, B. V., Reddy, G. R., Nair, G. K., Gandhi, S. C., Neelkantan, V., Reddy, K. G., 1964, *Ind. J. Soil Sci.,* **72,** 329.

583. Mendelsohn, F., 1961, *The Geology of the Northern Rhodesian Copper Belt,* McDonald, London.

584. Menezes de Sequeira, E., 1968, *Agronom. Lusit.,* **30,** 115.

585. Meyers, A. T., Hamilton, J. C., 1960, *Bull. Geol. Soc. Am.,* **71,** 1934.

586. Meyers, A. T., Hamilton, J. C., 1961, *U. S. Geol. Surv. Prof. Pap.,* **424B,** 286.

587. Miesch, A. T., Connor, J. J., 1968, *Computer Contribution,* No. 27, University of Kansas.

588. Miller, C. P., 1961, *Trans. Am. Inst. Min. Metall. Petrol. Eng.,* **220,** 255.

589. Miller, J. T., Byers, H. G., 1937, *J. Agr. Res.,* **55,** 58.

590. Miller, R. L., Goldberg, E. D., 1955, *Geochim. Cosmochim. Acta,* **8,** 53.

591. Millman, A. P., 1957, *Geochim. Cosmochim. Acta,* **12,** 85.

592. Minguzzi, C., Naldoni, K. M., 1950, *Mem. Atti. Soc. Tosc. Sci. Nat.,* **57,** 38.

593. Minguzzi, C., Vergnano, O., 1948, *Mem. Atti. Soc. Tosc. Sci. Nat.,* **55,** 49.

594. Mitchell, R. L., 1954, *Proc. Int. Peat Symp.,* Dublin, 3.

595. Mitchell, R. L., 1955, in *Chemistry of the Soil,* Reinhold, New York.

596. Mitchell, R. L., 1964, *Commonw. Bur. Soil Sci. Tech. Commn.,* **44A,** 225 pp.

597. Mitskevich, B. F., 1962, *Inf. Minist. Geol. Okhr. Nedr SSSR,* No. 1, 31.

598. Mizuno, N., 1967, *Hokkaido-ritsu Nogyo Shikensho Shuho,* No. 15, 48.

599. Moiseenko, U. I., 1959, *Geochemistry,* No. 1, 117.

600. Montgomery, J. H., Cochrane, D. R., Sinclair, A. J., 1974, *Proc. 3rd Int. Geochem. Explor. Symp.,* Toronto, No. 5, 47.

601. Montgomery, J. H., Cochrane, D. R., Sinclair, A. J., 1974, *Proc. 3rd Int. Geochem. Explor. Symp.,* No. 5, 85.

602. Moore, B. R., Wachs, T. C., Marshall, J. L., 1974, *Econ. Geol.,* **69,** 1184.

603. Morris, B. J., 1975, *Geol. Surv. S. Austral. Quart. Notes,* No. 56, 6.

604. Morrison, D. F., 1967, *Multivariate Statistical Methods,* McGraw-Hill, New York.

605. Morrison, R. S., Brooks, R. R., Reeves, R. D., Malaisse, F., 1979, *Pl. Soil,* **53,** 535.

606. Morton, F., Gams, H., 1925, *Cave Plants* (in Ger.), Hölzel, Vienna.

607. Mulder, D., 1953, *Proc. Nat. Conf. Fruitgrowing,* Montana di San Vicenti (Italy).

608. Mulder, E. G., Gerretson, F. C., 1952, *Adv. Agron.,* **4,** 222.

609. Murakami, Y., Fujiwara, S., Sato, M., Ohashi, S., 1958, *Proc. 2nd Int. Conf. Peacef. Uses At. Energ.,* Geneva, **2,** 131.

610. Muzaleva, L. D., Pershina, E. F., 1965, *Uchen. Zap. Petrozavod. Gos. Univ.,* **13,** 21.

611. Nash, J. T., Ward, F. N., 1977, *U. S. Geol. Surv. Open File Rep.,* **77-354,** 23 pp.

612. Nazarov, A. G., 1969, *Izv. Vses. Ucheb. Zavod. Geol. Razved.* No. 8, 57.

613. Nemec, B., Babička, J., Oborsky, A., 1936, *Bull. Int. Acad. Sci. Bohême,* **1,** 1.

614. Nemeryuk, G. E., 1970, *Fiziol. Rast.,* **17,** 673.

615. Nesvetalyova, N. G., 1961, *Int. Geol. Rev.,* **3,** 609.

616. Neumann, J. von, Morgenstern, O., 1953, *Theory of Games and Economic Behavior,* Princeton University Press, Princeton.

617. Nichol. I., Garrett, R. G., Webb, J. S., 1969, *Econ. Geol.,* **64,** 204, 3.

618. Nicolas, D. J., Brooks, R. R., 1969, *Proc. Australas. Inst. Min. Metall.,* No. 231, 59.

619. Nicolls, O. W., Provan, D. M. J., Cole, M. M., Tooms, J. S., 1966, *Trans. Inst. Min. Metall. Sec. B,* **75,** 307.

620. Nielsen, J. S., Brooks, R. R., Boswell, C. R., Marshall, N. J., 1973, *J. Appl. Ecol.,* **10,** 251.

621. Nikiforov, N. A., Fedorchuk, V. P., 1959, *Trudy Sred. Aziat. Politekh. Inst.,* No. 6, 178.

622. Nikonova, N. N., 1967, *Izv. Sib. Otd. Akad. Nauk SSSR Ser. Biol.-Med. Nauk,* No. 3, 25.

623. Nilsson, G., 1971, *Geol. För. Stockh. Förh.,* **93,** 725.

624. Nilsson, G., 1973, in *Prospecting in Areas of Glaciated Terrain,* Institution of Mining and Metallurgy, London.

625. Nilsson, I., 1972, *Oikos,* **23,** 132.

626. Nilsson, K. A., 1970, *Swed. Nat. Sci. Res. Counc. Ecol. Res. Commn. Bull.,* No. 5, 17.

627. Noguchi, A., 1956, *Kumamoto J. Sci. Ser. B,* **2,** 239.

628. Nuutilainen, J., Peuraniemi, V., 1977, in *Prospecting in Areas of Glaciated Terrain,* Institution of Mining and Metallurgy, London.

629. Obial, R. C., 1970, *Trans. Inst. Min. Metall. Sec. B,* **79,** 175.

630. Orlov, A. P., Robonen, V. I., Kirilenko, G. M., 1969, *Geological Prospecting with Ore-Searching Dogs* (in Russ.), Nedra Press, Moscow.

631. Ovchinnikov, L. N., Grigoryan, S. V., Garmash, A. A., 1967, *Z. Angew. Geol.,* **13,** 568.

632. Owen, D. B., 1962, *Handbook of Statistical Tables,* Addison-Wesley, Reading, Massachusetts.

633. Paarma, H., Raevaara, H., Talvitie, J., 1968, *Photogramm. J. Finl.,* **2,** 3.

634. Paarma, H., Vartiainen, H., Penninkilampi, J., 1977, *Prospecting in Areas of Glaciated Terrain,* Institution of Mining and Metallurgy, London.

635. Pagliuchi, F. D., 1925, *Eng. Min. J.,* **120,** 485.

636. Pulou, R., Gramont, X, de, Magny, J., Carles, J., 1965, *Bull. Soc. Hist. Nat. Toulouse,* **100,** 465.

637. Panteleev, V. M., Belaeva, Z. N., Volkov, G. A., Dorokhina, M. A., Serova, L. V., Tokareva, G. I., Galitsyn, M. S., 1979, *Issled. Inst. Gidrogeol. Inzh. Geol.,* No. 128, 80.

638. Pape, H., 1981, *Monogr. Ser. Min. Dep.,* No. 19, Borntraeger, Stuttgart.

639. Parfent'eva, N. S., 1955, *Possibility of Using Vegetation in the Exploration for Lead Deposits in Calcareous Rocks in the Central Karatau Range* (in Russ.), State University, Moscow.

640. Paribok, T. A., Alekseeva-Popova, N. V., 1966, *Bot. Zhur. SSSR,* **31,** 339.

641. Parks, J. M., 1966, *J. Geol.,* **74,** 703.

642. Passow, H., Rothstein, A., Clarkson, T. W., 1961, *Pharmac. Rev.,* **13,** 185.

643. Paul, P. F. M., 1953, Notes of discussion with H. Brauchli at Johns Hopkins University, Baltimore, 3 pp.

644. Pauli, W., 1968, *Proc. Can. Geol. Ass.,* **19,** 45.

645. Pearson, E. S., Hartley, H. D., 1966, *Biometrika Tables for Statisticians,* vol. 1, 3rd ed., Cambridge University Press, Cambridge.

646. Pease, R. W., Bowden, L. W., 1969, *Rem. Sens. Environ.,* **1,** 23.

647. Perel'man, A. I., 1966, *Landscape Geochemistry* (in Russ.), Izd. Vish. Shk., Moscow.

648. Perel'man, A. I., 1967, *Geochemistry of Epigenesis,* Plenum Press, New York.

649. Persson, H., 1948, *Rev. Bryol. Lichen.,* **17,** 76.

650. Persson, H., 1956, *J. Hattori Bot. Lab.,* **17,** 1.

651. Peterson, P. J., 1971, *Sci. Progr.,* **59,** 505.

652. Pinto da Silva, A. R., 1970, *Agronom. Lusit.,* **30,** 175.

653. Pirschle, K., 1930, *Jb. Wiss. Bot.,* **72,** 335.

654. Pirschle, K., 1932, *Jb. Wiss. Bot.,* **76,** 1.

655. Pius II, 1614, *Pii Secundi Pontificio Max. Commentarii Rerum Memorabilium Quae Temporibus Suis Contigerunt,* Francoforti.

656. Podkorytov, F. M., 1967, *Trudy Nauchno.-Issled. Inst. Sel'sk. Sev. Ukrain,* **14,** 277.

657. Polikarpochkin, V. V., Polikarpochkina, R. T., Abramov, I. I., 1965, *Vop. Geokhim. Sib. Otd. Akad. Nauk SSSR,* **1965,** 242.

658. Poluzerov, N. A., 1965, *Izv. Akad. Nauk Kazakh. SSR Ser. Biol. Nauk,* No. 2, 28.

659. Pomykala, J., 1975, *Tech. Poszukiwan,* **14,** 30.

660. Porutsky, G. V., Golovchenko, V. P., Cherednichenko, S. V., 1962, *Dokl. Akad. Nauk SSSR,* **146,** 1223.

661. Poskotin, D. L., 1963, *Trudy Sverdl. Gorn. Inst.,* **42,** 81.

662. Poskotin, D. L., Lyubimova, M. V., 1963, *Geochemistry,* No. 6, 620.

663. Prat, S., 1934, *Ber. Dt. Bot. Gesell.,* **52,** 65.

664. Pratt, P. F., Bair, F. L., McClean, G. W., 1967, *Trans. 8th Int. Congr., Soil Sci.,* Bucharest, **3,** 243.

665. Press, N. P., 1974, *Proc. 9th Int. Symp. Rem. Sens. Environ.,* Ann Arbor, 2027.

666. Prister, B. S., Prister, S. S., 1970, *Radiobiology,* **10,** 221.

667. Putnam, G. W., 1975, *Econ. Geol.,* **70,** 1225.

668. Puustjärvi, V., 1955, *Arch. Soc. Zool.-Bot. Fenn. "Vanamo",* **9,** 257.

669. Quin, B. F., Brooks, R. R., 1972, *N. Z. J. Sci.,* **15,** 308.

670. Quin, B. F., Brooks, R. R., 1974, *Pl. Soil,* **41,** 177.

671. Quin, B. F., Brooks, R. R., Painter, J. A. C., 1974, *J. Geochem. Explor.,* **3,** 43.

672. Quin, B. F., Brooks, R. R., Reay, P. F., 1972, *Pl. Soil,* **36,** 699.

673. Rackley, R. I., Shockey, P. N., Dahill, M. P., 1968, *Guidebk. Annual Fd. Conf. Wyoming Geol. Assoc.,* **20,** 115.

674. Ramsden, A. R., Ryall, W. R., 1979, *CSIRO Invest. Rep.,* **128,** 112p.

675. Rankama, K., 1940, *Compt. Rend. Soc. Géol. Finl.,* No. 14, 90.

676. Rankama, K., 1941, *Geol. Rundsch.,* **32,** 575.

677. Rasmussen, R. A., Went, F. W., 1965, *Proc. Nat. Acad. Sci.,* **53,** 215.

678. Rattigan, J. H., Gersteling, R. W., Tonkin, D. G., 1977, *J. Geochem. Explor.,* **8,** 203.

679. Ray, R. G., 1960, *U. S. Geol. Surv. Prof. Pap.,* **373,** 230 pp.

680. Razin, L. V. Rozhkov, I. S., 1966, *Geochemistry of Gold in the Crust of Weathering and in the Biosphere in the Gold Deposits of the Kuvanakh Type,* Nauka Press, Moscow.

681. Reeves, R. D., Brooks, R. R., 1978, *Trace Element Analysis of Geological Materials,* Wiley, New York.

682. Reeves, R. D., Brooks, R. R., 1983, *J. Geochem. Explor.* (in press).

683. Reeves, R. D., Brooks, R. R., Dudley, T. R., 1983, *Taxon* (in press).

684. Reeves, R. D., Brooks, R. R., McFarlane, R. M., 1981, *Am. J. Bot.,* **68,** 708.

685. Reeves, R. D., Brooks, R. R., McFarlane, R. M., 1983, *Am. J. Bot.* (in press).

686. Reeves, R. D., Brooks, R. R., Press, J. R., 1980, *Taxon,* **29,** 629.

687. Reilly, C., 1967, *Nature,* **215,** 667.

688. Remezov, N. B., 1938, *Sov. Bot.,* No. 6, 34.

689. Richardson, D. H. S., Beckett, P. J., Nieboer, E., 1980, in *Nickel in the Environment,* Wiley, New York.

690. Richardson, D. H. S., Nieboer, E., 1981, *Endeavour,* **5,** 127.

691. Riddell, J. E., 1952, *Que. Dep. Mines Prelim. Rep.,* **269,** 15 pp.

692. Roberts, B. A., 1980, *Abs. Proc. 80th Bot. Conf. Univ. Brit. Columbia,* 95.

693. Robertson, N., 1975, *Rri News* (Llandudno), **4,** 1.

694. Robinson, W. O., 1943, *Soil Sci.,* **56,** 1.

695. Robinson, W. O., 1951, *U. S. Dep. Agr. Leaflet,* 9 pp.

696. Robinson, W. O., Bastron, H., Murata, K. J., 1958, *Geochim. Cosmochim. Acta,* **14,** 55.

697. Robinson, W. O., Edgington, G., 1942, *Soil Sci.,* **53,** 309.

698. Robinson, W. O., Edgington, G., 1945, *Soil Sci.,* **60,** 15.

699. Robinson, W. O., Edgington, G., 1948, *Soil Sci.,* **66,** 197.

700. Robinson, W. O., Edgington, G., Byers, H. G., 1935, *U. S. Dep. Agr. Tech. Bull.,* **471,** 1.

701. Robinson, W. O., Lakin, H. W., Reichen, L. E., 1947, *Econ. Geol.,* **42,** 572.

702. Robinson, W. O., Whetstone, R. R., Edgington, G., 1950, *U. S. Dep. Agr. Tech. Bull.,* **1013,** 36 pp.

703. Robinson, W. O., Whetstone, R. R., Scribner, B. F., 1938, *Science,* **87,** 470.

704. Robyns, W., 1932, *Natuurw. Tijdschr.,* **14,** 101.

705. Romney, E. M., Childress, J. D., 1965, *Soil Sci.,* **100,** 210.

706. Rosenkvist, A. M., 1950, *Forh. Kong. Norsk. Videns. Sels.,* **22,** 82.

707. Rowntree, J. C., Mosher, D. V., 1976, *Proc. IAEA Symp. Explor. Uranium Ore Dep.,* Vienna, STI/PUB/434, **551** (IAEA-SM-208/58).

708. Roy, S., 1974, *Quart. J. Geol. Min. Metall. Soc. India,* **46,** 251.

709. Rühling, A., Tyler, G., 1970, *Oikos,* **21,** 92.

710. Rune, O., 1953, *Acta Phytogeogr. Suec.,* **31,** 1.

711. Russell, D. W., Moort, J. C. van, 1981, *Econ. Geol.,* **76,** 339.

712. Rusyayeva, G. G., Gapon, A. Ye., Karpov, I. et al., 1977, *Izv. Akad. Nauk (SUN),* Irkutsk.

713. Ryall, W. R., Nicholas, T., 1979, *J. Geol. Soc. Austral.,* **26,** 187.

714. Saagar, R., Esselaar, P. A., 1969, *Econ. Geol.,* **64,** 445.

715. Sabinin, D. A., 1955, *Physiological Principles of Plant Nutrition* (in Russ.), Akad. Nauk SSSR, Moscow.

716. Safronov, N. I., Polikarpochkin, V. V., Utgof, A. A., 1958, *Sborn. Metod. Tekh.-Geol. Rab.,* No. 1, 100.

717. Salmi, M., 1950, *Geotekh. Julk. Geol. Tutkimuslaitos,* **51,** 20 pp.

718. Salmi, M., 1955, *Bull. Commn. Géol. Finl.,* **169,** 5.

719. Salmi, M., 1956, *Bull. Commn. Géol. Finl.,* **175,** 22 pp.

720. Salmi, M., 1959, *Proc. Geochem. Explor. Symp. 20th Int. Geol. Congr.,* Mexico City, 243.

721. Sandell, E. B., 1959, *Colorimetric Determination of Traces of Metals,* 3rd ed., Interscience, New York.

722. Santos, G. G. Jr., 1973, *IAEA Tech. Rep. Ser. TRAEA2,* No. 144, 136.

723. Sarosiek, J., 1964, *Monograph. Bot.,* **18,** 1.

724. Schatz, A., 1955, *Bryologist,* **52,** 113.

725. Schiller, P., Cook, G. R., Beswick, C. K., 1971, *IAEA Panel Proc.,* 129.

726. Schiller, P., Skalova, A., 1975, *Chem. Zvesti.,* **29,** 745.

727. Schmidt-Collerus, J. J., 1979, *U. S. Dep. Energ. Rep.,* GJBX-130 (79), 291 pp.

728. Schofield, W. B., 1959, *Bryologist,* **62,** 248.

729. Schultz, A., 1912, *Jb. Westf. Prov. Ver. Wiss. Kunst,* **40,** 209.

730. Schwanitz, F., Hahn, H., 1954, *Z. Bot.,* **42,** 179.

731. Schwanitz, F., Hahn, H., 1954, *Z. Bot.,* **42,** 459.

732. Schwickerath, M., 1931, *Beitr. Naturdenkmalpflege,* **14,** 463.

733. Schwickerath, M., 1933, *Aachener Beitr. Heimatkunde,* **13,** 1.

734. Scott, R. A., 1961, *U. S. Geol. Surv. Res. Rep.,* **B-130.**

735. Se Sjue-Tzsin, Sjuj Ban-Lian, 1953, *Dichzhi Sjuozbao,* **32,** 360.

736. Selivanov, L. S., 1946, *Trudy Biogeokhim. Lab.,* No. 8, 5.

737. Serdyuchenko, D. P., Pavlov, V. A., 1967, *Redk. Element. Porod Razl. Metamorf. Fatsii,* 126.

738. Severne, B. C., 1972, Botanical methods for mineral exploration in Western Australia, PhD thesis, Massey University, New Zealand.

739. Severne, B. C., 1974, *Nature,* **248,** 807.

740. Severne, B. C., Brooks, R. R., 1972, *Planta,* **103,** 91.

741. Shacklette, H. T., 1960, *U. S. Geol. Surv. Prof. Pap.,* **400B,** 102.

742. Shacklette, H. T., 1961, *Bryologist,* **64,** 1.

743. Shacklette, H. T., 1962, *Can. Fd. Nat.,* **76,** 162.

744. Shacklette, H. T., 1964, *Can. Fd. Nat.,* **78,** 32.

745. Shacklette, H. T., 1965, *U. S. Geol. Surv. Bull.,* **1198C,** 18 pp.

746. Shacklette, H. T., 1965, *U. S. Geol. Surv. Bull.,* **1198D,** 21 pp.

747. Shacklette, H. T., 1967, *U. S. Geol. Surv. Bull.,* **1198G,** 17 pp.

748. Shacklette, H. T., 1972, *U. S. Geol. Surv. Bull.,* **1314G,** 28 pp.

749. Shacklette, H. T., Erdman, J. A., 1982, *J. Geochem. Explor.* (in press).

750. Shacklette, H. T., Lakin, H. W., Hubert, A. E., Curtin, G. C., 1970, *U. S. Geol. Surv. Bull.,* **1314B,** 1.

751. Shakhova, I. K., 1960, *Trudy Biogeokhim. Lab.,* No. 11, 232.

752. Shaw, D. M., 1961, *Geochim. Cosmochim. Acta,* **23,** 116.

753. Shchapova, G. F., 1938, *Bot. Zhur. SSSR,* **23,** 122.

754. Shewry, P. R., Woolhouse, H. W., Thompson, K., 1979, *Bot. J. Linn. Soc.,* **79,** 1.

755. Shiikawa, M., Tono, N., 1972, *Geochemical Exploration in the Region of the Kuroko Deposits,* Mining and Metallurgical Institute of Japan, Tokyo.

756. Shimwell, D. W., Laurie, P. E., 1972, *Environ. Pollut.,* **3,** 291.

757. Shnyukov, E. F., Usenk, V. P., Krasnozhina, Z. V., 1963, *Pit. Geokhim. Miner. Petrogr. Akad. Nauk Ukr. SSR,* **1963,** 107.

758. Shrift, A., 1969, *Ann. Rev. Pl. Physiol.,* **20,** 475.

759. Shvartsev, L. S., 1966, *Int. Geol. Rev.,* **8,** 1151.

760. Shvyryayeva, A. M., Mikhailova, G. A., 1965, in *Plant Indicators of Soils, Rocks, and Subsurface Waters,* Consultants Bureau, New York.

761. Siegel, S., 1956, *Non-parametric Statistics for the Behavioral Sciences,* McGraw-Hill, Tokyo.

762. Simmoneau, P., 1954, *Ann. Agronom Sér. A,* 91.

763. Sinclair, A. J., 1976, *Ass. Explor. Geochemists Spec. Vol.,* **4,** 95 pp.

764. Sizykh, V. I., Korneva, A. A., Tauson, L. V., in *Lithogeochemical and Biogeochemical Methods for Research in the Bom Gorkhonsk Deposits* (in Russ.), SUN, Novosibirsk.

765. Skarlygina-Ufimtseva, M. D., Berezkina, G. A., Chernyakhov, V. B., 1971, *Vest. Leningr. Univ. Geol. Geograf.,* No. 24, 134.

766. Skarlygina-Ufimtseva, M. D., Chernyakhov, V. B., Berezkina, G. A., 1976, *Izd. Leningr. Univ.,* 149 pp.

767. Skarlygina-Ufimtseva, M. D., Lobanova, A. B., Goncharov, G. N., Gorbenko, M. G., 1980, *Vest. Leningr. Univ. Geol. Geograf.,* No. 6, 85.

768. Skertchly, S. B. J., 1897, *Qld. Geol. Surv. Pub.,* No. 119, 51.

769. Smith, A. Y., 1960, *Geol. Surv. Can. Pap.,* **59-12,** 8 pp.

770. Smith, G. M., 1938, *Bryophytes and Pteridophytes* v. 2, McGraw-Hill, New York.

771. Sokolova, A. I., Khramova, V. V., 1961, *Trudy Sverdlovsk. Gorny Inst.*, **40**, 107.

772. Sokolova, V. Y., Yatsyuk, M. D., 1965, *Ukr. Bot. Zhur.*, **22**, 14.

773. Somers, E., 1960, *Nature*, **187**, 427.

774. Spix, J. B. von, Martius, C. F., 1824, *Travels in Brazil in the Years 1817–1820 Undertaken by Command of His Majesty the King of Bavaria*, Longmans, London.

775. Stafleu, F. A., 1974, *Index Herbariorum*, Scheltema and Holkema, Utrecht.

776. Starikov, V. S., Konovalov, B. T., Brushtein, I. M., 1964, *Geokhimiya*, No. 10, 1070.

777. Starr, C. C., 1949, *W. Miner*, **22**, 43.

778. Steed, G. M., Annels, A. E., Shrestha, P. L., Tater, P. S., 1976, *Trans. Inst. Min. Metall.*, **85**, 119.

779. Storozheva, M. M., 1954, *Trudy Biogeokhim. Lab.*, **10**, 64.

780. Stout, P. R., Hoagland, D. R., 1939, *Am. J. Bot.*, **26**, 320.

781. Streeter, D. J., 1970, *Sci. Progr.*, **58**, 419.

782. Szalay, A., 1958, *Proc. 2nd Int. Conf. Peacef. Uses At. Energ.*, Geneva, **2**, 182.

783. Szalay, A., 1968, in *Chemistry of the Earth's Crust* Vol. 2, Oldbourne, London, p. 456.

784. Szalay, A., Szilagyi, M., 1967, *Geochim. Cosmochim. Acta*, **31**, 1.

785. Talipov, R. M., 1964, *Geokhimiya*, No. 5, 457.

786. Talipov, R. M., 1964, in *Voprosi Genezia, Akad. Nauk Uzbek. SSR*, Tashkent, 95.

787. Talipov, R. M., 1965, *Mikroelementy Sel'sk. Akad. Nauk SSSR Otd. Khim.-Tekh. Biol. Nauk*, 342.

788. Talipov, R. M., 1966, *Biogeochemical Prospecting for Polymetallic and Copper Deposits under the Conditions of Uzbekistan* (in Russ.), FAN, Tashkent.

789. Talipov, R. M., Aripova, Kh., Karabaev, K. K., Khatamov, Sh., Akundkhodzhaeva, N., 1968, *Uzbek. Geol. Zhur.*, No. 5, 43.

790. Talipov, R. M., Glushchenko, V. M., Lezhneva, N. D., Nishanov, P. Kh., 1975, *Uzbek. Geol. Zhur.*, No. 4, 21.

791. Talipov, R. M., Khatamov, Sh., Karabaev, K. K., Aripova, Kh., Akunkhodzhaeva, N., 1967, *Uzbek. Geol. Zhur.*, **11**, 1977.

792. Talipov, R. M., Khatamov. Sh., 1974, *Uzbek. Geol. Zhur.*, No. 1, 278.

793. Talipov, R. M., Musin, R. A., Glushchenko, V. M., Nishanonov, P., Matchanov, D., Samigdzhanova, M. A., 1974, *Metallogenesis and Geochemistry of Uzbekistan* (in Russ.), FAN, Tashkent.

794. Talvitie, J., 1979, *Bull. Geol. Soc. Finl.*, **51**, 63.

795. Talvitie, J., Lehmuspelto, P., Vuotesi, T., 1981, *Geol. Surv. Finl. Rep.*, **50**, 13 pp.

796. Talvitie, J., Paarma, H., 1973, in *Prospecting in Areas of Glaciated Terrain*, Institution of Mining and Metallurgy, London.

797. Targioni-Tozzetti, G., 1774, *Account of Several Journeys Made in Different Parts of Tuscany in Order to Observe Natural Products and Ancient Monuments* (in Ital.), Firenze.

798. Taylor, L., 1975, Biogeochemical exploration for copper, lead, and zinc mineral deposits using juniper and sagebrush, Dugwan Range, Utah, MSc thesis, Brigham Young Univ., Provo.

799. Tchakirian, A., 1942, *Ann. Inst. Pasteur Paris*, **68**, 461.

800. Tennant, C. B., White, M. L., 1959, *Econ. Geol.*, **54**, 1281.

801. Thibault, D. H., Sheppard, M. I., 1980, *Rep. At. Energ. Can.*, **WNRE-495**, 20 pp.

802. Thornton, I., Webb, J. S., 1970, *J. Ind. Soc. Soil Sci.*, **18**, 357.

803. Thorpe, J., 1931, *Soil Sci.*, **32**, 283.

804. Thyssen, S. W., 1942, *Beitr. Angew. Geophys.*, **10**, 35.

805. Timperley, M. H., Brooks, R. R., Peterson, P. J., 1970, *Econ. Geol.*, **65**, 505.

806. Timperley, M. H., Brooks, R. R., Peterson, P. J., 1970, *J. Appl. Ecol.*, **7**, 429.

807. Timperley, M. H., Brooks, R. R., Peterson, P. J., 1972, *Econ. Geol.*, **67**, 669.

808. Timperley, M. H., Brooks, R. R., Peterson, P. J., 1972, *Proc. Australas. Inst. Min. Metall.*, No. 242, 25.

809. Timperley, M. H., Peterson, P. J., Brooks, R. R., 1973, *J. Exp. Bot.*, **24**, 889.

810. Titaeva, N. A., 1967, *Geokhimiya*, **12**, 1493.

811. Titaeva, N. A., Taskaev, A. I., Ovchenkov, V. Ya., Aleksakkin, R. M., Shuktomova, I. I., 1978, *Ekologiya* (Sverdlovsk), No. 4, 37.

812. Tkalich, S. M., 1938, *Vest. Dal'ne Vost. Fil. Akad. Nauk SSSR*, **32**, 3.

813. Tkalich, S. M., 1970, *Phytogeochemical Prospecting for Ore Deposits (in Russ.)*, Nedra Press, Moscow.

814. Tkalich, S. M., 1953, *Priroda*, No. 1, 93.

815. Tolgyesi, G., 1962, *Agrokem. Talajt.*, **11**, 203.

816. Tomassini, F. D., Puckett, K. J., Nieboer, E., Richardson, D. H. S., Grace, D., 1976, *Can. J. Bot.*, **54**, 1591.

817. Tomassini, F. D., 1976, The Measurement of photosynthetic C-14 fixation rates and potassium effects to assess the sensitivity of lichens to sulphur dioxide and the adaptation of x-ray fluorescence to determine the elemental content in lichens, MSc thesis, Laurentian University, Sudbury, Ontario.

818. Tooms, J. S., Jay, J. R., 1964, *Econ. Geol.*, **59**, 826.

819. Toverud, O., 1979, *Prospecting in Areas of Glaciated Terrain*, Institution of Mining and Metallurgy, London.

820. Townshend, J., 1824, *Geological and Mineralogical Research During a Period of More than Fifty Years in England, Scotland, Ireland, Switzerland, Holland, France, Flanders and Spain*, Gye, Bath.

821. Trelease, S. F., Beath, O. A., 1949, *Selenium: Its Geological Occurrence and its Biological Effects in Relation to Botany, Chemistry, Agriculture, Nutrition and Medicine*, Trelease and Beath, New York.

822. Trelease, S. F., Disomma, A. A., Jacobs, A. L., 1960, *Science*, **132**, 3427.

823. Tsung-Shan, K., 1957, *Sci. Sinica*, **6**, 1105.

824. Tsutsui, M., Levy, M. N., Nakamura, A., Ichikawa, M., Nori, K., 1970, *Introduction to Metal Pi-Complex Chemistry*, Plenum Press, New York.

825. Turrill, W. B., 1951, *J. Ecol.*, **39**, 14.

826. Tyuremnov, S. N., Barkhatova, D. I., Kulikova, G. G., 1968, *Vest. Mosk. Gos. Univ. Ser. 6*, **23**, 476.

827. Tyurina, G. I., Schibrik, V. I., 1962, *Trudy Zentr. Kazakh. Geol. Upr. Minist. Geol. Okhr. Nedr Kazakh. SSR*, No. 2, 44.

828. Tyutina, N. A., Aleskovsky, V. B., Vasil'ev, P. I., 1959, *Geokhimiya*, No. 6, 550.

829. Uglow, W. L., 1920, *Munit. Res. Commn. Can. Final Rep.*, 65.

830. Urbani, P. F., 1976, *Inf. Esc. Minas Lab. Univ. Cent. Venezuela*, No. 76-1, 77 pp.

831. Ure, A. M., 1975, *Anal. Chim. Acta*, **76**, 1.

832. Usik, L., 1969, *Geol. Surv. Can. Pap.*, **68-66**, 43 pp.

833. Uspensky, Y. Y., 1915, *Zhur. Opyt. Agron.*, **15**, 425.

834. Vakhromeev, G. S., 1962, *Zap. Vost. Sib. Otd. Vses. Miner. Obsch. Akad. Nauk SSSR*, No. 4, 196.

835. Velikova, F. I., Borovskaya, Y. B., Efendieva, N. G., 1963, *Izv. Akad. Nauk Azerb. SSR Ser. Geol.-Geograf. Nauk,* No. 4, 71.

836. Ventakesh, V., 1964, *Ind. Miner,* **18,** 101.

837. Ventakesh, V., 1966, *Steel Mines Rev.* (India), **6,** 3.

838. Vergnano, O., 1957, *Nuov. G. Bot. Ital.,* **64,** 652.

839. Vergnano, O., 1958, *Nuov. G. Bot. Ital.,* **65,** 133.

840. Vergnano Gambi, O., Brooks, R. R., Radford, C. C., 1979, *Webbia,* **33,** 269.

841. Vergnano Gambi, O., Cardini, F., 1967, *G. Bot. Ital.,* **101,** 63.

842. Vergnano Gambi, O., Gabbrielli, R., 1979, *Ofioliti,* **4,** 199.

843. Vergnano Gambi, O., Gabbrielli, R., Lotti, L., Polidori, V., 1971, *Webbia,* **25,** 353.

844. Vergnano Gambi, O., 1966, *Arch. Bot. Biogeograf. Ital.,* **42,** 1.

845. Verkhovskaya, I. N., Vavilov, P. P., Maslov, V. I., 1967, *Proc. Int. Symp. Radioecol. Concn. Processes,* Stockholm, 313.

846. Verma, K. K., Barman, G., 1977, Ind. Miner, 31, 69.

847. Vernadsky, V. I., 1934, *Outline of Geochemistry* (in Russ.), Russ. Izd., Moscow.

848. Viktorov, S. V., 1956, *Vest. Mosk. Gos. Univ.,* No. 5, 115.

849. Viktorov, S. V., Vostokova, Y. A., Vyshivkin, D. D., 1964, *Short Guide to Geobotanical Surveying,* Pergamon, Oxford.

850. Vine, J. D., Swanson, V. E., Bell, K. G., 1958, *Proc. 2nd Int. Conf. Peacef. Uses At. Energ.,* Geneva, **2,** 187.

851. Vinogradov, A. P., 1959, *The Geochemistry of Rare and Dispersed Chemical Elements in Soils,* Consultants Bureau, New York.

852. Vinogradov, A. P., 1964, *Int. Monogr. Earth Sci.,* No. 15, 317.

853. Virupaksha, T. K., Shrift, A., 1965, *Biochim. Biophys. Acta,* **107,** 69.

854. Vistelius, A. B., 1960, *J. Geol.,* **68,** 1.

855. Vogt, T., 1939, *Forh. Kong. Norsk. Videns. Sels.,* **12,** 82.

856. Vogt, T., 1942, *Forh. Kong. Norsk. Videns. Sels.,* **15,** 21.

857. Vogt, T., 1942, *Forh. Kong. Norsk. Videns. Sels.,* **15,** 5.

858. Vogt, T., Bergh, H., 1947, *Forh. Kong. Norsk. Videns. Sels.,* **19,** 76.

859. Vogt, T., Bergh, H., 1948, *Forh. Kong. Norsk. Videns. Sels.,* **20,** 100.

860. Vogt, T., Braadlie, O., 1942, *Forh. Kong. Norsk. Videns. Sels.,* **15,** 25.

861. Vogt, T., Braadlie, O., Bergh, H., 1943, *Forh. Kong. Norsk. Videns. Sels.,* **16,** 55.

862. Vogt, T., Bugge, J., 1943, *Forh. Kong. Norsk. Videns. Sels.,* **16,** 51.

863. Voitkevich, G. V., Alekseenko, V. A., Aleksenko, V. A., 1970, *Izv. Vuzov. Geol. Razved.,* No. 2, 64.

864. Vorob'yev, V. Ya., Gudoshnikov, V. V., 1967, *Sov. Geol.,* No. 3, 107.

865. Vostokova, Y. A., 1957, *Razv. Okhr. Nedr,* **23,** 33.

866. Vostokova, Y. A., Khdanova, G. I., 1961, *Int. Geol. Rev.,* **23,** 33.

867. Vostokova, Y. A., Vyshivkin, D. D., Kasynova, M. S., Nesvetaylova, N. G., Shvyryayeva, A. M., 1961, *Int. Geol. Rev.,* **3,** 598.

868. Wagner, K., 1936, *J. Prakt. Chem.,* **147,** 110.

869. Walker, N. C., 1976, *Summ. Sask. Geol. Surv. Invest.,* **1976,** 125.

870. Walker, N. C., 1979, *Proc. 7th Int. Geochem. Explor. Symp.,* Colorado, 361.

871. Walker, R. B., 1954, *Ecology,* **35,** 259.

872. Walker, R. B., Walker, H. M., Ashworth, P. R., 1955, *Pl. Physiol.,* **30,** 214.

873. Wallace, A., Bear, F. E., 1949, *Pl. Physiol.,* **24,** 664.

874. Wallace, T., 1951, *Int. Un. Biol. Sci. Colloq. No. 1 Ser. B*, 5.

875. Walsh, A., 1955, *Spectrochim. Acta*, **7**, 108.

876. Ward, F. N., Nakagawa, H. M., 1969, *Proc. 1st Int. Geochem. Symp.*, Colorado, 497.

877. Ward, J. V., 1973, *J. Fish. Res. Bd. Can.*, **30**, 841.

878. Ward, N. I., Brooks, R. R., 1978, *N. Z. J. Bot.*, **16**, 175.

879. Ward, N. I., Brooks, R. R., Reeves, R. D., 1976, *N. Z. J. Sci.*, **19**, 81.

880. Ward, N. I., Roberts, E., Brooks, R. R., 1977, *N.Z. J. Sci.*, **20**, 413.

881. Ward, N. I., Brooks, R. R., Roberts, E., 1977, *Bryologist*, **80**, 304.

882. Ward, N. I., Brooks, R. R., Roberts, E., Boswell, C. R., 1977, *Environ. Sci. Tech.*, **11**, 917.

883. Warren, H. V., 1962, *Trans. Roy. Soc. Can.*, **61**, 21.

884. Warren, H. V., Delavault, R. E., 1948, *Geophysics*, **13**, 609.

885. Warren, H. V., Delavault, R. E., 1949, *Bull. Geol. Soc. Am.*, **60**, 531.

886. Warren, H. V., Delavault, R. E., 1950, *Bull. Geol. Soc. Am.*, **61**, 123.

887. Warren, H. V., Delavault, R. E., 1950, *Trans. Can. Inst. Min. Metall.*, **53**, 236.

888. Warren, H. V., Delavault, R. E., 1950, *Trans. Roy. Soc. Can.*, **54**, 11.

889. Warren, H. V., Delavault, R. E., 1954, *Trans. Roy. Soc. Can.*, **47**, 71.

890. Warren, H. V., Delavault, R. E., 1955, *Can. Min. J.*, **76**, 49.

891. Warren, H. V., Delavault, R. E., 1955, *Trans. Roy. Soc. Can.*, **48**, 71.

892. Warren, H. V., Delavault, R. E., 1955, *Trans. Roy. Soc. Can.*, **49**, 111.

893. Warren, H. V., Delavault, R. E., 1957, *Tran. Roy. Soc. Can.*, **51**, 33.

894. Warren, H. V., Delavault, R. E., 1959, *Proc. Geochem. Explor. Symp. 20th Int. Geol. Congr.*, Mexico City, 255.

895. Warren, H. V., Delavault, R. E., 1960, *Trans. Roy. Soc. Can.*, **54**, 11.

896. Warren, H. V., Delavault, R. E., 1965, *W. Miner*, **38**, 64.

897. Warren, H. V., Delavault, R. E., Barakso, J., 1964, *Econ. Geol.*, **59**, 38.

898. Warren, H. V., Delavault, R. E., Barakso, J., 1966, *Econ. Geol.*, **61**, 1010.

899. Warren, H. V., Delavault, R. E., Cross, C. H., 1959, *Proc. 6th Commonw. Min. Metall. Congr.*, Montreal, 277.

900. Warren, H. V., Delavault, R. E., Cross, C. H., 1966, *W. Miner*, **339**, 36.

901. Warren, H. V., Delavault, R. E., Fletcher, K., Peterson, G. R., 1971, *Can. Inst. Min. Metall. Spec. Vol.*, **11**, 444.

902. Warren, H. V., Delavault, R. E., Fortescue, J. A. C., 1955, *Bull. Geol. Soc. Am.*, **66**, 229.

903. Warren, H. V., Delavault, R. E., Irish, R. I., 1949, *Trans. Roy. Soc. Can.*, **43**, 119.

904. Warren, H. V., Delavault, R. E., Irish, R. I., 1951, *Bull. Geol. Soc., Am.*, **62**, 919.

905. Warren, H. V., Delavault, R. E., Irish, R. I., 1952, *Bull. Geol. Soc. Am.*, **63**, 435.

906. Warren, H. V., Delavault, R. E., Irish, R. I., 1952, *Econ. Geol.*, **47**, 131.

907. Warren, H. V., Delavault, R. E., Routley, D. G., 1953, *Trans. Roy. Soc.*, **47**, 71.

908. Warren, H. V., Hajek, J. H., 1973, *W. Miner*, **46**, 124.

909. Warren, H. V., Howatson, C. H., 1947, *Bull. Geol. Soc. Am.*, **58**, 803.

910. Warrington, K., 1937, *Ann. Appl. Biol.*, **24**, 475.

911. Watson, J. P., 1970, *Trans. Inst. Min. Metall. Sec. B*, **79**, 53.

912. Watson, J. P., 1972, *Soil Sci.*, **113**, 317.

913. Webb, J. S., 1964, *New Scientist*, **23**, 504.

914. Webb, J. S., Millman, A. P., 1951, *Trans. Inst. Min. Metall.*, **60**, 473.

915. Webb, J. S., Nichol, I., Thornton, I., 1968, *Proc. 23rd Int. Geol. Congr.*, **6**, 131.

916. Webb, J. S., Thornton, I., Fletcher, K., 1968, *Nature*, **217**, 1010.

917. Weiss, O., 1971, *Can. Inst. Min. Metall., Spec. Vol.*, **11**, 502.

918. Weiss, O., 1967, U. S. Pat. 3309518.

919. Wenrich-Verbeek, K. J., 1979, *Abs. Progr. Geol. Soc. Am.*, **11**, 538.

920. Wenrich-Verbeek, K. J., 1980, *U. S. Geol. Surv. Prof. Pap.*, **1175**, 56.

921. West, F. W., 1965, *Chamber of Mines J.* (Zimbabwe), **7**, 40.

922. West, F. W., 1970, *Chamber of Mines J.* (Zimbabwe), **12**, 32.

923. West, T. S., Williams, X. K., 1969 *Anal. Chim. Acta*, **45**, 27.

924. White, W. H., 1950, *Trans. Can. Inst. Min. Metall.*, **53**, 368.

925. Whitehead, N. E., Brooks, R. R., 1969, *Bryologist*, **72**, 501.

926. Whitehead, N. E., Brooks, R. R., 1969, *Econ. Geol.*, **64**, 50.

927. Whitehead, N. E., Brooks, R. R., 1969, *J. Appl. Ecol*, **6**, 301.

928. Whitehead, N. E., Brooks, R. R., Coote, G. E., 1971, *N. Z. J. Sci.*, **14**, 66.

929. Wild, H., 1968, *Kirkia*, **7**, 1.

930. Wild, H., 1968, *Kirkia*, **8**, 1.

931. Wild, H., 1970, *Kirkia Suppl.*, **7**, 62 pp.

932. Wild, H., 1971, *Mitt. Bot. Staatssamml. Munchen*, **10**, 266.

933. Wild, H., 1974, *Kirkia*, **9**, 233.

934. Wild, H., 1974, *Kirkia*, **9**, 243.

935. Wild, H., 1974, *Kirkia*, **9**, 265.

936. Wild, H., 1975, *Trans. Rhod. Sci. Ass.*, **57**, 1.

937. Wild, H., 1978, in *Biogeography and Ecology of Southern Africa*, Junk, The Hague.

938. Wild, H., 1970, *Chamber Mines J.* (Zimbabwe), **12**, 30.

939. Will, O. M., 1955, *Nature*, **176**, 1180.

940. Williams, D. E., Vlamis, J., *Pl. Soil*, **8**, 183.

941. Williams, G. J., 1934, *N. Z. J. Sci. Technol.*, **15**, 344.

942. Williams, K. T., 1938, *U. S. Dep. Agr. Yrb.*, **1938**, 831.

943. Williams, X. K., 1967, *N. Z. J. Geol. Geophys.*, **10**, 771.

944. Wither, E. D., Biogeochemical studies in Southeast Asia by use of herbarium material, MSc thesis, Massey University, New Zealand.

945. Wither, E. D., Brooks, R. R., 1977, *J. Geochem. Explor.*, **8**, 579.

946. Wodzicki, A., 1959, *N. Z. J. Geol. Geophys.*, **2**, 602.

947. Wolfe, W. J., 1971, *Can. Inst. Min. Metall. Spec. Vol.*, **11**, 72.

948. Wolfe, W. J., 1971, *Can. Min. Metall. Bull.*, No. 715, 72.

949. Wolfe, W. J., 1974, *J. Geochem. Explor.*, **3**, 25.

950. Wolff, E. T. von, 1871, 1880, *Aschenanalysen von Wirtschaftlichen Produkten Fabrikabfallen und Wildwachsenden Pflanzen* (2 v. in one), Wiegland and Hempel, Berlin.

951. Worthington, J. E., 1955, *Econ. Geol.*, **50**, 420.

952. Yakovleva, M. N., 1963, *Vop. Priklad. Radiogeol.*, No. 1, 271.

953. Yakovleva, M. N., 1964, *Byull. Nauchno.-Tekh. Inst. Gos. Geol. Kom. SSSR Otd. Vses. Nauchno.-Issled. Inst. Miner. Syr'ya*, No. 3, 13.

954. Yarovaya, G. A., 1960, *Trudy Biogeokhim. Lab.*, No. 2, 208.

955. Yates, T. E., Brooks, R. R., Boswell, C. R., 1974, *J. Appl. Ecol.*, **11**, 563.

956. Yates, T. E., Brooks, R. R., Boswell, C. R., 1974, *N. Z. J. Sci.*, **17**, 151.

957. Yezdakov, V. I., Yezdakova, L. A., 1968, *Vses. Ucheb. Zavod. Izv. Geol. Razved.*, No. 3, 89.

958. Yezdakova, L. A., 1966, *Bot. Zhur. SSSR,* **51,** 727.

959. Yliruokanen, I., 1975, *Ann. Acad. Sci. Fenn. Ser. A,* **176,** 471.

960. Yliruokanen, I., 1975, *Ann. Acad. Sci. Fenn. Ser. A,* **176,** 5.

961. Yliruokanen, I., 1975, *Bull. Geol. Soc. Finl.,* **47,** 71.

962. Yost, E., Wenderoth, S., 1971, *Proc. 7th Int. Symp. Rem. Sens. Environ.,* Ann Arbor, 269.

963. Zagoshkin, V. A., Shimansky, A. A., in *Secondary Lithogeochemical and Biogeochemical Dispersion Haloes in Rare Earth Pegmatite Deposits,* Akad. Nauk, Moscow.

964. Zajic, J. E., 1969, *Microbial Biogeochemistry,* Academic Press, New York.

965. Zalashkova, N. E., Lizunov, N. V., Sitnin, A. A., 1958, *Razv. Okhr. Nedr,* **24,** 9.

966. Ziauddin, M., Roy, S., 1974, *Geol. Surv. India News,* **5,** 4.

APPENDIXES

APPENDIX I Table of Values for *F* Test

Probability (P)	ϕ_2	\multicolumn{12}{c}{ϕ_1 (corresponding to greater mean square)}											
		1	2	3	4	5	6	7	8	9	10	15	∞
0.10	1	39.9	49.5	53.6	55.8	57.2	58.2	58.9	59.4	59.9	60.2	61.2	63.3
0.05		161	199	216	225	230	234	237	239	241	242	246	254
0.01		4,052	4,999	5,403	5,625	5,764	5,859	5,928	5,982	6,022	6,056	6,157	6,366
0.10	2	8.53	9.00	9.16	9.24	9.29	9.33	9.35	9.38	9.39	9.39	9.42	9.49
0.05		18.5	19.0	19.2	19.2	19.3	19.3	19.4	19.4	19.4	19.4	19.4	19.5
0.01		98.5	99.0	99.2	99.2	99.3	99.3	99.4	99.4	99.4	99.4	99.4	99.5
0.10	3	5.54	5.46	5.39	5.34	5.31	5.28	5.27	5.25	5.24	5.23	5.20	5.13
0.05		10.1	9.55	9.28	9.12	9.01	8.94	8.89	8.85	8.81	8.79	8.70	8.53
0.01		34.1	30.8	29.5	28.7	28.2	27.9	27.7	27.5	27.3	27.2	26.9	26.1
0.10	4	4.54	4.32	4.19	4.11	4.05	4.01	3.98	3.95	3.94	3.92	3.87	3.76
0.05		7.71	6.94	6.59	6.39	6.26	6.16	6.09	6.04	6.00	5.96	5.86	5.63
0.01		21.2	18.0	16.7	16.0	15.5	15.2	15.0	14.8	14.7	14.5	14.2	13.5
0.10	5	4.06	3.78	3.62	3.52	3.45	3.40	3.37	3.34	3.32	3.30	3.24	3.10
0.05		6.61	5.79	5.41	5.19	5.05	4.95	4.88	4.82	4.77	4.74	4.62	4.36
0.01		16.3	13.3	12.1	11.4	11.0	10.7	10.5	10.3	10.2	10.1	9.72	9.02
0.10	6	3.78	3.46	3.29	3.18	3.11	3.05	3.01	2.98	2.96	2.94	2.87	2.72
0.05		5.99	5.14	4.76	4.53	4.39	4.28	4.21	4.15	4.10	4.06	3.94	3.67
0.01		13.7	10.9	9.78	9.15	8.75	8.47	8.26	8.10	7.98	7.87	7.56	6.88
0.10	7	3.59	3.26	3.07	2.96	2.88	2.83	2.78	2.75	2.72	2.70	2.63	2.47
0.05		5.59	4.74	4.35	4.12	3.97	3.87	3.79	3.73	3.68	3.64	3.51	3.23
0.01		12.2	9.55	8.45	7.85	7.46	7.19	6.99	6.84	6.72	6.62	6.31	5.65

α	df												
0.10	8	3.46	3.11	2.92	2.81	2.73	2.67	2.62	2.59	2.56	2.54	2.46	2.29
0.05		5.32	4.46	4.07	3.84	3.69	3.58	3.50	3.44	3.39	3.35	3.22	2.93
0.01		11.3	8.65	7.59	7.01	6.63	6.37	6.18	6.03	5.91	5.81	5.52	4.86
0.10	9	3.36	3.01	2.81	2.69	2.61	2.55	2.51	2.47	2.44	2.42	2.34	2.16
0.05		5.12	4.26	3.86	3.63	3.48	3.37	3.29	3.23	3.18	3.14	3.01	2.71
0.01		10.6	8.02	6.99	6.42	6.06	5.80	5.61	5.47	5.35	5.26	4.96	4.31
0.10	10	3.29	2.92	2.73	2.61	2.52	2.46	2.41	2.38	2.35	2.32	2.24	2.06
0.05		4.96	4.10	3.71	3.48	3.33	3.22	3.14	3.07	3.02	2.98	2.85	2.54
0.01		10.0	7.56	6.55	5.99	5.64	5.39	5.20	5.06	4.94	4.85	4.56	3.91
0.10	12	3.18	2.81	2.61	2.48	2.39	2.33	2.28	2.24	2.21	2.19	2.10	1.90
0.05		4.75	3.89	3.49	3.26	3.11	3.00	2.91	2.85	2.80	2.75	2.62	2.30
0.01		9.33	6.93	5.95	5.41	5.06	4.82	4.64	4.50	4.39	4.30	4.01	3.36
0.10	15	3.07	2.70	2.49	2.36	2.27	2.21	2.16	2.12	2.09	2.06	1.97	1.76
0.05		4.54	3.68	3.29	3.06	2.90	2.79	2.71	2.64	2.59	2.54	2.40	2.07
0.01		8.68	6.36	5.42	4.89	4.56	4.32	4.14	4.00	3.89	3.80	3.52	2.87
0.10	16	3.05	2.67	2.46	2.33	2.24	2.18	2.13	2.09	2.06	2.03	1.94	1.72
0.05		4.49	3.63	3.24	3.01	2.85	2.74	2.66	2.59	2.54	2.49	2.35	2.01
0.01		8.53	6.23	5.29	4.77	4.44	4.20	4.03	3.89	3.78	3.69	3.41	2.75
0.10	24	2.93	2.54	2.33	2.19	2.10	2.04	1.98	1.94	1.91	1.88	1.78	1.53
0.05		4.26	3.40	3.01	2.78	2.62	2.51	2.42	2.36	2.30	2.25	2.11	1.73
0.01		7.82	5.61	4.72	4.22	3.90	3.67	3.50	3.36	3.26	3.17	2.89	2.21
0.10	60	2.79	2.39	2.18	2.04	1.95	1.87	1.82	1.77	1.74	1.71	1.60	1.29
0.05		4.00	3.15	2.76	2.53	2.37	2.25	2.17	2.10	2.04	1.99	1.84	1.39
0.01		7.08	4.98	4.13	3.65	3.34	3.12	2.95	2.82	2.72	2.63	2.35	1.60
0.10	∞	2.71	2.30	2.08	1.94	1.85	1.77	1.72	1.67	1.63	1.60	1.49	1.00
0.05		3.84	3.00	2.60	2.37	2.21	2.10	2.01	1.94	1.88	1.83	1.67	1.00
0.01		6.63	4.61	3.78	3.32	3.02	2.80	2.64	2.51	2.41	2.32	2.04	1.00

	Probability (P)							
ϕ	0.5	0.2	0.1	0.05	0.02	0.01	0.005	0.001
1	1.00	3.08	6.31	12.7	31.8	63.7	127	637
2	0.816	1.89	2.92	4.30	6.97	9.92	14.1	31.6
3	0.765	1.64	2.35	3.18	4.54	5.84	7.45	12.9
4	0.741	1.53	2.13	2.78	3.75	4.60	5.60	8.61
5	0.727	1.48	2.01	2.57	3.37	4.03	4.77	6.87
6	0.718	1.44	1.94	2.45	3.14	3.71	4.32	5.96
7	0.711	1.42	1.89	2.36	3.00	3.50	4.03	5.41
8	0.706	1.40	1.86	2.31	2.90	3.36	3.83	5.04
9	0.703	1.38	1.83	2.26	2.82	3.25	3.69	4.78
10	0.700	1.37	1.81	2.23	2.76	3.17	3.58	4.59
11	0.697	1.36	1.80	2.20	2.72	3.11	3.50	4.44
12	0.695	1.36	1.78	2.18	2.68	3.05	3.43	4.32
13	0.694	1.35	1.77	2.16	2.65	3.01	3.37	4.22
14	0.692	1.35	1.76	2.14	2.62	2.98	3.33	4.14
15	0.691	1.34	1.75	2.13	2.60	2.95	3.29	4.07
16	0.690	1.34	1.75	2.12	2.58	2.92	3.25	4.01
17	0.689	1.33	1.74	2.11	2.57	2.90	3.22	3.96
18	0.688	1.33	1.73	2.10	2.55	2.88	3.20	3.92
19	0.688	1.33	1.73	2.09	2.54	2.86	3.17	3.88
20	0.687	1.33	1.72	2.09	2.53	2.85	3.15	3.85
21	0.686	1.32	1.72	2.08	2.52	2.83	3.14	3.82
22	0.686	1.32	1.72	2.07	2.51	2.82	3.12	3.79
23	0.685	1.32	1.71	2.07	2.50	2.81	3.10	3.77
24	0.685	1.32	1.71	2.06	2.49	2.80	3.09	3.74
25	0.684	1.32	1.71	2.06	2.49	2.79	3.08	3.72
26	0.684	1.32	1.71	2.06	2.48	2.78	3.07	3.71
27	0.684	1.31	1.70	2.05	2.47	2.77	3.06	3.69
28	0.683	1.31	1.70	2.05	2.47	2.76	3.05	3.67
29	0.683	1.31	1.70	2.05	2.46	2.76	3.04	3.66
30	0.683	1.31	1.70	2.04	2.46	2.75	3.03	3.65
40	0.681	1.30	1.68	2.02	2.42	2.70	2.97	3.55
60	0.679	1.30	1.67	2.00	2.39	2.66	2.91	3.46
120	0.677	1.29	1.66	1.98	2.36	2.62	2.86	3.37
∞	0.674	1.28	1.64	1.96	2.33	2.58	2.81	3.29

The table gives the probability that the absolute value of *t* will exceed the tabulated value.

APPENDIX III   Table of Values for Chi² Test

Probability (P)

ϕ	0.99	0.975	0.95	0.90	0.50	0.10	0.05	0.025	0.02	0.001
1	0.000	0.001	0.004	0.016	0.455	2.71	3.84	5.02	6.63	10.83
2	0.020	0.051	0.103	0.211	1.39	4.61	5.99	7.38	9.21	13.82
3	0.115	0.216	0.352	0.584	2.37	6.25	7.81	9.35	11.34	16.27
4	0.297	0.484	0.711	1.06	3.36	7.78	9.49	11.14	13.28	18.47
5	0.554	0.831	1.15	1.61	4.35	9.24	11.07	12.83	15.09	20.52
6	0.872	1.24	1.64	2.20	5.35	10.64	12.59	14.45	16.81	22.46
7	1.24	1.69	2.17	2.83	6.35	12.02	14.07	16.01	18.48	24.32
8	1.65	2.18	2.73	3.49	7.34	13.36	15.51	17.53	20.09	26.13
9	2.09	2.70	3.33	4.17	8.34	14.68	16.92	19.02	21.67	27.88
10	2.56	3.25	3.94	4.87	9.34	15.99	18.31	20.48	23.21	29.59
11	3.05	3.82	4.57	5.58	10.34	17.28	19.68	21.92	24.73	31.26
12	3.57	4.40	5.23	6.30	11.34	18.55	21.03	23.34	26.22	32.91
13	4.11	5.01	5.89	7.04	12.34	19.81	22.36	24.74	27.69	34.53
14	4.66	5.63	6.57	7.79	13.34	21.06	23.68	26.12	29.14	36.12
15	5.23	6.26	7.26	8.55	14.34	22.31	25.00	27.49	30.58	37.70
16	5.81	6.91	7.96	9.31	15.34	23.54	26.30	28.85	32.00	39.25
17	6.41	7.56	8.67	10.09	16.34	24.77	27.59	30.19	33.41	40.79
18	7.01	8.23	9.39	10.86	17.34	25.99	28.87	31.53	34.81	42.31
19	7.63	8.91	10.12	11.65	18.34	27.20	30.14	32.85	36.19	43.82
20	8.26	9.59	10.85	12.44	19.34	28.41	31.41	34.17	37.57	45.32
21	8.90	10.28	11.59	13.24	20.34	29.62	32.67	35.48	38.93	46.80
22	9.54	10.98	12.34	14.04	21.34	30.81	33.92	36.78	40.29	48.27
23	10.20	11.69	13.09	14.85	22.34	32.01	35.17	38.08	41.64	49.73
24	10.86	12.40	13.85	15.66	23.34	33.20	36.42	39.36	42.98	51.18
25	11.52	13.12	14.61	16.47	24.34	34.38	37.65	40.65	44.31	52.62
26	12.20	13.84	15.38	17.29	25.34	35.56	38.89	41.92	45.64	54.05
27	12.88	14.57	16.15	18.11	26.34	36.74	40.11	43.19	46.96	55.48
28	13.56	15.31	16.93	18.94	27.34	37.92	41.34	44.46	48.28	56.89
29	14.26	16.05	17.71	19.77	28.34	39.09	42.56	45.72	49.59	58.30
30	14.95	16.79	18.49	20.60	29.34	40.26	43.77	46.98	50.89	59.70

APPENDIX IV   Table of Significance Values for r

	Probability (P)				
ϕ	0.10	0.05	0.02	0.01	0.001
1	0.988	0.997	0.999	1.000	1.000
2	0.900	0.950	0.980	0.990	0.999
3	0.805	0.878	0.934	0.959	0.992
4	0.729	0.811	0.882	0.917	0.974
5	0.669	0.754	0.833	0.874	0.951
6	0.621	0.707	0.789	0.834	0.925
7	0.582	0.666	0.750	0.798	0.898
8	0.549	0.632	0.716	0.765	0.872
9	0.521	0.602	0.685	0.735	0.847
10	0.497	0.576	0.658	0.708	0.823
11	0.476	0.553	0.634	0.684	0.801
12	0.457	0.532	0.612	0.661	0.780
13	0.441	0.514	0.592	0.641	0.760
14	0.426	0.497	0.574	0.623	0.742
15	0.412	0.482	0.558	0.606	0.725
16	0.400	0.468	0.543	0.590	0.708
17	0.389	0.456	0.528	0.575	0.693
18	0.378	0.444	0.516	0.561	0.679
19	0.369	0.433	0.503	0.549	0.665
20	0.360	0.423	0.492	0.537	0.652
25	0.323	0.381	0.445	0.487	0.597
30	0.296	0.349	0.409	0.449	0.554
35	0.275	0.325	0.381	0.418	0.519
40	0.257	0.304	0.358	0.393	0.490
45	0.243	0.287	0.338	0.372	0.465
50	0.231	0.273	0.322	0.354	0.443
60	0.211	0.250	0.295	0.325	0.408
70	0.195	0.232	0.274	0.302	0.380
80	0.183	0.217	0.256	0.283	0.357
90	0.173	0.205	0.242	0.267	0.337
100	0.164	0.195	0.230	0.254	0.321

INDEX OF BOTANICAL AND ZOOLOGICAL NAMES

PLANTS

Lasiochlamys sp. (Euphorbiac.), 257
Lecanora polytropa (Ehrh.) Rabenh. (Lichen),
   47
Lecidea decipiens (Hedw.) Ach. (Lichen), 47
L. dicksonii, 47
Ledum palustre L. (Ericac.), 140, 159
Leptospermum scoparium J. R. et G. Forst.
   (Myrtac.), 57, 129, 130
Leucanthemum sp. (Compos.), 141
Limonium suffruticosum O. K. (Plumbaginac.),
   35, 53
Linaria sp. (Scrophulariac.), 22
Lindernia sp. (Scrophulariac.), 257
L. damblonii Duvig., 35
L. perennis Duvig., 35
Lonicera sp. (Caprifoliac.), 140
L. caerulea L., 159
L. periclymemum L., 141
Lophocolea planiuscula (Tayl.) G. L. et N.
   (Bryophyte), 181
Loudetia simplex (Nees) C. E. Hubb.
   (Gramin.), 26
L. superba De Not (Gramin), 26
Lychnis alpina L. (Caryophyllac.), 31, 35, 38,
   39, 41, 132, 133, 251, 255
L. alpina var. *serpentinicola*, 36, 41, 132, 133

Macadamia sp. (Proteac.), 256
Macaranga hispida Muell.—Arg.
   (Euphorbiac.), 253
Malvastrum coccineum A. Gray (Malvac.), 130
Maytenus sp. (Celastrac.), 256
M. bureauvianus (Loes.) Loes., 36, 40
Melaleuca sheatheana W. V. Fitzg. (Myrtac.),
   18
Melandrium dioica (L.) Coss. et Germ.
   (Caryophyllac.), 251
Melicytus ramiflorus J. R. et G. Forst
   (Violac.), 229, 230
Merceya sp. (Bryophyte), 47
M. latifolia Kindb. ex Mac. (Bryophyte), 35,
   47
Mielichhoferia sp. (Bryophyte), 47
M. elongata Hornsch., 47
M. macrocarpa (Hook. ex Drumm.) Bruch. et
   Schimp. ex Jaeg., 47
M. mielichhoferi (Funck. ex Hook.) Loeske,
   35, 47
Minuartia verna (L.) Hiern. (Caryophyllac.),
   35, 37
M. verna ssp. *hercynica*, 42
Mohria caffrorum (L.) Desv. (Schizaeac.), 27
Monadenium sp. (Orchidac.), 27
Monocymbidium ceresiiforme (Nees) Stapf.
   (Orichidac.), 44

Moraea carsonii Bak. (Iridac.), 26
Mora excelsa Benth. (Legumin.), 99
Moringa oleifera Lamk. (Moringac.), 253
Munroa squarrosa Torr. (Gramin.), 130
Myosotis monroi Cheesem. (Boraginac.), 24, 25
Myristica sp. (Myristicac.), 257
Myrsine divaricata A. Cunn. (Myrsinac.), 130

Neptunia amplexicaulis Domin. (Legumin.), 23
Noccaea sp. (Crucif.), 256
Nothofagus fusca (Hook f.) Oerst. (Fagac.),
   135, 151, 223, 227
Nyssa sp. (Nyssac.), 250
N. sylvatica Marsh., 38, 250

Ochna schweinfurthiana F. Hoffm. (Ochnac.),
   26
Ocimum(=*Becium*) *homblei* De Wild. (Labiat.),
   81
Olax obtusifolia de Wild. (Olacac.), 44
Olearia muelleri (Sond.) Benth. (Compos.), 18
O. rani (A. Cunn.) Druce (Compos.), 137
Oligotrichum hercynicum (Hedw.) Lam. et DC
   (Bryophyte), 35
Oncotheca sp. (Oncothecac.), 257
Oryzopsis sp. (Gramin.), 23

Pancheria sp. (Cunoniac.), 257
Pandiaka sp. (Amaranthac.), 256
P. metallorum Duvig. et Van Bockstal
   (Amaranthac.), 26, 36
Papaver commutatum (Fisch.) Mey et Trautw.
   (Papaverac.), 57, 58
Pearsonia metallifera Wild (Legumin.), 131
Pellaea goudotii (Pteridophyte), 27
Peltaria sp. (Crucif.), 256
Peltigera apthosa (L.) Willd. (Lichen), 140
Pennisetum sp. (Gramin.), 44
Peperomia pellucida H. B. K. (Piperac.), 253
Peraphyllum ramossissima Nutt. ex Torr.
   (Rosac.), 57, 58
Philonotis fontana (Hedw.) Brid. (Bryophyte),
   178
Phragmanthera rufescens (DC) Balle var.
   cornetii (Dewevre) Balle (Loranthac.), 26
Phyllanthus sp. (Euphorbiac.), 249, 257
P. serpentinus Moore, 249
Picea sp. (Pinac.), 76
P. engelmanni (Parry) Engelm., 235
P. excelsa Link., 148
P. glauca Hort. ex Beissn., 76
P. mariana (Mill.) B. S. P., 76, 261
P. rubens Sarg., 174
Pimelea suteri Kirk (Thymelac.), 24, 25, 41
Pinus contorta Dougl. (Pinac.), 140, 225

BACTERIA

SEAWEEDS

BIRD

BIVALVE

FISH

INSECTS

SUBJECT INDEX